Ribosomal Proteins in Ribosome Assembly

Ribosomal Proteins in Ribosome Assembly

Guest Editors

Brigitte Pertschy
Ingrid Zierler

Basel • Beijing • Wuhan • Barcelona • Belgrade • Novi Sad • Cluj • Manchester

Guest Editors

Brigitte Pertschy
Institute of Molecular
Biosciences
University of Graz
Graz
Austria

Ingrid Zierler
Institute of Molecular
Biosciences
University of Graz
Graz
Austria

Editorial Office
MDPI AG
Grosspeteranlage 5
4052 Basel, Switzerland

This is a reprint of the Special Issue, published open access by the journal *Biomolecules* (ISSN 2218-273X), freely accessible at: www.mdpi.com/journal/biomolecules/special_issues/Ribosomal_Proteins.

For citation purposes, cite each article independently as indicated on the article page online and using the guide below:

Lastname, A.A.; Lastname, B.B. Article Title. *Journal Name* **Year**, *Volume Number*, Page Range.

ISBN 978-3-7258-3218-7 (Hbk)
ISBN 978-3-7258-3217-0 (PDF)
https://doi.org/10.3390/books978-3-7258-3217-0

© 2025 by the authors. Articles in this book are Open Access and distributed under the Creative Commons Attribution (CC BY) license. The book as a whole is distributed by MDPI under the terms and conditions of the Creative Commons Attribution-NonCommercial-NoDerivs (CC BY-NC-ND) license (https://creativecommons.org/licenses/by-nc-nd/4.0/).

Contents

About the Editors vii

Brigitte Pertschy and Ingrid Zierler
Ribosomal Proteins in Ribosome Assembly
Reprinted from: Biomolecules 2024, 15, 13, https://doi.org/10.3390/biom15010013 1

Kavan Gor and Olivier Duss
Emerging Quantitative Biochemical, Structural, and Biophysical Methods for Studying Ribosome and Protein–RNA Complex Assembly
Reprinted from: Biomolecules 2023, 13, 866, https://doi.org/10.3390/biom13050866 5

Chiara Di Vona, Laura Barba, Roberto Ferrari and Susana de la Luna
Loss of the DYRK1A Protein Kinase Results in the Reduction in Ribosomal Protein Gene Expression, Ribosome Mass and Reduced Translation
Reprinted from: Biomolecules 2023, 14, 31, https://doi.org/10.3390/biom14010031 31

Andreas Steiner, Sébastien Favre, Maximilian Mack, Annika Hausharter, Benjamin Pillet, Jutta Hafner, Valentin Mitterer, Dieter Kressler, Brigitte Pertschy and Ingrid Zierler
Dissecting the Nuclear Import of the Ribosomal Protein Rps2 (uS5)
Reprinted from: Biomolecules 2023, 13, 1127, https://doi.org/10.3390/biom13071127 54

Anne-Marie Landry-Voyer, Zabih Mir Hassani, Mariano Avino and François Bachand
Ribosomal Protein uS5 and Friends: Protein–Protein Interactions Involved in Ribosome Assembly and Beyond
Reprinted from: Biomolecules 2023, 13, 853, https://doi.org/10.3390/biom13050853 75

Matthew John Eastham, Andria Pelava, Graeme Raymond Wells, Nicholas James Watkins and Claudia Schneider
RPS27a and RPL40, Which Are Produced as Ubiquitin Fusion Proteins, Are Not Essential for p53 Signalling
Reprinted from: Biomolecules 2023, 13, 898, https://doi.org/10.3390/biom13060898 93

Margaret L. Rodgers, Yunsheng Sun and Sarah A. Woodson
Ribosomal Protein S12 Hastens Nucleation of Co-Transcriptional Ribosome Assembly
Reprinted from: Biomolecules 2023, 13, 951, https://doi.org/10.3390/biom13060951 111

Taylor N. Ayers and John L. Woolford
Putting It All Together: The Roles of Ribosomal Proteins in Nucleolar Stages of 60S Ribosomal Assembly in the Yeast *Saccharomyces cerevisiae*
Reprinted from: Biomolecules 2024, 14, 975, https://doi.org/10.3390/biom14080975 124

Sara Martín-Villanueva, Carla V. Galmozzi, Carmen Ruger-Herreros, Dieter Kressler and Jesús de la Cruz
The Beak of Eukaryotic Ribosomes: Life, Work and Miracles
Reprinted from: Biomolecules 2024, 14, 882, https://doi.org/10.3390/biom14070882 150

Ting Yu, Junyi Jiang, Qianxi Yu, Xin Li and Fuxing Zeng
Structural Insights into the Distortion of the Ribosomal Small Subunit at Different Magnesium Concentrations
Reprinted from: Biomolecules 2023, 13, 566, https://doi.org/10.3390/biom13030566 177

Zainab Fakih, Mélodie B. Plourde and Hugo Germain
Differential Participation of Plant Ribosomal Proteins from the Small Ribosomal Subunit in Protein Translation under Stress
Reprinted from: *Biomolecules* **2023**, *13*, 1160, https://doi.org/10.3390/biom13071160 . . . . **193**

About the Editors

Brigitte Pertschy

Brigitte Pertschy heads a research group at the Institute of Molecular Biosciences at the University of Graz. Her research covers various aspects of ribosome assembly, including the assembly-promoting role of snoRNPs, post-translational modifications of ribosomal proteins and ribosome assembly factors, and the nuclear import and chaperoning of ribosomal proteins.

Ingrid Zierler

Ingrid Zierler is a senior postdoc at the Institute of Molecular Biosciences at the University of Graz. Her research focuses on the involvement of ribosomal proteins in ribosome biogenesis, including the function of dedicated ribosomal protein chaperones, ribosomal protein import, and the function of ribosomal proteins in ribosome assembly. In a recent independent project funded by her own grant, she investigated the role of disease-relevant mutations of ribosomal protein S15 in the late steps of 40S ribosomal subunit maturation.

Editorial

Ribosomal Proteins in Ribosome Assembly

Brigitte Pertschy [1,2,*] and Ingrid Zierler [1,2,*]

1. Institute of Molecular Biosciences, University of Graz, Humboldtstrasse 50, 8010 Graz, Austria
2. BioTechMed-Graz, Mozartgasse 12/II, 8010 Graz, Austria
* Correspondence: brigitte.pertschy@uni-graz.at (B.P.); ingrid.zierler@uni-graz.at (I.Z.)

Ribosomes are the cellular machinery responsible for translating mRNA into proteins, a process fundamental to all domains of life from bacteria to eukaryotes. The synthesis of new ribosomes is a highly orchestrated process involving the assembly of ribosomal proteins with ribosomal RNA (rRNA) into the complex three-dimensional structure of mature ribosomes. This Special Issue focusses on the multifaceted roles of ribosomal proteins in this intricate process, covering topics from their synthesis and transport to their assembly with rRNA, the consequences of their absence, and extra-ribosomal functions.

The current understanding of the role of ribosomal proteins in ribosome assembly is the result of decades of research and technological advances. The review by Gor and Duss (Emerging Quantitative Biochemical, Structural, and Biophysical Methods for Studying Ribosome and Protein–RNA Complex Assembly) provides an overview of the state-of-the-art methods used to study ribosome assembly, focusing on bacteria, and covers some of the methods applied in the research articles within this Special Issue.

But let us move to the beginning (of ribosome synthesis)—in particular, the production of the components for new ribosomes. As well as rRNA, ribosomal protein mRNA also has to be transcribed, and once translated, newly synthesized ribosomal proteins have to be safely transported into the nucleus where the assembly with the rRNA takes place. These early steps of ribosome synthesis are addressed by several articles in this Special Issue.

The study by Di Vona et al. (Loss of the DYRK1A Protein Kinase Results in the Reduction in Ribosomal Protein Gene Expression, Ribosome Mass and Reduced Translation) provides insights into the regulation of ribosomal protein gene transcription in human cells. This study uncovers a novel motif present in a subset of ribosomal protein gene promoters that is bound by the DYRK1A kinase. The presence of DYRK1A enhances the recruitment of the TATA-binding protein, promoting ribosomal protein gene expression. The importance of this novel regulatory pathway is reflected by the observation that the depletion of DYRK1A leads to a global reduction in ribosomal protein gene mRNA, with a concomitant reduction of ribosomal proteins and, consequently, fewer ribosomes.

Received: 25 November 2024
Accepted: 27 November 2024
Published: 26 December 2024

Citation: Pertschy, B.; Zierler, I. Ribosomal Proteins in Ribosome Assembly. *Biomolecules* **2025**, *15*, 13. https://doi.org/10.3390/biom15010013

Copyright: © 2024 by the authors. Licensee MDPI, Basel, Switzerland. This article is an open access article distributed under the terms and conditions of the Creative Commons Attribution (CC BY) license (https://creativecommons.org/licenses/by/4.0/).

While translation of ribosomal protein mRNA occurs in the cytoplasm, most newly synthesized ribosomal proteins in eukaryotes must be transported into the nucleus. The research article by Steiner et al. (Dissecting the Nuclear Import of the Ribosomal Protein Rps2 (uS5)) provides insights into how the efficient nuclear import of the ribosomal protein uS5 is achieved in yeast cells. The findings of this study reveal that uS5 possesses at least two nuclear localization sequences (NLSs): an internal NLS recognized by the importin-β Pse1 and an N-terminal NLS. Interestingly, the N-terminal NLS overlaps with the binding site for the uS5-specific chaperone Tsr4, raising the question how the chaperoning and nuclear import of uS5 are coordinated.

Beyond importins and Tsr4, uS5 interacts with multiple additional proteins outside the ribosome. Landry-Voyer et al. (Ribosomal Protein uS5 and Friends: Protein–Protein Interactions Involved in Ribosome Assembly and Beyond) provide an overview of the interaction network of uS5, with a primary focus on the human protein. Human uS5 interacts with the Tsr4 ortholog PDCD2, a dedicated chaperone that protects uS5 from aggregation. Its paralog, PDCD2L, appears to have acquired an independent function, potentially acting as an export adaptor for pre-40S subunits. Furthermore, uS5 is bound by an arginine methyltransferase and a zinc finger protein, although the precise functions of these interactions remain unclear. While some of uS5's binding partners are involved in ribosome synthesis, others may direct uS5 toward as-yet-uncharacterized extra-ribosomal functions.

An important extra-ribosomal function of many ribosomal proteins is the regulation of p53 signaling in response to impaired ribosome synthesis, a topic studied by Eastham et al. (RPS27a and RPL40, Which Are Produced as Ubiquitin Fusion Proteins, Are Not Essential for p53 Signalling) in human cell lines. RPS27a (eS31) and RPL40 (eL40) are unique among ribosomal proteins as they are synthesized as ubiquitin fusion proteins that require cleavage to release the functional ribosomal protein and ubiquitin. This study investigates whether these two ribosomal proteins, as well as serving as ubiquitin providers, also contribute to p53 signaling. Using several different cell lines, the authors reveal that knockdown of eS31 and eL40 stabilizes p53 in only some of them, suggesting that the involvement of eS31 and eL40 in p53 signaling is cell type specific.

After synthesis of ribosomal proteins, the next critical step is their assembly with the ribosomal RNA as it is transcribed and begins to fold. The article by Rodgers et al. (Ribosomal Protein S12 Hastens Nucleation of Co-Transcriptional Ribosome Assembly) addresses the intricate mechanisms by which one of the early binding ribosomal proteins, S4 (uS4), associates with the ribosomal RNA during transcription. In particular, the authors reveal that one of the later binding proteins, S12 (uS12), accelerates uS4 binding during pre-16S rRNA transcription. This study suggests that uS12 functions as an rRNA chaperone, interacting transiently with a region near the uS4 binding site to facilitate rapid uS4 assembly. It is tempting to speculate that other late binding ribosomal proteins may also play early roles as RNA chaperones.

Beyond RNA folding, the absence of a ribosomal protein can have various effects on ribosome biogenesis, including the inhibition of pre-rRNA processing, or the delayed assembly of other ribosomal proteins or ribosome assembly factors. The review article by Ayers and Woolford (Putting It All Together: The Roles of Ribosomal Proteins in Nucleolar Stages of 60S Ribosomal Assembly in the Yeast *Saccharomyces cerevisiae*) addresses the roles of various ribosomal proteins during the nucleolar stages of 60S ribosomal subunit assembly, focusing on studies of yeast. Taking cryo-EM structural data into consideration, the authors propose several mechanisms for the assembly of ribosomal proteins into nascent ribosomes. They highlight the interplay between neighboring ribosomal proteins and the structural changes of rRNA upon ribosomal protein binding.

The role of ribosomal proteins in the assembly of ribosomal subunits in eukaryotes is also addressed in the review article by Martín-Villanueva et al. (The Beak of Eukaryotic Ribosomes: Life, Work and Miracles), this time focusing on the role of ribosomal proteins that are part of the protrusion of the small subunit termed the "beak structure". The authors discuss how the composition and structure of the beak have evolved from an all-rRNA structure in bacteria to a more complex arrangement involving multiple ribosomal proteins in eukaryotes and how this structure is assembled.

Having covered the various steps of ribosome assembly in this Special Issue, the last two articles examine how alterations in ribosomal protein composition can influence the structure and function of mature ribosomes. Yu et al. (Structural Insights into the Distortion of the Ribosomal Small Subunit at Different Magnesium Concentrations) investigate the structural dependence of bacterial ribosomes on magnesium ions. A reduction in Mg^{2+} concentration leads to distortions of the decoding center and the dissociation of ribosomal protein S12 (uS12), and it eliminates the interactions of S16 (bS16) with h17, h10, and h15. These findings underscore the critical role of magnesium in maintaining ribosomal integrity, a role that is likely also relevant during ribosome assembly.

Last but not least, the article by Fakih et al. (Differential Participation of Plant Ribosomal Proteins from the Small Ribosomal Subunit in Translation under Stress) addresses how different ribosomal protein isoforms affect translation in plants. This study reveals that specific ribosomal protein isoforms are relevant for the efficient translation of a specific set of mRNAs encoding proteins functioning in pathogen defense. These findings support the hypothesis that plants may utilize functionally specialized ribosomes, incorporating particular ribosomal protein variants to respond effectively to stress conditions.

The research and review articles compiled in this Special Issue offer insights into the multifaceted roles of ribosomal proteins in ribosome biogenesis and beyond. From the regulation of ribosomal protein synthesis and nuclear import to their assembly with rRNA and participation in extra-ribosomal functions, this collection emphasizes the necessity of ribosomal proteins in ribosome biogenesis and function. We hope that these contributions will stimulate further research in the field, advancing our understanding of ribosome biogenesis and its implications for human health and disease.

Conflicts of Interest: The authors declare no conflict of interest.

List of Contributions:

1. Gor, K.; Duss, O. Emerging Quantitative Biochemical, Structural, and Biophysical Methods for Studying Ribosome and Protein-RNA Complex Assembly. *Biomolecules* **2023**, *13*, 866. https://doi.org/10.3390/biom13050866.
2. Di Vona, C.; Barba, L.; Ferrari, R.; de la Luna, S. Loss of the DYRK1A Protein Kinase Results in the Reduction in Ribosomal Protein Gene Expression, Ribosome Mass and Reduced Translation. *Biomolecules* **2023**, *14*, 31. https://doi.org/10.3390/biom14010031.
3. Steiner, A.; Favre, S.; Mack, M.; Hausharter, A.; Pillet, B.; Hafner, J.; Mitterer, V.; Kressler, D.; Pertschy, B.; Zierler, I. Dissecting the Nuclear Import of the Ribosomal Protein Rps2 (uS5). *Biomolecules* **2023**, *13*, 1127. https://doi.org/10.3390/biom13071127.
4. Landry-Voyer, A.-M.; Mir Hassani, Z.; Avino, M.; Bachand, F. Ribosomal Protein uS5 and Friends: Protein-Protein Interactions Involved in Ribosome Assembly and Beyond. *Biomolecules* **2023**, *13*, 853. https://doi.org/10.3390/biom13050853.
5. Eastham, M.J.; Pelava, A.; Wells, G.R.; Watkins, N.J.; Schneider, C. RPS27a and RPL40, Which Are Produced as Ubiquitin Fusion Proteins, Are Not Essential for p53 Signalling. *Biomolecules* **2023**, *13*, 898. https://doi.org/10.3390/biom13060898.
6. Rodgers, M.L.; Sun, Y.; Woodson, S.A. Ribosomal Protein S12 Hastens Nucleation of Co-Transcriptional Ribosome Assembly. *Biomolecules* **2023**, *13*, 951. https://doi.org/10.3390/biom13060951.
7. Ayers, T.N.; Woolford, J.L. Putting It All Together: The Roles of Ribosomal Proteins in Nucleolar Stages of 60S Ribosomal Assembly in the Yeast *Saccharomyces cerevisiae*. *Biomolecules* **2024**, *14*, 975. https://doi.org/10.3390/biom14080975.
8. Martín-Villanueva, S.; Galmozzi, C.V.; Ruger-Herreros, C.; Kressler, D.; de la Cruz, J. The Beak of Eukaryotic Ribosomes: Life, Work and Miracles. *Biomolecules* **2024**, *14*, 882. https://doi.org/10.3390/biom14070882.
9. Yu, T.; Jiang, J.; Yu, Q.; Li, X.; Zeng, F. Structural Insights into the Distortion of the Ribosomal Small Subunit at Different Magnesium Concentrations. *Biomolecules* **2023**, *13*, 566. https://doi.org/10.3390/biom13030566.

10. Fakih, Z.; Plourde, M.B.; Germain, H. Differential Participation of Plant Ribosomal Proteins from the Small Ribosomal Subunit in Protein Translation under Stress. *Biomolecules* **2023**, *13*, 1160. https://doi.org/10.3390/biom13071160.

Disclaimer/Publisher's Note: The statements, opinions and data contained in all publications are solely those of the individual author(s) and contributor(s) and not of MDPI and/or the editor(s). MDPI and/or the editor(s) disclaim responsibility for any injury to people or property resulting from any ideas, methods, instructions or products referred to in the content.

Review

Emerging Quantitative Biochemical, Structural, and Biophysical Methods for Studying Ribosome and Protein–RNA Complex Assembly

Kavan Gor [1,2] and Olivier Duss [1,*]

1 Structural and Computational Biology Unit, European Molecular Biology Laboratory (EMBL), 69117 Heidelberg, Germany; kavan.gor@embl.de
2 Faculty of Biosciences, Collaboration for Joint PhD Degree between EMBL and Heidelberg University, 69117 Heidelberg, Germany
* Correspondence: olivier.duss@embl.de

Citation: Gor, K.; Duss, O. Emerging Quantitative Biochemical, Structural, and Biophysical Methods for Studying Ribosome and Protein–RNA Complex Assembly. *Biomolecules* 2023, *13*, 866. https://doi.org/10.3390/biom13050866

Academic Editors: Brigitte Pertschy and Ingrid Rössler

Received: 11 April 2023
Revised: 5 May 2023
Accepted: 9 May 2023
Published: 19 May 2023

Copyright: © 2023 by the authors. Licensee MDPI, Basel, Switzerland. This article is an open access article distributed under the terms and conditions of the Creative Commons Attribution (CC BY) license (https:// creativecommons.org/licenses/by/ 4.0/).

Abstract: Ribosome assembly is one of the most fundamental processes of gene expression and has served as a playground for investigating the molecular mechanisms of how protein–RNA complexes (RNPs) assemble. A bacterial ribosome is composed of around 50 ribosomal proteins, several of which are co-transcriptionally assembled on a ~4500-nucleotide-long pre-rRNA transcript that is further processed and modified during transcription, the entire process taking around 2 min in vivo and being assisted by dozens of assembly factors. How this complex molecular process works so efficiently to produce an active ribosome has been investigated over decades, resulting in the development of a plethora of novel approaches that can also be used to study the assembly of other RNPs in prokaryotes and eukaryotes. Here, we review biochemical, structural, and biophysical methods that have been developed and integrated to provide a detailed and quantitative understanding of the complex and intricate molecular process of bacterial ribosome assembly. We also discuss emerging, cutting-edge approaches that could be used in the future to study how transcription, rRNA processing, cellular factors, and the native cellular environment shape ribosome assembly and RNP assembly at large.

Keywords: RNP assembly; ribosome assembly; protein–RNA interactions; RNA folding; assembly intermediates; in vitro reconstitutions; mass spectrometry; single-molecule fluorescence microscopy; cryo–electron microscopy; RNA structure probing

1. Introduction

Ribosomes are responsible for protein synthesis and are some of the largest and most complex macromolecular machines in a cell. Prokaryotic ribosomes are made up of a large subunit (LSU or 50S) and a small subunit (SSU or 30S). The *Escherichia coli* (*E. coli*) LSU consists of 23S and 5S ribosomal RNAs (rRNA) bound by 33 ribosomal proteins (r-proteins), while the SSU consists of 16S rRNA and 21 r-proteins [1]. The assembly of a ribosome is a very complex and multistep process that consumes about 40% of a cell's energy [2]. Assembly is initiated with the transcription of a primary rRNA transcript containing ~4500 nucleotides. Transcription is assisted by the rRNA transcription antitermination complex (rrnTAC), which reduces transcription pausing and prevents early termination [3–5]. The primary transcript is co-transcriptionally processed by multiple specific RNases to form the three rRNA fragments (16S, 23S and 5S rRNAs) [6–9] that simultaneously fold into secondary and tertiary RNA structures [10–12]. Co-transcriptional rRNA folding follows the vectorial (5′ to 3′) direction and allows for the sequential binding of r-proteins [13–18]. Co-transcriptional rRNA processing, rRNA folding, and r-protein binding is accompanied by the introduction of base modifications, such as pseudouridinylations and methylations [19,20]. Furthermore, these processes are assisted by multiple assembly factors, such as GTPases, helicases, and maturation factors [1,21]. Remarkably, it takes only about 2 min for the cell to assemble a functional bacterial ribosome [22]. Consequently, the assembly intermediates of this process are short-lived and contribute to only ~2% of the total ribosome population [23], making them difficult to study.

Ribosome assembly, and RNP assembly in general, is very difficult to investigate. Apart from the complexity of the process and the low abundance of assembly intermediates, many of the biomolecular interactions that form during assembly are transient and dynamic in nature and therefore difficult to capture biochemically and structurally. Furthermore, the assembly processes are often very heterogeneous and consist of multiple parallel assembly pathways.

Ribosome assembly has been studied over many decades and, despite its complexity and technical limitations, various aspects of the process are well understood. There are several reviews that provide detailed overviews of various aspects of the assembly process [1,20,21,24–30]. Here, we aim to provide a methodological perspective on studying ribosome assembly and the assembly of other RNPs, such as the spliceosome, various mRNPs, and large non-coding RNPs. We summarise the various biochemical, structural, and biophysical methods employed over the years for studying different facets of the ribosome-assembly mechanism, with a focus on bacterial ribosome assembly. This review highlights the exciting parallel between the evolution of our understanding of ribosome assembly and the technological advancements that have led to the development of new methods (Figure 1). We start by discussing in vitro reconstitutions that employ a bottom-up approach using minimal components to understand the assembly process in a very controlled manner. Time-resolved mass spectrometry, RNA structural probing, and cryo–electron microscopy have provided information on the kinetics of assembly and have permitted the structural visualisation of the assembly process at high resolution. Single-molecule experiments have become instrumental in understanding how the different processes are functionally coupled with each other as they allow us to follow complicated, multistep processes in real-time. We conclude our review by discussing approaches that we think will be required in the future to understand how the ribosome and other complex protein–RNA machineries are assembled so fast and efficiently in vivo.

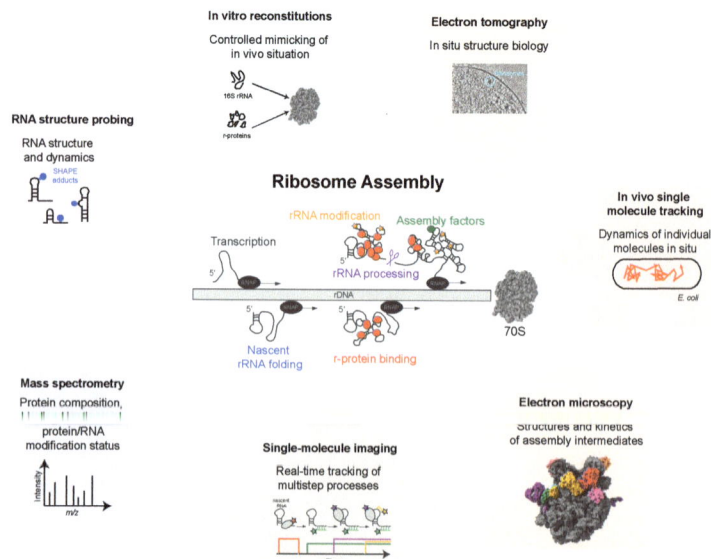

Figure 1. Overview of the biochemical, structural, and biophysical methods for studying ribosome and RNP assembly. Top right tomogram adapted and reproduced from [31], 70S (PDB: 4V6G) and 50S intermediate (PDB: 7BL5).

2. Biochemical Reconstitutions

2.1. In Vitro Reconstitutions

In the early days of studying ribosome assembly, it was evident that a ribosome is a very complex machinery composed of multiple r-proteins interacting with rRNA. In order to understand its assembly, the two subunits of the ribosome were studied separately. In vitro reconstitution/omission experiments were performed by mixing purified rRNA with different sets of r-proteins and then purifying the resulting assembly intermediates via the ultracentrifugation of sucrose gradients [32]. Initial attempts to reconstitute these subunits indicated that the 30S can be reconstituted in a single step [17], while several heating steps and various Mg^{2+} concentrations were required to reconstitute the 50S [33,34]. The reconstituted ribosomes were tested for their ability to read polyU templates [35,36], form peptide bonds [37], or bind tRNA [38], suggesting that these in vitro reconstitutions provide active ribosomes. Reconstitution experiments indicated that the binding of r-proteins occurred in a sequential order and allowed for the organisation of the ribosome assembly into assembly maps (Nomura map for 30S and Nierhaus map for 50S) containing the thermodynamic binding dependencies of the various r-proteins [16–18]. In vivo experiments using cold-sensitive mutant strains and strains lacking r-proteins validated the assembly maps derived from the in vitro reconstitution methods [39].

2.2. In Vivo Mimicry

While these reconstitution efforts were successful in describing the in vitro thermodynamic assembly pathway, assembly was much less efficient and required unphysiological heating steps and buffer conditions. Furthermore, the reconstituted ribosomes were not tested for their ability to translate a complete mRNA [40]. Importantly, these experimental conditions did not properly mimic the in vivo situation. Inside cells, the rRNA is efficiently transcribed and co-transcriptionally processed, modified, and bound by r-proteins simultaneously [41–44]. This entire process is assisted by multiple assembly factors. Developments in the field of cell-free systems spearheaded by the Jewett Lab have been used to reconstitute ribosomes with high activity in near-native assembly conditions. An integrative ribosome synthesis, assembly, and translation (iSAT) assay combines co-transcriptional ribosome assembly and the subsequent translation of mRNA via the assembled ribosome in a single reaction, with GFP as a readout for the successful assembly of an active ribosome (Figure 2A–C) [40]. The iSAT reaction consists of a plasmid containing the entire rRNA operon initiated by a T7 promoter sequence, T7 RNA polymerase, all r-proteins purified from native ribosomes (TP70), a second plasmid coding for the reporter mRNA sequence (GFP), and cell extract (S150) containing all the cellular factors required for ribosome assembly and translation (Figure 2A). The cell extract allows for the correct processing [45] and modification [46] of rRNA. Since all the key components required in ribosome assembly (as well as other components, such as assembly factors that assist ribosome assembly) are then present, the assembly of the ribosome is expected to proceed in a native way i.e., the processes of transcription, rRNA processing, r-protein binding, and base modifications are expected to occur simultaneously and assisted by assembly factors. While earlier iSAT reactions had translational efficiencies of 20% when compared to in vivo-purified ribosomes [45], their efficiency can be improved to 70% by adding crowding and reducing agents to the iSAT reactions [47]. iSAT reactions have been further extended to include the synthesis of individual r-proteins [48], yet the assay still needs to be further developed to allow all r-proteins to be synthesised in the same reaction. Of note, iSAT reactions work efficiently despite using T7 RNAP instead of the native *E. coli* RNAP. Better mimicking the in vivo situation, future adaptions of iSAT should include the native *E. coli* RNAP in order to also properly reproduce the native rRNA transcription speed and pausing behaviour, which is assisted by the rrnTAC.

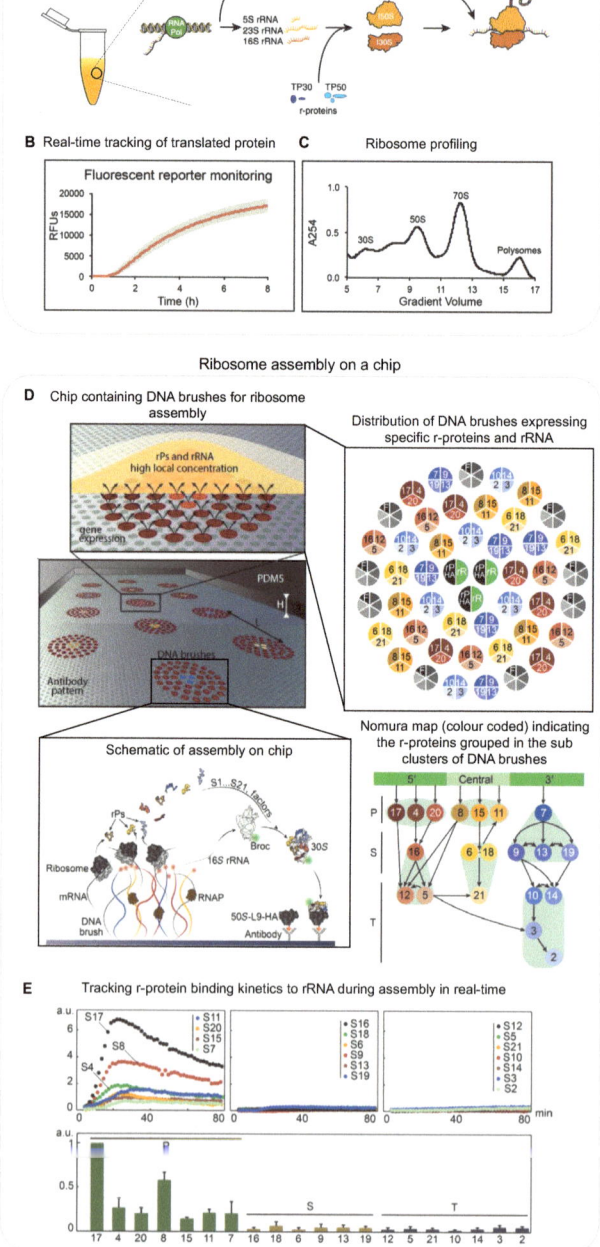

Figure 2. (**A–C**) Integrated ribosome synthesis, assembly, and translation (iSAT): (**A**) schematic of the one-pot iSAT reaction for the synthesis and assembly of ribosomes and the translation of a reporter protein; (**B**) real-time monitoring of fluorescence intensity as a reporter for translation activity of ribosomes

produced in iSAT reaction; (**C**) ribosome profiling of the iSAT reaction. (**D,E**) **Ribosome assembly on a chip:** (**D**) schematic of chip surface and the distribution of DNA brushes (centre), zoomed-in schematic of one DNA cluster (top), distribution of DNA brushes (right) encoding for rRNA (black) and r-proteins-HA (green), assembly factors (grey) and other r-proteins colour-coded as in the Nomura map (bottom right), and schematic of ribosome assembly on chip (bottom centre); (**E**) time traces of primary, secondary, and tertiary r-proteins binding to rRNA during assembly on a chip (top: left to right) and normalised maximum signal from primary (green), secondary (yellow), and tertiary (grey) r-protein-HA (bottom: left to right). (**A**) is reproduced with permission from [40]. (**B,C**) are reproduced with minor adaptations with permission from [47]. (**D,E**) are reproduced from [49].

Inspired by the lab-on-a-chip approach, the one-pot iSAT reaction assay has also been performed on a chip to reconstitute 30S subunits in near-native conditions (Figure 2D) [49]. Genes encoding for r-proteins and rRNA were immobilised on a chip surface as DNA brushes along with anti-HA antibodies (Figure 2D, right panel). One r-protein at a time was designed as a fusion protein with a HA tag, and the rRNA was modified to include a Broccoli aptamer. All genes were transcribed, and r-proteins were translated locally at the surface. The resultant increase in fluorescence signal from the broccoli aptamer on the regions of the chip coated with anti-HA antibodies indicated that the rRNA was bound by the HA-tagged r-protein and all upstream binding r-proteins according to the Nomura assembly map (Figure 2D bottom, right panel). Using this approach, they could recapitulate the r-protein binding dependencies (Nomura map) and their binding kinetics (Figure 2E). They were also able to monitor the late stages of 30S assembly, including the binding of the mature 30S to the 50S.

In summary, biochemical reconstitutions are a powerful tool for investigating the intricate details of a specific process using a minimalistic system. Recent ribosome reconstitutions that mimic native conditions have enabled researchers to study mutant ribosomes [50], incorporate non-canonical amino acids [51], and investigate the process of evolution in the context of ribosome assembly and function [52]. Furthermore, these methods can enable the investigation of the role of various assembly factors in wild-type versus mutant ribosomes and the engineering of new ribosomes with specific functions.

3. Mass Spectrometry

While the in vitro reconstitution/omission experiments allowed for the construction of ribosome assembly maps that summarise thermodynamic protein-binding dependencies, they do not contain any information on its protein-binding kinetics during assembly. By combining quantitative mass spectrometry (qMS) with pulse-chase experiments using stable isotope labelling, it became possible to complement the thermodynamic r-protein binding dependencies with r-protein binding rates.

3.1. In Vitro Mass Spectrometry for r-Proteins

Pulse-chase qMS (PC-qMS) allows for the tracking of binding rates for all r-proteins to the rRNA in a single experiment [15]. The rRNA is incubated with heavy isotope-labelled r-proteins for a specific amount of time, and then chased with an excess of light isotope-labelled r-proteins to complete the assembly (Figure 3A, left panel). The completely assembled subunits are then isolated on a sucrose gradient and the value of the heavy-to-total protein ratio for each protein is determined using mass spectrometry and plotted as a function of time (Figure 3A, centre panel). The resulting binding curves provide the average binding rates for the individual r-proteins (Figure 3A, centre and right panels). By repeating these experiments at different protein concentrations and temperatures, the authors demonstrated that (1) RNA folding and protein binding occur at similar rates, (2) the rate-limiting steps for different proteins is similar at low or high temperatures, and (3) the final steps of 30S synthesis are limited by many different transitions. Similar experiments were performed with a pre-folded 16S rRNA that was pre-bound with a subset of r-proteins [53]. They observed multiphase binding kinetics of r-proteins, suggesting further complexity in the assembly pathway. Their observations also indicate the presence of multiple assembly pathways and a delicate interplay between thermodynamic dependency and kinetic cooperativity. PC-qMS was also used to investigate the influence of assembly factors for the assembly of the 30S, showing, for example, that RimP allows for the faster binding of S9 and S19 but prevents the binding of S12 and S13, potentially by blocking their binding sites [54].

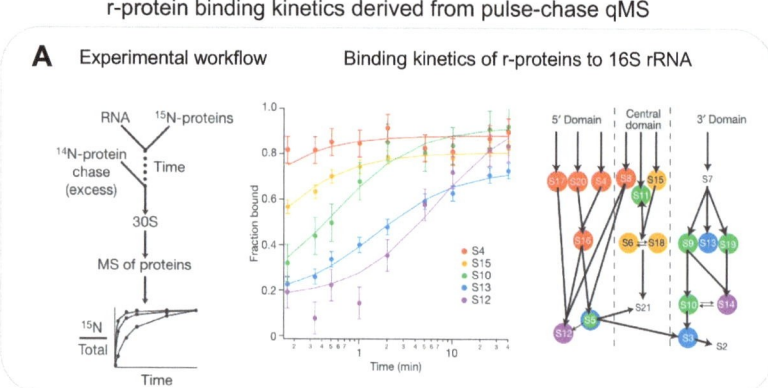

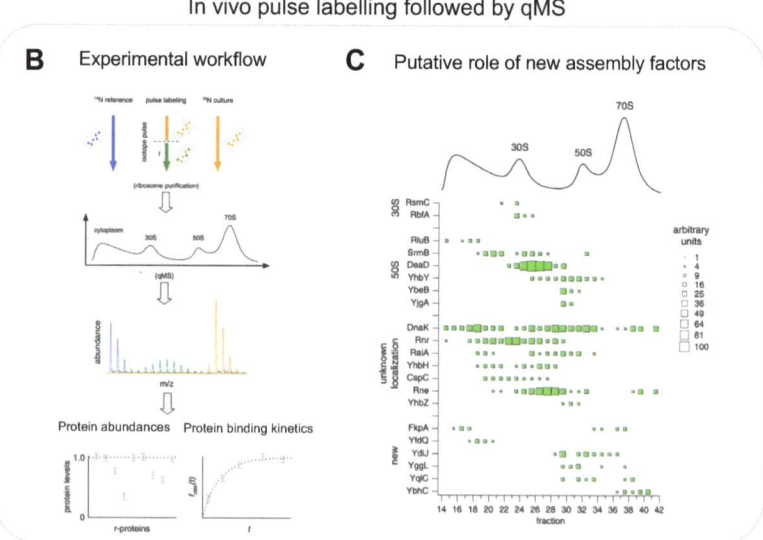

Figure 3. (**A**) In vitro pulse-chase qMS to determine protein binding kinetics: schematic of pulse-chase qMS workflow (left), r-protein binding curves to 16S rRNA (centre), Nomura assembly map coloured according to binding rates derived from pulse-chase qMS (right). (**B**,**C**) **In vivo pulse labelling to determine protein binding kinetics and discovery of new assembly factors bound to the ribosome-assembly intermediates:** (**B**) experimental workflow of in vivo pulse labelling and corresponding quantification by MS. (**C**) qMS-based identification and discovery of assembly factors and their potential role in assembly of specific subunits (right). (**A**) is reproduced with permission from [15], and (**B**) is adapted and reproduced and (**C**) is reproduced with permission from [22].

3.2. In Vivo Mass Spectrometry for r-Proteins and Assembly Factors

qMS-based methods were also applied to recapitulate the assembly pathway in vivo and for the identification of multiple assembly factors [22] (Figure 3B,C). The authors used an in vivo stable isotope pulse-labelling approach to characterise the exact r-protein composition of various populations of intermediates (Figure 3B). The cells were grown in heavy isotope media and pulse labelled with light isotope media. Various fractions from the sucrose gradient corresponding to assembly intermediates were digested by trypsin and subjected to qMS. The resultant in vivo data validated the presence of four assembly intermediates of 30S particles, as observed by Mulder et al. using in vitro reconstitutions. The 50S assembly was more continuous in cells and revealed six assembly intermediates, which indicated a general pathway where the 50S assembly starts opposite to the peptidyl transferase centre, forms intermediates where r-proteins are added globally to the whole structure, and ends with the formation of the central protrusion. Likewise, subjecting the fractions of a sucrose gradient to qMS analysis led to the identification of 15 known and 6 unknown assembly factors that co-occurred with specific assembly intermediates, indicating their role in that particular stage of assembly (Figure 3C).

MS was also used to understand the effects of cellular knockouts of assembly factors on the composition of ribosome-assembly intermediates. Experiments with strains lacking specific assembly factors showed a slower growth rate and an accumulation of assembly intermediates [54–57]. An investigation into in vivo assembly intermediates from mutant strains that lacked the assembly factors LepA or RsgA, for example, showed reduced levels of late-binding r-proteins, suggesting the role of these assembly factors in the late stages of assembly [58].

Apart from quantifying the composition of assembly intermediates, MS can also be used to investigate the post-translational modifications of r-proteins during assembly. For example, the Woodson Lab used MS to understand the extent of S5 and S18 acetylation during in vivo ribosome assembly and its effect on the formation of specific rRNA contacts [59].

qMS methods were also applied to study eukaryotic ribosome assembly. For example, Sailer et al. used multiple different affinity-tagged assembly factors to pull down and crosslink different intermediates of pre-60S particles from *Saccharomyces cerevisiae* [60]. A mass spectrometry analysis of the crosslinked peptides produced a protein–protein interaction map that identified the localisation of 22 unmapped assembly factors. The association based on relative abundances between the newly mapped assembly factors and specific intermediates indicated the approximate time at which they act in the assembly pathway.

3.3. In Vivo Mass Spectrometry for RNA Modifications

rRNA is modified by methylations as well as pseudouridinylations [19]. These modifications are deposited site-specifically by multiple different modification enzymes during the course of the assembly process. Traditionally, modifications are detected using reverse transcriptase primer extension techniques [61], or P1 nuclease digestion followed by thin-layer chromatography (TLC) or high-performance liquid chromatography (HPLC) [62,63]. Although these are very sensitive methods, they are tedious as they allow for the observation of only one modification at a time and are suitable for detecting only specific modifications. qMS analyses of RNA enable the detection of multiple site-specific modifications simultaneously. Typically, isotope-based labelling is used to detect the fraction of RNA molecules that is site-specifically methylated. However, the accurate quantification of lowly abundant modifications can be challenging. Furthermore, since pseudouridine is a structural isomer of uridine, it cannot be detected. Popova et al. used a metabolic labelling approach to validate methylations and detect pseudouridinylated residues [64]. CD_3-methionine (the precursor to SAM) leads to a +3 Da mass shift that can be distinctly and confidently annotated. Similarly, 5,6-D-uracil leads to a -1 Da mass shift for a pseudouridinylated residue. Using this approach on assembly intermediates purified from cells, the authors were able to characterise the stages at which each residue is modified during the assembly process. For example, most of the modifications on the 23S rRNA occur early during assembly, as opposed to the 16S where the modifications are incorporated from a 5′ to 3′ direction, in agreement with a co-transcriptional rRNA modification process. Another study used qMS on S-adenosylmethionine (SAM; methyl donor used by methyltransferases)-depleted cells to study the importance of RNA modification on ribosome assembly [65].

Overall, mass spectrometry is a highly sensitive and quantitative method for determining the binding kinetics of r-proteins to rRNA as well as for studying when multiple r-protein or rRNA chemical modifications are introduced during assembly.

4. Electron Microscopy

Electron microscopy (EM) has been proven instrumental in providing high-resolution structural information on ribosome-assembly intermediates. Both for negative-stain and cryogenic EM, ribosome-assembly intermediates either from in vitro reconstitutions or purified from cells are applied to a grid for imaging. Optimally, the individual particles meant to be imaged are present in multiple different orientations to reconstruct a 3D image [66]. Seminal work by the Williamson Lab in 2010 demonstrated the potential of using structural information derived from a heterogenous population of assembly intermediates for understanding the mechanisms of ribosome assembly [67]. They performed time-resolved low-resolution negative-stain EM after mixing 16S rRNA with all 30S r-proteins and then freezing them at different time points. They were able to visualise 14 different assembly intermediates, which were classified into 4 major groups (Figure 4A). The population of the first group, representing the smallest assembly intermediate, decreased over time. The second group peaked at several minutes, while the third and fourth groups appeared only at later time points. In combination with PC-qMS, they were able to reconstruct a detailed assembly pathway for the 30S subunit in vitro, demonstrating multiple parallel assembly pathways (Figure 4A).

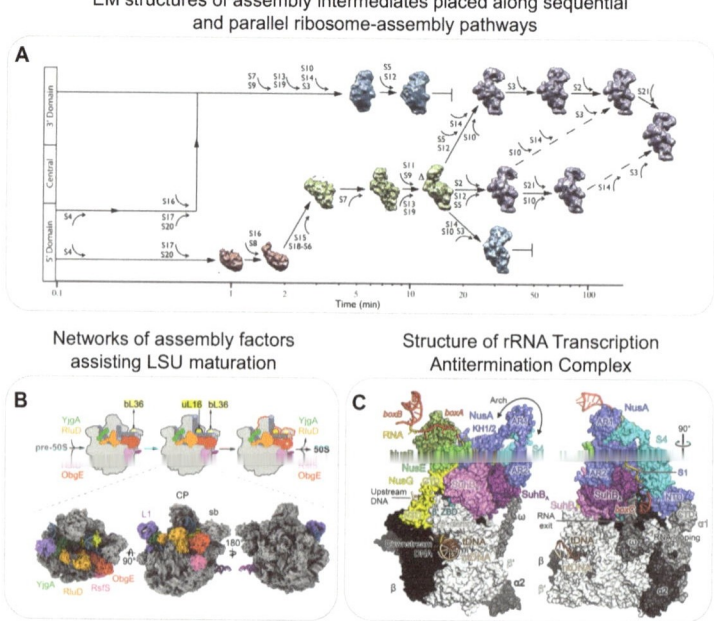

Figure 4. Electron microscopy: (**A**) Assembly intermediates populating parallel 30S assembly pathways visualised by negative-stain EM. (**B**) Model of how late-stage 50S maturation is guided by assembly

factors (top); model can be constructed using several high-resolution cryoEM structures of pre-50S bound by assembly factors YjgA, RluD, RsfS and ObgE. (**C**) High-resolution cryoEM structures of the complete rRNA transcription antitermination complex (rrnTAC) responsible for efficient transcription of rRNA. Reproduced with permission: (**A**) from [67], (**B**) from [68], and (**C**) from [69].

The resolution revolution in 2013 led to significant improvements in electron detection technology and reconstruction algorithms [70,71]. This enabled its use in investigating more heterogeneous populations of complexes present in the same sample, providing the basis for imaging multiple assembly intermediates that populate the 50S assembly pathway both in vitro and in vivo.

The in vitro reconstitution of 50S is a two-step process that leads to the activation of 50S [72]. High-resolution cryoEM of the two-step reconstitution process displayed five main classes (subpopulations) resulting from the first step and a mature 50S structure resulting from the second. 50S assembly initiates at its core, followed by the L1 protuberance and the central protuberance (CP). Interestingly, the main difference between the last class from step one and the fully mature 50S is a structural rearrangement of the rRNA that leads to the maturation of the peptidyl transferase centre.

In order to perform experiments in native conditions and gain perspective on ribosome assembly in vivo, Davis et al. used high-resolution cryoEM of assembly intermediates isolated from a bL17 (r-protein of 50S)-depleted strain to enrich intermediates [73]. Subpopulation averaging revealed that, similar to that of the in vitro experiments, the in vivo 50S assembly consists of a heterogeneous ensemble of intermediates. The different subpopulations that progressively evolved into a more mature complex could be further grouped together, thus providing structural evidence of parallel pathways of 50S assembly. Interestingly, a reanalysis of this compositionally and conformationally heterogenous data using a neural network-based framework called CryoDRGN revealed a previously unreported assembly intermediate [74]. CryoDRGN is a powerful tool that enables the automated classification of various states, which is typically performed using multiple manual and expert-guided rounds of hierarchical 3D classification.

A correlative analysis using qMS and cryoEM data from bL17 depleted cells also indicated that the unidentified densities in subpopulations from one of the assembly pathways corresponds to assembly factor YjgA [73]. Putative YjgA binding blocked the docking of a helix crucial for inter-subunit bridge formation, suggesting that YjgA acts as a late-stage assembly factor for maturation. Recent evidence suggests that the presence of assembly factors in vivo directly affects the order of maturation of specific regions. For example, in contrast to those of in vitro assembly [72], the core and the central protuberance formations were suggested to be interdependent in vivo [68,75]. Another detailed characterisation of pre-50S assembly intermediates revealed a network of assembly factors, such as ObgE, RsfS, YjgA, RldU, and YhbY, that orchestrate 50S maturation (Figure 4B) [68]. Several other studies have used cryo-EM to determine structures of bacterial ribosome-assembly intermediates to understand the function of assembly factors but are not further reviewed here [76–83].

Apart from structurally characterising later assembly intermediates that are formed once transcription is already completed and the majority of the rRNA is already processed, recent structural work has also provided information on the process of early rRNA transcription by the rRNA transcription antitermination complex and on the mechanism of initial rRNA processing. The rrnTAC is the macromolecular machinery responsible for the efficient transcription of rRNA in the cell [3–5]. The rrnTAC assembles on RNA polymerase (RNAP) and reduces NusA-mediated transcriptional pausing, R-loop formation, polymerase backtracking, and intrinsic as well as Rho-dependent termination. It enables chaperone-mediated rRNA folding and the formation of long-range rRNA–rRNA interactions. The high-resolution cryoEM structures of an in vitro-reconstituted rrnTAC-associated transcription complex revealed the presence of NusA, NusB, NusE, NusG, SuhB, and S4 (Figure 4C) [69]. Interestingly, in the rrnTAC, NusA is repositioned to prevent pausing caused by hairpin stabilisation as well as intrinsic termination. Similarly, the presence of NusG in the rrnTAC suppresses RNAP backtracking. The interactions of NusA, NusE, and SuhB with the C-terminus of NusG prevent it from recruiting Rho. Furthermore, the formation of a ringlike structure made by SuhB and S4 around the *E. coli* polymerase exit channel prevents Rho from directly interacting with the exit channel, therefore preventing Rho-dependent termination. Finally, the authors demonstrated that the 5′ end of RNA is bound by S4, and the emerging 3′ end of RNA is bound by Nus factors along with SuhB on the ring. This brings the distal regions of RNA close in space to form the long-range interactions that are required for creating a substrate for rRNA processing by RNase III [84].

The 4500-nucleotide-long primary rRNA transcript is initially processed by dsRNA-specific RNases to generate the pre-rRNA fragments that further mature into 16S, 23S, and 5S [1,8,9]. In *B. subtilis*, a mature 23S is obtained by Mini-III and a mature 5S by M5 processing [85,86]. The structural characterisation of these RNases with their respective substrates revealed that the Mini-III binds pre-23S ds-rRNA as a dimer, where each subunit cleaves one of the strands [87]. In contrast, the N-terminal domain of M5 binds to the 3′ strand of ds-rRNA and cleaves it. This then leads to structural rearrangements enabling the C-terminal domain of M5 to bind and cleave the 5′ strand. Mini-III and M5 are assisted by r-proteins, such as uL3 and uL18, that bind to the respective substrates and keep them in a conformation that can be recognised by the enzymes.

The above examples highlight the role of cryoEM as a powerful structural method for increasing our understanding of the structural and mechanistic details of ribosome assembly and for providing time-resolved information.

5. RNA Structure Probing

Multiple different studies have indicated the role of rRNA secondary and tertiary structures in the binding of r-proteins. The simple chemical or enzymatic probing of rRNA structures is a powerful method but has limited time-resolution and throughput [88]. The structural determination of rRNA during assembly is also difficult due to the presence of a heterogeneous set of conformations and the difficulty in resolving flexible regions. Hydroxyl radical footprinting (HRF) and DMS/SHAPE probing are two complementary methods that provide information on RNA tertiary and secondary structures, respectively. These methods overcome some of the above-mentioned limitations and have successfully been used to obtain more-detailed and better-resolved information on rRNA folding.

5.1. In Vitro RNA Structure Probing

RNA secondary structure information can be obtained using chemical reagents such as DMS and many other probes (e.g., glyoxal, 1m7, 1m6, NMIA, BzCN, NAI), which react specifically with single-stranded RNA but do not chemically modify dsRNA [89]. The introduced adducts result in a stop when read by a reverse transcriptase. The resulting fragments are detected using primer extension. This approach has been used to predict 16S rRNA secondary structures with 97% accuracy [90]. In order to increase throughput, an alternate strategy, termed SHAPE-MaP, uses manganese ions during the reverse transcription step, which causes reverse transcriptase to introduce a mutation into cDNA rather than stop at the modified sites [91]. The cDNA is sequenced and the percentage of the underlying mutations is used to generate a reactivity profile for predicting the secondary structure (Figure 5A). SHAPE-MaP was used to track the rRNA structure during ribosome assembly. For example, the SHAPE-MaP-based structural probing of 23S rRNA in the presence and absence of r-proteins showed very similar reactivity profiles, suggesting that the 23S rRNA assumes its secondary structure even in the absence of r-proteins [73].

While most of the chemical reagents introduced above require seconds to several minutes to react with their RNA substrate, and therefore limit the time resolution, hydroxyl radical footprinting (HRF) provides information at few-milliseconds resolution and therefore allows for the study of very early rRNA folding events and of the formation of protein–RNA interactions [92,93]. In HRF, rRNA is exposed to short pulses of hydroxyl radicals that are generated by X-rays. These hydroxyl radicals react with the unprotected RNA backbone and thereby cleave the RNA into smaller fragments. A site-specific primer extension is used to amplify these fragments. The probability of cleavage depends on solvent accessibility, and therefore reports on the RNA tertiary structure and/or its interaction with proteins. HRF experiments were performed in a time-resolved manner by mixing 16S rRNA with all r-proteins and exposing the reaction to an X-ray pulse at different time points after mixing, thus providing the first time point as early as 20 ms after mixing (Figure 5B–D) [92]. Apart from validating the kinetics of early and late r-protein binding as determined using PC-qMS, these experiments showed that 30S assembly nucleates from different points along rRNA (Figure 5C). Additionally, the authors observed that the initial encounter complexes refold during assembly. For example, S7 initially binds in a non-native conformation (protecting only H43 within 20–50 ms) and adapts a native conformation (protecting H29, H37, and H41) only much later in assembly (Figure 5D). These experiments demonstrate the potential for HRF to provide information on RNA structural changes at a nucleotide resolution and at a millisecond-to-second timescale.

rRNA folds into a very heterogeneous set of conformations during ribosome assembly [10,11,94,95]. Therefore, RNA structural probing will provide an average of all conformations present in the sample to be probed. While this has not yet been performed in studying rRNA folding during ribosome assembly, recent analysis pipelines have shown the potential for dissecting RNA heterogeneity using the property of DMS to achieve multi-hit kinetics and using single-molecule sequencing as a read-out [96–100].

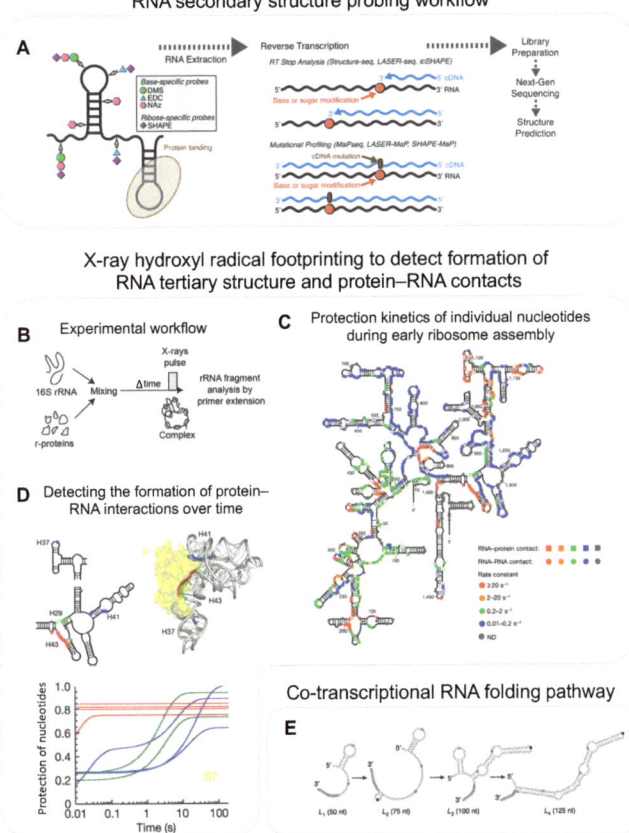

Figure 5. RNA secondary structure determination using chemical probing and high-throughput

sequencing: (A) general workflow of RNA secondary structure probing. (B–D) **Tertiary structure and protein–RNA interaction determination using hydroxyl radical footprinting (HRF):** (B) experimental setup of in vitro time-resolved HRF. (C) Protection rates of individual residues of the 16S rRNA representing formation of RNA–RNA tertiary contacts as well as RNA–protein contacts. (D) Kinetics of rRNA backbone protection as a result of S7 binding represented on secondary (top left) and 3D (top right) structures and normalised fitted curves indicating the protection of residues (y-axis) as a function of time (x-axis) (bottom). Colour codes for Figure 5D are the same as indicated in Figure 5C. **Co-transcriptional RNA folding intermediates:** (E) model for co-transcriptional folding of the SRP RNA as determined by co-transcriptional SHAPE-seq. (A) is reproduced and adapted with permission from [101]. Reproduced with permission: (C,D) from [92], and (E) from [102].

5.2. Co-Transcriptional RNA Structure Probing

RNA probing assays have been performed on pre-transcribed rRNA, but rRNA folds co-transcriptionally in vivo and is affected by the speed of transcription [103]. While it has been shown that co-transcriptional rRNA folding is different from the folding of a pre-transcribed RNA [10,11,95], the structural probing of rRNA has not been performed in the context of transcription yet. However, co-transcriptional probing has been employed to study relatively simpler systems that undergo ligand-induced conformational changes, such as the fluoride riboswitch and the SRP RNA [102]. For this, DMS/SHAPE-based structural probing was adapted by designing roadblocks on the 3′ end of a DNA transcription template. The roadblocks prevent polymerase from transcribing further. The reactivity profiles of RNA fragments that were transcribed from different lengths of DNA templates (made by placing roadblocks at different positions) allowed the authors to mimic co-transcriptional RNA folding pathways, simulating, however, an infinitesimally slow transcription rate (Figure 5E).

5.3. In Vivo RNA Structure Probing

RNA structural probing has also been performed in vivo to understand how rRNA folds in the native cellular environment and in the natural context of rRNA transcription. Soper et al. used HRF to study how assembly factors affect rRNA structure formation during assembly [59]. They compared protection resulting from mutant strains that lacked assembly factors, such as RbfA and RimM, to wild-type strains to determine the putative binding site of these assembly factors. Furthermore, closely analysing the assembly intermediates of mutant strains helped discover the role of these assembly factors in the assembly pathway. These assembly factors lead to global structural changes at late time points in assembly, binding to the 50S inter-subunit interface and thus acting as a checkpoint for quality control.

In order to obtain information on early co-transcriptional rRNA folding in vivo, a protocol was developed that uses the metabolic labelling of cells to separate newly transcribing rRNA intermediates from the total pool of rRNA [104]. Transcriptionally inactive cells (in nutrient-poor media) are labelled with 4-thiouridine (4sU) right before feeding with rich media. This allows for the isolation of nascent-transcribed rRNA, which can be probed by DMS or HRF. The results of the in vivo DMS probing of nascent 16S rRNA recapitulated the general vectorial folding pathway and specifically provided information on rRNA interactions at 30 seconds resolution. The results of the in vivo HRF probing of nascent rRNA are expected to provide exciting insights into rRNA tertiary structure formation at milliseconds resolution. Furthermore, this approach can potentially be extended by developing cell-compatible and faster acting probes to achieve millisecond-to-second resolutions for secondary structure probing [105].

6. Single-Molecule Methods

The so-far discussed ensemble biochemical, biophysical, and structural methods have led to the detailed characterisation of the mechanism of ribosome assembly, including the order and kinetics of r-protein binding, the dynamics of RNA structure formation, and the structural characterisation of assembly intermediates formed at various stages of assembly. However, these ensemble methods provide an average over the individual molecules. The need for averaging leads to the following major challenges: (1) the heterogeneity of the ribosome-assembly process cannot be sufficiently resolved, i.e., it is not possible to separately monitor the trajectories along the reaction coordinates of the individual assembly pathways; (2) it is not possible to dissect how different molecular processes, such as transcription progression, RNA folding, and protein binding, are functionally coupled to each other; and (3) dynamic structural changes may not be resolved.

Single-molecule methods instead allow for tracking the activity of individual molecules over long time scales at high temporal resolutions, thereby directly following multistep processes in real-time and dissecting the heterogeneity. To observe a single-molecule for minutes to hours, molecules of interest are immobilised on the chemically functionalised surface of a glass coverslip, typically using a biotin–streptavidin/neutravidin interaction (Figure 6A) [106]. Fluorescently labelling the molecule on the surface or the ligands that can bind to the surface-immobilised molecules allows for monitoring the conformational changes, binding events, and enzymatic activities of the molecules in real-time using total internal reflection fluorescence (TIRF) microscopy to reduce the fluorescence background. A fluorescent resonance energy transfer (FRET) can be used to directly measure the distance changes between a donor and one or several acceptor dyes [107] and thereby, for example, inform on conformational changes as they happen in real-time [108].

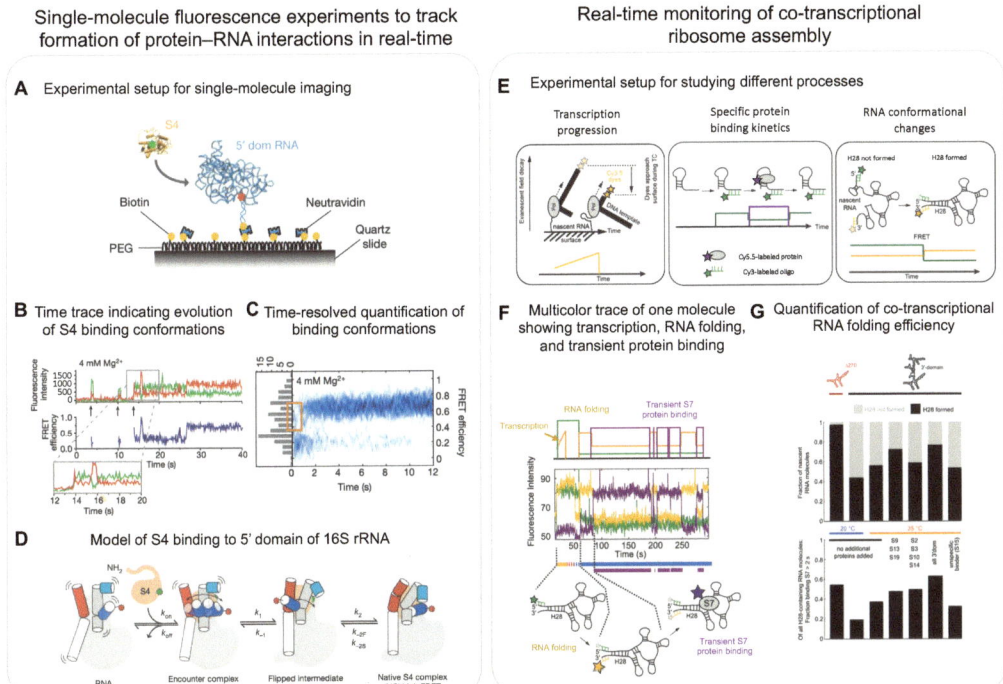

Figure 6. (**A–D**) **Single molecule fluorescence microscopy experiments for tracking changes to protein–RNA interactions in real-time:** (**A**) experimental setup of typical single-molecule experiments: schematic shows specific binding of S4 to the 5′ domain of 16S rRNA using single-molecule FRET. S4 was labelled with a donor dye (Cy3 in green) and the immobilised RNA by an acceptor dye (Cy5 in red). (**B**) Single-molecule trace of S4 binding to the 5′ domain of the 16S rRNA, leading to anti-correlated changes to the Cy3 and Cy5 channels over time. (**C**) Ensemble FRET efficiency

plot highlighting a non-native intermediate state of S4 binding (orange box). (**D**) Proposed model of rRNA rearrangements upon S4 binding (bottom panel). (**E–G**) **Real-time tracking of multiple processes occurring during co-transcriptional ribosome assembly:** (**E**) experimental setups for simultaneously detecting transcription progression (left), specific protein binding kinetics (centre), and RNA conformational changes (right). (**F**) Multicolour single-molecule trace showing real-time transcription progression, long-range rRNA helix-28 (H28) formation, and transient binding of r-protein S7. (**G**) Quantification of single-molecule data from experiments shown in (**E,F**) under different conditions: the plots show the efficiency of H28 formation (top) and the efficiency of S7 binding to the subset of molecules that have H28 formed (bottom). (**A–D**) are reproduced with permission from [94] and (**E–G**) are reproduced and (**F**) is adapted with permission from [11].

6.1. Multicolour Single-Molecule Fluorescence Microscopy

Some of the initial single-molecule experiments investigated the folding of the H20–H21–H22 three-way junction of 16S rRNA on S15 binding [109]. The three-way junction was immobilised using one of the helices, and the other two helices (H22 and H21) were labelled with a donor and acceptor dye, respectively. In the absence of S15, the three helices adapt a planar conformation that results in the limited transfer of energy from donor to acceptor (low FRET). However, in the presence of S15, the helices form a non-planar tertiary structure that brings the two dyes closer and leads to high FRET efficiency. Further, using a fast buffer-exchanging system, the authors titrated the levels of Mg^{2+} ions to determine that the three-way junction reacts instantaneously to Mg^{2+} ion levels.

One and a half decades later, more sophisticated multicolour experiments allowed for the visualisation of multiple processes at the same time, specifically the simultaneous tracking of rRNA folding and r-protein binding. Kim et al. investigated the binding of S4 (primary binding r-protein) to a 5-way junction (5WJ) in the 5′ domain of 16S rRNA (Figure 6A) [94]. They used a similar helix labelling system as described above for H3 and H16, and additionally labelled the r-protein S4 with another acceptor. Using this approach, they showed that S4 initially binds in a low FRET state (non-native conformation) and then later transitions into a high FRET state (native conformation) (Figure 6B,C). Performing similar experiments on the 5WJ indicated that helix H3 initially adopts a flipped conformation that recruits S4. This then enables H3 to dock onto S4 and assume a native conformation, suggesting that S4 guides rRNA folding (Figure 6D).

Interestingly, similar experiments applied to the initial binding of S15 to the central domain H20–H21–H22 junction showed that the binding of S15 leads directly into a high FRET state that does not change over time [95]. This suggests that, unlike that of S4, the S15 binding site immediately folds into its native conformation upon recruitment of the primary binding protein S15.

Further multicolour experiments on the 5′-domain system highlighted that r-proteins can efficiently change the rRNA folding landscape [110]. Monitoring the recruitment of S4, S20, and S16 showed that S16 can be stably recruited to a complex consisting of S4 and S20. The stable recruitment of S16 leads to conformational changes that enable H12 to interact with H3, which prevents H3 from flipping out and stabilising the native conformation. Overall, these experiments showed that r-protein binding changes the energy landscape such that only certain barriers can be crossed and, thus, limits the conformational search space.

6.2. Co-Transcriptional Single-Molecule Imaging

Ribosome assembly occurs co-transcriptionally, and, thus, the processes of rRNA folding and r-protein binding are linked to transcription [41–44,111]. Duss et al. developed a method to simultaneously monitor the process of transcription elongation and r-protein binding to a nascent rRNA directly emerging from the RNAP [11,95]. To this end, a stalled transcription elongation complex was formed that consists of a DNA template labelled with dyes at the 3′ end, native *E. coli* RNAP, and nascent rRNA (Figure 6E, left panel). This stalled complex was obtained by initiating transcription using only three out of the four

NTPs on a sequence missing the fourth nucleotide. The stalled transcription complex was then immobilised to the imaging surface through the 5′ end of its nascent RNA using a complementary biotinylated probe. The experiment was initiated by the addition of all four NTPs. The progression of transcription brings the fluorescently labelled 3′ end of the DNA template closer to the surface, which leads to an exponential increase in signal intensity (Figure 6E, left panel) as a result of an exponential increase in excitation in the evanescent field generated by total internal reflection when moving closer to the surface. A plateau in fluorescence intensity during transcription termination demonstrated that RNAP can stall for a few seconds before dissociating from the DNA template, which is identified as a sudden intensity drop in its signal to zero [95]. The authors then monitored, using real-time transcription, the elongation of the 16S rRNA H20–H21–H23 three-way junction and, simultaneously, the binding kinetics of S15 to the nascent RNA (Figure 6E, centre panel). They found that S15 can only bind once the full-length three-way junction RNA has been transcribed. A detailed characterisation of the S15 binding events revealed three populations of nascent RNA molecules: (1) natively folded RNA molecules that stably bound S15 immediately upon transcription of the full-length three-way junction, (2) partially folded RNA molecules that bound S15 transiently, and (3) misfolded RNA molecules that did not bind S15 at all. They further showed that pre-transcribed RNA has distinct properties compared to co-transcriptionally folded RNA [95].

While this study indirectly reported on RNA folding using protein binding kinetics as a read-out, direct information on rRNA folding was missing. In a follow up study, the authors developed an approach that allows for the simultaneous tracking of (1) transcription elongation, (2) the co-transcriptional folding of nascent RNA, and (3) the binding of one or two proteins to nascent RNA (Figure 6E) [11]. Studying the 3′ domain of 16S rRNA showed that the primary binding r-protein S7 first engages transiently with nascent RNA before becoming stably incorporated, which happens upon binding of the secondary and tertiary binding proteins. Furthermore, the authors observed that the binding of S7 was more efficient on smaller constructs as opposed to the full-length 3′ domain, indicating a higher tendency of longer rRNA to misfold and thereby preventing r-protein binding. Four-colour experiments then showed that the binding of S7 directly depends on the formation of a long-range helix (H28), which forms more efficiently if less RNA needs to be transcribed before the 5′ and 3′ halves of this helix can meet to form the long-range helix (Figure 6F,G). This directly demonstrated that the formation of long-range RNA interactions are impeded by the 5′ to 3′ directional process of transcription [112]. Remarkably, rRNA folding efficiency increased in the presence of the 3′-domain binding r-proteins, indicating that r-proteins can chaperone rRNA folding and guide the energy landscape of ribosome assembly.

A similar study on the 5′ domain of 16S rRNA showed that the primary binding r-protein S4 binds transiently to the transcribing rRNA, whereas S4 could bind stably to pre-transcribed rRNA [10]. This suggests that structures formed early during transcription are not competent to stably recruit S4. They also found that the addition of secondary binding r-proteins led to longer-lived S4 binding events. These studies together suggest that r-protein binding-based rRNA remodelling is a general mechanism of ribosome assembly.

Other approaches to track co-transcriptional RNA folding have also been developed but have not been applied yet to study co-transcriptional ribosome assembly. For example, forming an artificial transcription bubble, the authors of one study were able to introduce two different fluorescent labels site-specifically into nascent RNA [113] to study the co-transcriptional folding of a thiamine pyrophosphate (TPP) riboswitch. A FRET signal was used to study different conformational states of the aptamer assumed during transcription, in the presence and absence of the TPP ligand. In a similar approach, an azido UTP was site-specifically introduced into RNA and linked to a dye using copper-free click chemistry [114]. This approach revealed an inverse relationship between transcription speed and the metabolite-dependent folding of TPP riboswitch.

In another elegant study, a superhelicase was used to simulate and study the co-transcriptional folding of an RNA ribozyme [115]. First, a fully transcribed RNA, site-specifically labelled with two dyes, was hybridised with a complementary strand of DNA. This RNA–DNA hybrid was then immobilised to the surface for single-molecule imaging in the presence of the superhelicase. Transcription was mimicked using the addition of ATP, which triggered helicase activity to make the RNA single-stranded in the direction from the 5′ to 3′. They were able to investigate the RNA transitioning from a single-stranded state (low FRET) to a secondary folded (intermediate FRET) and a tertiary folded state (high FRET). Helicase activity can potentially be matched to transcription speed, but it still lacks the native transcriptional pausing that can directly influence RNA folding.

6.3. Optical Tweezers

While single-molecule fluorescence microscopy studies are powerful for tracking co-transcriptional RNA folding and the binding of proteins simultaneously at relatively high throughput, they lack the ability for tracking transcription elongation at single-nucleotide resolution. Optical tweezers, instead, can trap biomolecules—for example, transcription complexes between two beads—and allow for the observation of transcription progression [116] and RNAP pausing at single-nucleotide resolution [117]. Optical tweezers have been used to characterise real-time co-transcriptional RNA folding to understand the switching function of the adenine riboswitch and the resultant changes in RNA conformation upon ligand binding [118]. Optical tweezers also provide information on the forces exerted by biomolecules. For example, to understand how r-proteins stabilise rRNA structures, they mechanically unfolded and folded an irregular stem in domain II of 23S rRNA [119] in the presence and absence of r-protein L20. They found that L20 made the rRNA more resistant to mechanical unfolding by acting as a clamp around both strands of the rRNA stem.

Overall, single-molecule methods are very sensitive and provide direct and quantitative information. They inherently resolve biological heterogeneity and provide high temporal resolutions for tracking small and fast conformational changes of flexible regions that are averaged-out by conventional structural methods. Importantly, they provide information on how several different processes are functionally coupled with each other and how different assembly intermediates are placed along a reaction coordinate.

7. Integrative Methods

Multiple different biochemical, structural, and biophysical methods have been employed in studying the complex, multistep process of ribosome assembly. Yet, none of the methods can independently provide information on the entire process. Here, we highlight a few selected examples that integrate various methods.

In order to study the assembly mechanism of the bacterial 50S subunit in vivo, Davis et al. used a depleted bL17 strain to accumulate 50S assembly intermediates [73]. High-resolution cryoEM was used to determine the structures of 13 assembly intermediates. However, missing densities in the structures of these immature particles precludes the ability to obtain information on RNA structure and the associated proteins in these presumably dynamic regions. They used SHAPE-MaP-based chemical structural probing data to determine that, in these assembly intermediates, the 23S rRNA had a native secondary structure. Interestingly, sequencing reads also showed that some of the rRNA was not completely processed in the assembly intermediates. This is in agreement with previous reports that suggest final rRNA maturation occurs very late in assembly [1,120,121]. In order to provide information on the protein composition of the structural blocks that were missing in the cryoEM maps, they performed qMS and showed that the majority of r-proteins are already bound to these dynamic regions, and these blocks just need to be docked to the rest of the subunit to become a mature 50S subunit. Finally, one of the major drawbacks of structural methods is their inability to give direct information on function. In this case, to determine if assembly intermediates are capable of maturing into functional subunits, Davis et al. pulse-labelled bL17-depleted cells with heavy labelled media and simultaneously induced bL17 production. As expected, the peak in the sucrose gradient of the bL17-depleted assembly intermediates disappeared completely and the native 70S peak increased in intensity. This native 70S peak had heavy labelled bL17 incorporated, indicating that the addition of bL17 can rescue the intermediate and complete the maturation process.

In another study, Soper et al. used a combination of hydroxyl radical footprinting and qMS to understand the role of cellular factors in RNA folding and ribosome-assembly quality control [59]. Hydroxyl radical footprinting experiments showed how the assembly factor RimM reduces the misfolding of 16S head during transcription in vivo. Instead, qMS allowed them to confirm that, in absence of RimM/RbfA, some tertiary r-proteins are missing from the assembly intermediates. Further, they observed that the acetylation state of S18 directly correlates with the folding of rRNA and the formation of specific RNA–protein contacts during assembly.

A more recent study used native co-transcriptional in vitro reconstitutions in cell extract (iSAT) and characterised 50S assembly intermediates using time-resolved cryoEM and qMS to quantify both r-protein composition and the status of rRNA modifications during assembly [46]. The structures derived from the iSAT reaction were highly heterogenous. Thirteen structures were classified, spanning from one of the smallest known assembly intermediates detected to date (made of 600 nts and 3 r-proteins) to the latest stages of assembly with a nearly complete 50S subunit. Remarkably, studies that perform in vitro reconstitutions from purified components [72,122], co-transcriptional in vitro reconstitutions with cell extract [46], and characterising intermediates in vivo [73] show similar assembly intermediates, providing a general consensus on the mechanism of 50S assembly.

Overall, integrating multiple methods is very powerful and crucial for mechanistically understanding ribosome assembly and the assembly of other RNPs in detail.

8. Future Methods

The combination of different biochemical, biophysical, and structural approaches has allowed us to understand in great detail how the very complex process of ribosome assembly works at the molecular level. Moving forward, the major challenges to solve are (1) understanding how different processes in ribosome assembly are functionally coupled with each other and (2) visualising the structure and dynamics of ribosome assembly in the dense native cellular environment. In the following section, we will discuss emerging methods that we think will help to address these challenges.

8.1. Multicolour and Multiscale Single-Molecule Methods

Single-molecule methods are uniquely suited for understanding how different processes are functionally coupled with each other. The multicolour single-molecule fluorescence microscopy approaches discussed above demonstrate the potential to track multiple processes simultaneously. For example, they allow us to understand how transcription, RNA folding, and protein binding are directly interconnected [11]. Moving forward, more-complex in vitro reconstitutions that include more factors and processes will become accessible. Furthermore, experiments in cell extracts that contain all cellular factors will bridge the gap with in vivo experiments.

Apart from developing more-complex multicolour single-molecule fluorescence experiments, the future will also include combining single-molecule experiments with force experiments such as optical tweezers. For example, combining the two single-molecule modalities may allow for the tracking of transcription elongation and RNA folding at single-nucleotide resolution and, in addition, correlate the binding of one or two proteins to co-transcriptionally folding rRNA. In a recent study, the authors used force changes as a readout to monitor the individual codon translocation of ribosomes on mRNA or the unwinding of mRNA secondary structures by ribosomes, and simultaneously monitored the binding of fluorescently labelled elongation factor EF-G (Figure 7A) [123]. As a further extension of this technology, LUMICKS has extended the imaging part from single-colour to multicolour fluorescent microscopy [124]. However, despite its power in studying multiple processes simultaneously, this method lacks throughput. The optical tweezer technology can only study one complex at a time. To study very complex and heterogenous systems, such as ribosome assembly, efforts will be required to increase its throughput and automation, such as the commercial introduction of microfluids by LUMICKS.

Mass photometry imaging is another single-molecule method that uses interferometric scattering to determine the mass of individual molecules [125]. This, in combination with other methods, could be useful for studying the size distribution of assembly intermediates during different stages of assembly.

Recent advancements in direct RNA single-molecule nanopore sequencing may provide new opportunities for understanding how and when RNA modifications are introduced during ribosome assembly. In this technique, voltage is applied to a pore located in a membrane so the resulting ionic current can be detected [126]. When RNA passes through the pore, the detected current changes depending on which nucleotide is passing. Similarly, modified nucleotides also lead to a change in current that is specific to each RNA modification. In principle, this allows for the direct detection of all modifications present on a single molecule of RNA. The direct sequencing of 16S rRNA successfully detected the presence of m^7G and pseudouridine at the population level [127]. Current advances in data analysis methods have allowed for the study of multiple other modifications, such as and not limited to m^6A, m^5C, m^1G, m^62A, I, Nm, and $2'$-OMe [128–130]. Recent developments have highlighted the potential of nanopore sequencing for detecting multiple RNA modifications on the same molecule at single transcript resolution [128,131,132]. This opens up an avenue to investigate if there is a specific order in which RNA modifications are introduced. The chemical probing of RNA followed by direct RNA nanopore sequencing was used to predict RNA secondary structures [133]. Combining base modification detection with chemical probing-based RNA structure determination could allow for an investigation on how RNA modification and RNA structure formation are functionally coupled in RNP assembly [134].

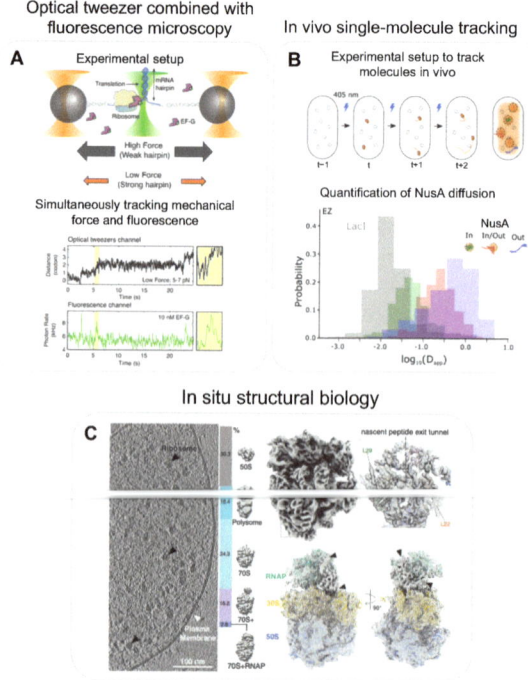

Figure 7. Multiscale single-molecule methods for studying RNP dynamics at nucleotide resolution: (**A**) experimental setup of optical tweezers combined with fluorescence microscopy for studying

mRNA unwinding during translation (top); time traces indicating change in distance upon one codon translation (centre) and changes in fluorescence intensity upon elongation factor binding (bottom). **In vivo single-molecule tracking to study spatial localisation and dynamics:** (**B**) experimental setup of in vivo single-molecule tracking experiment (top); quantification of tracking data by plotting the distribution of the apparent diffusion coefficients, indicating dynamic movements of transcription factor NusA within and outside the presumable transcription condensates. **In situ structural biology:** (**C**) representative tomographic slice of an *M. pneumoniae* cell and quantitative classification of ribosome subtomograms (left); resultant structures of 70S (top right), and RNAP-ribosome supercomplex (bottom right). Adapted and reproduced with permission: (**A**) from [123] and (**C**) from [135]. (**B**) is adapted and reproduced from [136].

8.2. In Vivo Single-Molecule Tracking

For in vivo single-molecule tracking, individual molecules are not tethered to the coverslip, but molecules of interest are endogenously tagged with a fluorescent reporter and tracked in real time while they are moving within a cell [137]. The majority of molecules are much too abundant in the cell to be tracked all at once as a result of the diffraction limit of light. Therefore, a small subset of the molecules can be photoactivated first and then excited with a different wavelength for tracking. One common endogenous tag, which can be linked to the protein of interest, is mMaple3 [138]. This photoconvertible protein is activated by illuminating at 405 nm and can then be imaged by exciting the protein at 561 nm. Some initial studies looked at the clustering of RNA polymerase (RNAP) using in vivo single-molecule localisation to characterise the RNAP organisation inside cells. Interestingly, RNAP localisation experiments showed that the spatial clustering of RNAP is independent of rRNA transcription activity, as opposed to what was suggested earlier, but rather dependent on the underlying nucleoid structure [139]. Pushing this further, the transcription factor NusA, which is part of the rrnTAC involved in early ribosome assembly, was tracked in vivo [136] (Figure 7B, top panel). By evaluating the different single-molecule tracks and converting them to apparent diffusion coefficients, the authors found that NusA diffuses in three states: slow-moving molecules were assigned to the NusA molecules associated with the transcription complex, fast-moving molecules as freely diffusing, and a third class with intermediate mobility was assigned to the NusA molecules present in a transcription condensate, which likely forms by liquid-liquid phase separation. The individual components can freely diffuse in and out of these clusters, indicating that the droplets are dynamic (Figure 7B, bottom panel). These studies provided evidence that not only does eukaryotic ribosome assembly occur in a biomolecule condensate (nucleolus), but that a similar condensed state may also organise bacterial ribosome assembly. Such a mechanism could explain the much higher ribosome-assembly efficiency in vivo compared to in vitro reconstitutions.

Similar experiments were also applied for studying eukaryotic ribosome assembly (which occurs in both the nucleolus and cytoplasm [140]), for example, to track the export of pre-60S particles from the nucleolus to the cytoplasm through the nuclear pore complex [141]. The authors observed that transport is a single rate-limiting step and takes about 24 ms on average. Furthermore, the quantification of exports from single pores revealed that only one third of export attempts are successful, and the overall mass flux can be as high as 125 MDa per second.

Similar experiments could, in the future, allow us to track the dynamics of individual r-proteins or assembly factors to gain a better understanding of ribosome assembly in vivo. While single-molecule tracking can be extended to more than one colour, and recent break-throughs with the MINFLUX technology have maximised spatiotemporal resolution to nanometre spatial and submillisecond temporal resolutions [142–144], the requirement for the stochastic activation of single fluorophores in an ocean of otherwise unlabelled molecules makes it very unlikely that two differently labelled molecules would interact with each other. Therefore, directly tracking individual protein–RNA interactions or macromolecular conformational changes in vivo will require new technologies to be developed.

8.3. Cryo–Electron Tomography

Cryo–electron tomography (cryoET) is an emerging method for gaining structural understanding directly in native cellular contexts. CryoET uses the same basic idea as single-particle cryoEM to reconstruct 3D images. The main difference is that, in tomography, an image is acquired by tilting the sample at multiple different angles [145]. This provides images of the sample at multiple different orientations, which can be used to reconstruct a 3D image for each individual particle. This is in contrast with single-particle cryoEM, which typically uses averaged information from hundreds of thousands of particles present in different orientations [66]. Thus, cryoET can be used to look at individual complexes inside whole cells or sections of cells, thereby preserving their native structure.

For example, Xue et al. were able to identify *Mycoplasma pneumoniae* ribosomes during various stages of translation and provide a detailed map of the translation elongation cycle within a single cell [31]. Importantly, they were able to identify the specific translation state for each ribosome in the cell, providing spatial functional information on its translation status. They were able to quantitatively show that 26% of all ribosomes in their study were polysomes and determine the orientation of each ribosome in the polysome with respect to each other and their overall packing density. By comparing the individual ribosomes within a polysome, they could determine that the r-protein L9 of the leading ribosome adopts an extended conformation, protruding into the binding site of the translation elongation factors of the trailing ribosome and thereby providing a mechanism for preventing ribosome collisions. Applying similar approaches to the study of bacterial ribosome assembly in cellular contexts will be challenging due to the low abundance of ribosome-assembly intermediates compared to fully assembled ribosomes. Imaging cells treated with antibiotics to accumulate ribosome-assembly intermediates could be the first step to tackle this challenging problem.

In another study from the Mahamid Lab, structures of an RNAP–ribosome supercomplex, termed expressome, were visualised in situ by combining cryoET with cross-linking mass spectrometry (Figure 7C, left panel) [135]. The structures showed for the first time how transcription–translation coupling is structurally organised in vivo (Figure 7C, right panel). They showed that the transcription factor NusA mediates coupling by physically linking the RNAP with the ribosome in *M. pneumoniae*. Furthermore, they visualised in high-resolution a state in which the ribosome has collided with the RNAP in the presence of an antibiotic that stalls the RNAP. Similar approaches could be used to visualise how bacterial ribosome assembly is coupled with transcription.

Eukaryotic ribosome assembly is separated from translation and takes place inside the nucleolus, which is a multiphasic biomolecular condensate that spatially organises maturing ribosome-assembly intermediates [140]. The Baumeister Lab used cryoET on native nucleoli of *Chlamydomonas reinhardtii* to show that pre-60S (LSU precursor) and SSU processome (SSU precursor) have different spatial localisation patterns. Furthermore, they classified three low-resolution structural assembly intermediates for each pre-60S and SSU processome. The maturation of these intermediates followed a gradient from the inside to the outside of the granular component [146].

Overall, these pioneering studies provide a starting point and demonstrate the potential for studying the complex process of ribosome assembly at high resolution in a native cellular context. Studying in vivo ribosome assembly could potentially answer questions such as the number of alternate pathways present in the assembly process and quantify the percentage flux in each of these pathways.

9. Conclusions

The assembly of a ribosome is a very complicated process involving the transcription, folding, modification, and processing of rRNA and the binding of dozens of r-proteins to nascent rRNA, assisted by dozens of assembly factors. Remarkably, the entire assembly process is completed within 2 min in the dense cellular environment. A plethora of biochemical, biophysical, and structural methods have helped further our understanding of this process in a quantitative manner: Sophisticated in vitro reconstitution systems in cell extracts that closely mimic the native process have been developed to bridge the gap between in vitro reconstitution from purified components and assembly in vivo. The use of pulse-chase quantitative mass spectrometry, time-resolved cryo–electron microscopy, and time-resolved RNA structure probing approaches has provided compositional and high-resolution structural data for understanding the kinetics of ribosome assembly and is instrumental in characterising multiple assembly intermediates along parallel assembly pathways. Recent multicolour single-molecule fluorescence experiments have shown the potential to follow how individual RNAs transcribe, simultaneously fold, and start to assemble into protein–RNA complexes in real time, providing information on how multiple different processes are functionally coupled with each other. Moving forward, in vivo single-molecule tracking, as well as cryo–electron tomography, will provide us with a much-needed understanding of how ribosomes assemble in their dense native cellular environment. Combining our efforts toward developing bottom-up reconstitutions of active systems that exhibit ever-increasing complexity with biophysical and structural approaches for visualising systems in vivo will bring us closer to understanding and, importantly, generating predictive models of how complex cellular processes work in a living cell [147].

Author Contributions: Conceptualisation, O.D. and K.G.; writing—original draft preparation, review, and editing, K.G. and O.D.; visualisation, K.G.; supervision, O.D.; funding acquisition, O.D. All authors have read and agreed to the published version of the manuscript.

Funding: This research was funded by the European Molecular Biology Laboratory.

Institutional Review Board Statement: Not applicable.

Informed Consent Statement: Not applicable.

Data Availability Statement: Not applicable.

Acknowledgments: We thank the entire Duss Lab for their helpful discussions.

Conflicts of Interest: The authors declare no conflict of interest.

References

1. Shajani, Z.; Sykes, M.T.; Williamson, J.R. Assembly of bacterial ribosomes. *Annu. Rev. Biochem.* **2011**, *80*, 501–526. [CrossRef] [PubMed]
2. Nierhaus, K.H. The assembly of prokaryotic ribosomes. *Biochimie* **1991**, *73*, 739–755. [CrossRef] [PubMed]
3. Huang, Y.H.; Said, N.; Loll, B.; Wahl, M.C. Structural basis for the function of SuhB as a transcription factor in ribosomal RNA synthesis. *Nucleic Acids Res.* **2019**, *47*, 6488–6503. [CrossRef] [PubMed]
4. Singh, N.; Bubunenko, M.; Smith, C.; Abbott, D.M.; Stringer, A.M.; Shi, R.; Court, D.L.; Wade, J.T. SuhB Associates with Nus Factors To Facilitate 30S Ribosome Biogenesis in *Escherichia coli*. *MBio* **2016**, *7*, e00114. [CrossRef]
5. Torres, M.; Condon, C.; Balada, J.M.; Squires, C.; Squires, C.L. Ribosomal protein S4 is a transcription factor with properties remarkably similar to NusA, a protein involved in both non-ribosomal and ribosomal RNA antitermination. *EMBO J.* **2001**, *20*, 3811–3820. [CrossRef]
6. Li, Z.; Pandit, S.; Deutscher, M.P. Maturation of 23S ribosomal RNA requires the exoribonuclease RNase T. *RNA* **1999**, *5*, 139–146. [CrossRef]
7. Li, Z.; Pandit, S.; Deutscher, M.P. RNase G (CafA protein) and RNase E are both required for the 5′ maturation of 16S ribosomal RNA. *EMBO J.* **1999**, *18*, 2878–2885. [CrossRef]
8. Ginsburg, D.; Steitz, J.A. The 30 S ribosomal precursor RNA from *Escherichia coli*. A primary transcript containing 23 S, 16 S, and 5 S sequences. *J. Biol. Chem.* **1975**, *250*, 5647–5654. [CrossRef]
9. King, T.C.; Schlessinger, D. S1 nuclease mapping analysis of ribosomal RNA processing in wild type and processing deficient *Escherichia coli*. *J. Biol. Chem.* **1983**, *258*, 12034–12042. [CrossRef]

10. Rodgers, M.L.; Woodson, S.A. Transcription Increases the Cooperativity of Ribonucleoprotein Assembly. *Cell* **2019**, *179*, 1370–1381.e1312. [CrossRef]
11. Duss, O.; Stepanyuk, G.A.; Puglisi, J.D.; Williamson, J.R. Transient Protein-RNA Interactions Guide Nascent Ribosomal RNA Folding. *Cell* **2019**, *179*, 1357–1369.e1316. [CrossRef] [PubMed]
12. Herschlag, D. RNA chaperones and the RNA folding problem. *J. Biol. Chem.* **1995**, *270*, 20871–20874. [CrossRef] [PubMed]
13. Adilakshmi, T.; Ramaswamy, P.; Woodson, S.A. Protein-independent folding pathway of the 16S rRNA 5′ domain. *J. Mol. Biol.* **2005**, *351*, 508–519. [CrossRef] [PubMed]
14. Powers, T.; Daubresse, G.; Noller, H.F. Dynamics of in vitro assembly of 16 S rRNA into 30 S ribosomal subunits. *J. Mol. Biol.* **1993**, *232*, 362–374. [CrossRef]
15. Talkington, M.W.; Siuzdak, G.; Williamson, J.R. An assembly landscape for the 30S ribosomal subunit. *Nature* **2005**, *438*, 628–632. [CrossRef]
16. Held, W.A.; Ballou, B.; Mizushima, S.; Nomura, M. Assembly mapping of 30 S ribosomal proteins from *Escherichia coli*. Further studies. *J. Biol. Chem.* **1974**, *249*, 3103–3111. [CrossRef]
17. Mizushima, S.; Nomura, M. Assembly mapping of 30S ribosomal proteins from *E. coli*. *Nature* **1970**, *226*, 1214. [CrossRef]
18. Rohl, R.; Nierhaus, K.H. Assembly map of the large subunit (50S) of *Escherichia coli* ribosomes. *Proc. Natl. Acad. Sci. USA* **1982**, *79*, 729–733. [CrossRef]
19. Decatur, W.A.; Fournier, M.J. rRNA modifications and ribosome function. *Trends Biochem. Sci.* **2002**, *27*, 344–351. [CrossRef]
20. Rodgers, M.L.; Woodson, S.A. A roadmap for rRNA folding and assembly during transcription. *Trends Biochem. Sci.* **2021**, *46*, 889–901. [CrossRef]
21. Britton, R.A. Role of GTPases in bacterial ribosome assembly. *Annu. Rev. Microbiol.* **2009**, *63*, 155–176. [CrossRef] [PubMed]
22. Chen, S.S.; Williamson, J.R. Characterization of the ribosome biogenesis landscape in *E. coli* using quantitative mass spectrometry. *J. Mol. Biol.* **2013**, *425*, 767–779. [CrossRef] [PubMed]
23. Lindahl, L. Intermediates and time kinetics of the in vivo assembly of *Escherichia coli* ribosomes. *J. Mol. Biol.* **1975**, *92*, 15–37. [CrossRef] [PubMed]
24. Davis, J.H.; Williamson, J.R. Structure and dynamics of bacterial ribosome biogenesis. *Philos. Trans. R. Soc. Lond. B Biol. Sci.* **2017**, *372*, 20160181. [CrossRef]
25. Naganathan, A.; Culver, G.M. Interdependency and Redundancy Add Complexity and Resilience to Biogenesis of Bacterial Ribosomes. *Annu. Rev. Microbiol.* **2022**, *76*, 193–210. [CrossRef]
26. Oborska-Oplova, M.; Gerhardy, S.; Panse, V.G. Orchestrating ribosomal RNA folding during ribosome assembly. *Bioessays* **2022**, *44*, e2200066. [CrossRef]
27. Sykes, M.T.; Williamson, J.R. A complex assembly landscape for the 30S ribosomal subunit. *Annu. Rev. Biophys.* **2009**, *38*, 197–215. [CrossRef]
28. Williamson, J.R. Biophysical studies of bacterial ribosome assembly. *Curr. Opin. Struct. Biol.* **2008**, *18*, 299–304. [CrossRef]
29. Woodson, S.A. RNA folding and ribosome assembly. *Curr. Opin. Chem. Biol.* **2008**, *12*, 667–673. [CrossRef]
30. Woodson, S.A. RNA folding pathways and the self-assembly of ribosomes. *Acc. Chem. Res.* **2011**, *44*, 1312–1319. [CrossRef]
31. Xue, L.; Lenz, S.; Zimmermann-Kogadeeva, M.; Tegunov, D.; Cramer, P.; Bork, P.; Rappsilber, J.; Mahamid, J. Visualizing translation dynamics at atomic detail inside a bacterial cell. *Nature* **2022**, *610*, 205–211. [CrossRef] [PubMed]
32. Culver, G.M.; Noller, H.F. Efficient reconstitution of functional *Escherichia coli* 30S ribosomal subunits from a complete set of recombinant small subunit ribosomal proteins. *RNA* **1999**, *5*, 832–843. [CrossRef] [PubMed]
33. Aoyama, R.; Masuda, K.; Shimojo, M.; Kanamori, T.; Ueda, T.; Shimizu, Y. In vitro reconstitution of the *Escherichia coli* 70S ribosome with a full set of recombinant ribosomal proteins. *J. Biochem.* **2022**, *171*, 227–237. [CrossRef] [PubMed]
34. Nierhaus, K.H.; Dohme, F. Total reconstitution of functionally active 50S ribosomal subunits from *Escherichia coli*. *Proc. Natl. Acad. Sci. USA* **1974**, *71*, 4713–4717. [CrossRef]
35. Rheinberger, H.J.; Nierhaus, K.H. Partial release of AcPhe-Phe-tRNA from ribosomes during poly(U)-dependent poly(Phe) synthesis and the effects of chloramphenicol. *Eur. J. Biochem.* **1990**, *193*, 643–650. [CrossRef]
36. Tamaru, D.; Amikura, K.; Shimizu, Y.; Nierhaus, K.H.; Ueda, T. Reconstitution of 30S ribosomal subunits in vitro using ribosome biogenesis factors. *RNA* **2018**, *24*, 1512–1519. [CrossRef]
37. Semrad, K.; Green, R. Osmolytes stimulate the reconstitution of functional 50S ribosomes from in vitro transcripts of *Escherichia coli* 23S rRNA. *RNA* **2002**, *8*, 401–411. [CrossRef]
38.
39. Nomura, M.; Traub, P.; Guthrie, C.; Nashimoto, H. The assembly of ribosomes. *J. Cell Physiol.* **1969**, *74* (Suppl. S1), 241. [CrossRef]
40. Jewett, M.C.; Fritz, B.R.; Timmerman, L.E.; Church, G.M. In vitro integration of ribosomal RNA synthesis, ribosome assembly, and translation. *Mol. Syst. Biol.* **2013**, *9*, 678. [CrossRef]
41. Miller, O.L., Jr.; Hamkalo, B.A.; Thomas, C.A., Jr. Visualization of bacterial genes in action. *Science* **1970**, *169*, 392–395. [CrossRef] [PubMed]
42. Miller, O.L., Jr.; Beatty, B.R. Visualization of nucleolar genes. *Science* **1969**, *164*, 955–957. [CrossRef]

43. Hofmann, S.; Miller, O.L., Jr. Visualization of ribosomal ribonucleic acid synthesis in a ribonuclease III-Deficient strain of *Escherichia coli*. *J. Bacteriol.* **1977**, *132*, 718–722. [CrossRef] [PubMed]
44. Gotta, S.L.; Miller, O.L., Jr.; French, S.L. rRNA transcription rate in *Escherichia coli*. *J. Bacteriol.* **1991**, *173*, 6647–6649. [CrossRef] [PubMed]
45. Fritz, B.R.; Jewett, M.C. The impact of transcriptional tuning on in vitro integrated rRNA transcription and ribosome construction. *Nucleic Acids Res.* **2014**, *42*, 6774–6785. [CrossRef]
46. Dong, X.; Doerfel, L.K.; Sheng, K.; Rabuck-Gibbons, J.N.; Popova, A.M.; Lyumkis, D.; Williamson, J.R. Near-physiological in vitro assembly of 50S ribosomes involves parallel pathways. *Nucleic Acids Res.* **2023**, *51*, 2862–2876. [CrossRef]
47. Fritz, B.R.; Jamil, O.K.; Jewett, M.C. Implications of macromolecular crowding and reducing conditions for in vitro ribosome construction. *Nucleic Acids Res.* **2015**, *43*, 4774–4784. [CrossRef]
48. Shimojo, M.; Amikura, K.; Masuda, K.; Kanamori, T.; Ueda, T.; Shimizu, Y. In vitro reconstitution of functional small ribosomal subunit assembly for comprehensive analysis of ribosomal elements in *E. coli*. *Commun. Biol.* **2020**, *3*, 142. [CrossRef]
49. Levy, M.; Falkovich, R.; Daube, S.S.; Bar-Ziv, R.H. Autonomous synthesis and assembly of a ribosomal subunit on a chip. *Sci. Adv.* **2020**, *6*, eaaz6020. [CrossRef]
50. d'Aquino, A.E.; Azim, T.; Aleksashin, N.A.; Hockenberry, A.J.; Kruger, A.; Jewett, M.C. Mutational characterization and mapping of the 70S ribosome active site. *Nucleic Acids Res.* **2020**, *48*, 2777–2789. [CrossRef]
51. Liu, Y.; Davis, R.G.; Thomas, P.M.; Kelleher, N.L.; Jewett, M.C. In vitro-Constructed Ribosomes Enable Multi-site Incorporation of Noncanonical Amino Acids into Proteins. *Biochemistry* **2021**, *60*, 161–169. [CrossRef] [PubMed]
52. Hammerling, M.J.; Fritz, B.R.; Yoesep, D.J.; Kim, D.S.; Carlson, E.D.; Jewett, M.C. In vitro ribosome synthesis and evolution through ribosome display. *Nat. Commun.* **2020**, *11*, 1108. [CrossRef] [PubMed]
53. Bunner, A.E.; Beck, A.H.; Williamson, J.R. Kinetic cooperativity in *Escherichia coli* 30S ribosomal subunit reconstitution reveals additional complexity in the assembly landscape. *Proc. Natl. Acad. Sci. USA* **2010**, *107*, 5417–5422. [CrossRef] [PubMed]
54. Bunner, A.E.; Nord, S.; Wikstrom, P.M.; Williamson, J.R. The effect of ribosome assembly cofactors on in vitro 30S subunit reconstitution. *J. Mol. Biol.* **2010**, *398*, 1–7. [CrossRef]
55. Nord, S.; Bylund, G.O.; Lovgren, J.M.; Wikstrom, P.M. The RimP protein is important for maturation of the 30S ribosomal subunit. *J. Mol. Biol.* **2009**, *386*, 742–753. [CrossRef]
56. Bylund, G.O.; Persson, B.C.; Lundberg, L.A.; Wikstrom, P.M. A novel ribosome-associated protein is important for efficient translation in *Escherichia coli*. *J. Bacteriol.* **1997**, *179*, 4567–4574. [CrossRef]
57. Inoue, K.; Alsina, J.; Chen, J.; Inouye, M. Suppression of defective ribosome assembly in a rbfA deletion mutant by overexpression of Era, an essential GTPase in *Escherichia coli*. *Mol. Microbiol.* **2003**, *48*, 1005–1016. [CrossRef]
58. Gibbs, M.R.; Moon, K.M.; Chen, M.; Balakrishnan, R.; Foster, L.J.; Fredrick, K. Conserved GTPase LepA (Elongation Factor 4) functions in biogenesis of the 30S subunit of the 70S ribosome. *Proc. Natl. Acad. Sci. USA* **2017**, *114*, 980–985. [CrossRef]
59. Clatterbuck Soper, S.F.; Dator, R.P.; Limbach, P.A.; Woodson, S.A. In vivo X-ray footprinting of pre-30S ribosomes reveals chaperone-dependent remodeling of late assembly intermediates. *Mol. Cell* **2013**, *52*, 506–516. [CrossRef]
60. Sailer, C.; Jansen, J.; Sekulski, K.; Cruz, V.E.; Erzberger, J.P.; Stengel, F. A comprehensive landscape of 60S ribosome biogenesis factors. *Cell Rep.* **2022**, *38*, 110353. [CrossRef]
61. Motorin, Y.; Muller, S.; Behm-Ansmant, I.; Branlant, C. Identification of modified residues in RNAs by reverse transcription-based methods. *Methods Enzymol.* **2007**, *425*, 21–53. [CrossRef] [PubMed]
62. Siibak, T.; Remme, J. Subribosomal particle analysis reveals the stages of bacterial ribosome assembly at which rRNA nucleotides are modified. *RNA* **2010**, *16*, 2023–2032. [CrossRef] [PubMed]
63. Kellner, S.; Ochel, A.; Thuring, K.; Spenkuch, F.; Neumann, J.; Sharma, S.; Entian, K.D.; Schneider, D.; Helm, M. Absolute and relative quantification of RNA modifications via biosynthetic isotopomers. *Nucleic Acids Res.* **2014**, *42*, e142. [CrossRef] [PubMed]
64. Popova, A.M.; Williamson, J.R. Quantitative analysis of rRNA modifications using stable isotope labeling and mass spectrometry. *J. Am. Chem. Soc.* **2014**, *136*, 2058–2069. [CrossRef] [PubMed]
65. Ishiguro, K.; Arai, T.; Suzuki, T. Depletion of S-adenosylmethionine impacts on ribosome biogenesis through hypomodification of a single rRNA methylation. *Nucleic Acids Res.* **2019**, *47*, 4226–4239. [CrossRef] [PubMed]
66. Murata, K.; Wolf, M. Cryo-electron microscopy for structural analysis of dynamic biological macromolecules. *Biochim. Biophys. Acta Gen. Subj.* **2018**, *1862*, 324–334. [CrossRef] [PubMed]
67. Mulder, A.M.; Yoshioka, C.; Beck, A.H.; Bunner, A.E.; Milligan, R.A.; Potter, C.S.; Carragher, B.; Williamson, J.R. Visualizing ribosome biogenesis: Parallel assembly pathways for the 30S subunit. *Science* **2010**, *330*, 673–677. [CrossRef]
68. Nikolay, R.; Hilal, T.; Schmidt, S.; Qin, B.; Schwefel, D.; Vieira-Vieira, C.H.; Mielke, T.; Burger, J.; Loerke, J.; Amikura, K.; et al. Snapshots of native pre-50S ribosomes reveal a biogenesis factor network and evolutionary specialization. *Mol. Cell* **2021**, *81*, 1200–1215.e1209. [CrossRef]
69. Huang, Y.H.; Hilal, T.; Loll, B.; Burger, J.; Mielke, T.; Bottcher, C.; Said, N.; Wahl, M.C. Structure-Based Mechanisms of a Molecular RNA Polymerase/Chaperone Machine Required for Ribosome Biosynthesis. *Mol. Cell* **2020**, *79*, 1024–1036.e1025. [CrossRef]
70. Kuhlbrandt, W. Biochemistry. The resolution revolution. *Science* **2014**, *343*, 1443–1444. [CrossRef]

71. Callaway, E. 'It opens up a whole new universe': Revolutionary microscopy technique sees individual atoms for first time. *Nature* **2020**, *582*, 156–157. [CrossRef] [PubMed]
72. Nikolay, R.; Hilal, T.; Qin, B.; Mielke, T.; Burger, J.; Loerke, J.; Textoris-Taube, K.; Nierhaus, K.H.; Spahn, C.M.T. Structural Visualization of the Formation and Activation of the 50S Ribosomal Subunit during In Vitro Reconstitution. *Mol. Cell* **2018**, *70*, 881–893.e883. [CrossRef] [PubMed]
73. Davis, J.H.; Tan, Y.Z.; Carragher, B.; Potter, C.S.; Lyumkis, D.; Williamson, J.R. Modular Assembly of the Bacterial Large Ribosomal Subunit. *Cell* **2016**, *167*, 1610–1622.e1615. [CrossRef] [PubMed]
74. Zhong, E.D.; Bepler, T.; Berger, B.; Davis, J.H. CryoDRGN: Reconstruction of heterogeneous cryo-EM structures using neural networks. *Nat. Methods* **2021**, *18*, 176–185. [CrossRef] [PubMed]
75. Wang, W.; Li, W.; Ge, X.; Yan, K.; Mandava, C.S.; Sanyal, S.; Gao, N. Loss of a single methylation in 23S rRNA delays 50S assembly at multiple late stages and impairs translation initiation and elongation. *Proc. Natl. Acad. Sci. USA* **2020**, *117*, 15609–15619. [CrossRef] [PubMed]
76. Rabuck-Gibbons, J.N.; Popova, A.M.; Greene, E.M.; Cervantes, C.F.; Lyumkis, D.; Williamson, J.R. SrmB Rescues Trapped Ribosome Assembly Intermediates. *J. Mol. Biol.* **2020**, *432*, 978–990. [CrossRef] [PubMed]
77. Razi, A.; Davis, J.H.; Hao, Y.; Jahagirdar, D.; Thurlow, B.; Basu, K.; Jain, N.; Gomez-Blanco, J.; Britton, R.A.; Vargas, J.; et al. Role of Era in assembly and homeostasis of the ribosomal small subunit. *Nucleic Acids Res.* **2019**, *47*, 8301–8317. [CrossRef]
78. Thurlow, B.; Davis, J.H.; Leong, V.; Moraes, T.F.; Williamson, J.R.; Ortega, J. Binding properties of YjeQ (RsgA), RbfA, RimM and Era to assembly intermediates of the 30S subunit. *Nucleic Acids Res.* **2016**, *44*, 9918–9932. [CrossRef] [PubMed]
79. Guo, Q.; Goto, S.; Chen, Y.; Feng, B.; Xu, Y.; Muto, A.; Himeno, H.; Deng, H.; Lei, J.; Gao, N. Dissecting the in vivo assembly of the 30S ribosomal subunit reveals the role of RimM and general features of the assembly process. *Nucleic Acids Res.* **2013**, *41*, 2609–2620. [CrossRef]
80. Jomaa, A.; Stewart, G.; Martin-Benito, J.; Zielke, R.; Campbell, T.L.; Maddock, J.R.; Brown, E.D.; Ortega, J. Understanding ribosome assembly: The structure of in vivo assembled immature 30S subunits revealed by cryo-electron microscopy. *RNA* **2011**, *17*, 697–709. [CrossRef] [PubMed]
81. Leong, V.; Kent, M.; Jomaa, A.; Ortega, J. Escherichia coli rimM and yjeQ null strains accumulate immature 30S subunits of similar structure and protein complement. *RNA* **2013**, *19*, 789–802. [CrossRef] [PubMed]
82. Boehringer, D.; O'Farrell, H.C.; Rife, J.P.; Ban, N. Structural insights into methyltransferase KsgA function in 30S ribosomal subunit biogenesis. *J. Biol. Chem.* **2012**, *287*, 10453–10459. [CrossRef] [PubMed]
83. Zhang, X.; Yan, K.; Zhang, Y.; Li, N.; Ma, C.; Li, Z.; Zhang, Y.; Feng, B.; Liu, J.; Sun, Y.; et al. Structural insights into the function of a unique tandem GTPase EngA in bacterial ribosome assembly. *Nucleic Acids Res.* **2014**, *42*, 13430–13439. [CrossRef] [PubMed]
84. Bubunenko, M.; Court, D.L.; Al Refaii, A.; Saxena, S.; Korepanov, A.; Friedman, D.I.; Gottesman, M.E.; Alix, J.H. Nus transcription elongation factors and RNase III modulate small ribosomal subunit biogenesis in *Escherichia coli*. *Mol. Microbiol.* **2013**, *87*, 382–393. [CrossRef]
85. Redko, Y.; Condon, C. Ribosomal protein L3 bound to 23S precursor rRNA stimulates its maturation by Mini-III ribonuclease. *Mol. Microbiol.* **2009**, *71*, 1145–1154. [CrossRef]
86. Stahl, D.A.; Pace, B.; Marsh, T.; Pace, N.R. The ribonucleoprotein substrate for a ribosomal RNA-processing nuclease. *J. Biol. Chem.* **1984**, *259*, 11448–11453. [CrossRef]
87. Oerum, S.; Dendooven, T.; Catala, M.; Gilet, L.; Degut, C.; Trinquier, A.; Bourguet, M.; Barraud, P.; Cianferani, S.; Luisi, B.F.; et al. Structures of B. subtilis Maturation RNases Captured on 50S Ribosome with Pre-rRNAs. *Mol. Cell* **2020**, *80*, 227–236.e225. [CrossRef]
88. Stiegler, P.; Carbon, P.; Zuker, M.; Ebel, J.P.; Ehresmann, C. Structural organization of the 16S ribosomal RNA from *E. coli*. Topography and secondary structure. *Nucleic Acids Res.* **1981**, *9*, 2153–2172. [CrossRef]
89. Spitale, R.C.; Incarnato, D. Probing the dynamic RNA structurome and its functions. *Nat. Rev. Genet.* **2023**, *24*, 178–196. [CrossRef]
90. Deigan, K.E.; Li, T.W.; Mathews, D.H.; Weeks, K.M. Accurate SHAPE-directed RNA structure determination. *Proc. Natl. Acad. Sci. USA* **2009**, *106*, 97–102. [CrossRef]
91. Siegfried, N.; Busan, S.; Rice, G.M.; Nelson, J.A.; Weeks, K.M. RNA motif discovery by SHAPE and mutational profiling (SHAPE-MaP). *Nat. Methods* **2014**, *11*, 959–965. [CrossRef] [PubMed]
92. Adilakshmi, T.; Bellur, D.L.; Woodson, S.A. Concurrent nucleation of 16S folding and induced fit in 30S ribosome assembly. *Nature* **2008**, *455*, 1268–1272. [CrossRef] [PubMed]
93. 16S rRNA. *Science* **1989**, *244*, 783–790. [CrossRef] [PubMed]
94. Kim, H.; Abeysirigunawarden, S.C.; Chen, K.; Mayerle, M.; Ragunathan, K.; Luthey-Schulten, Z.; Ha, T.; Woodson, S.A. Protein-guided RNA dynamics during early ribosome assembly. *Nature* **2014**, *506*, 334–338. [CrossRef]
95. Duss, O.; Stepanyuk, G.A.; Grot, A.; O'Leary, S.E.; Puglisi, J.D.; Williamson, J.R. Real-time assembly of ribonucleoprotein complexes on nascent RNA transcripts. *Nat. Commun.* **2018**, *9*, 5087. [CrossRef]
96. Olson, S.W.; Turner, A.W.; Arney, J.W.; Saleem, I.; Weidmann, C.A.; Margolis, D.M.; Weeks, K.M.; Mustoe, A.M. Discovery of a large-scale, cell-state-responsive allosteric switch in the 7SK RNA using DANCE-MaP. *Mol. Cell* **2022**, *82*, 1708–1723.e1710. [CrossRef]

97. Morandi, E.; Manfredonia, I.; Simon, L.M.; Anselmi, F.; van Hemert, M.J.; Oliviero, S.; Incarnato, D. Genome-scale deconvolution of RNA structure ensembles. *Nat. Methods* **2021**, *18*, 249–252. [CrossRef]
98. Tomezsko, P.J.; Corbin, V.D.A.; Gupta, P.; Swaminathan, H.; Glasgow, M.; Persad, S.; Edwards, M.D.; McIntosh, L.; Papenfuss, A.T.; Emery, A.; et al. Determination of RNA structural diversity and its role in HIV-1 RNA splicing. *Nature* **2020**, *582*, 438–442. [CrossRef]
99. Homan, P.J.; Favorov, O.V.; Lavender, C.A.; Kursun, O.; Ge, X.; Busan, S.; Dokholyan, N.V.; Weeks, K.M. Single-molecule correlated chemical probing of RNA. *Proc. Natl. Acad. Sci. USA* **2014**, *111*, 13858–13863. [CrossRef]
100. Khoroshkin, M.; Asarnow, D.; Navickas, A.; Winters, A.; Yu, J.; Zhou, S.K.; Zhou, S.; Palka, C.; Fish, L.; Ansel, K.M.; et al. A systematic search for RNA structural switches across the human transcriptome. *bioRxiv* **2023**. [CrossRef]
101. Mitchell, D., 3rd; Assmann, S.M.; Bevilacqua, P.C. Probing RNA structure in vivo. *Curr. Opin. Struct. Biol.* **2019**, *59*, 151–158. [CrossRef] [PubMed]
102. Watters, K.E.; Strobel, E.J.; Yu, A.M.; Lis, J.T.; Lucks, J.B. Cotranscriptional folding of a riboswitch at nucleotide resolution. *Nat. Struct. Mol. Biol.* **2016**, *23*, 1124–1131. [CrossRef] [PubMed]
103. Lewicki, B.T.; Margus, T.; Remme, J.; Nierhaus, K.H. Coupling of rRNA transcription and ribosomal assembly in vivo. Formation of active ribosomal subunits in *Escherichia coli* requires transcription of rRNA genes by host RNA polymerase which cannot be replaced by bacteriophage T7 RNA polymerase. *J. Mol. Biol.* **1993**, *231*, 581–593. [CrossRef] [PubMed]
104. Hulscher, R.M.; Bohon, J.; Rappe, M.C.; Gupta, S.; D'Mello, R.; Sullivan, M.; Ralston, C.Y.; Chance, M.R.; Woodson, S.A. Probing the structure of ribosome assembly intermediates in vivo using DMS and hydroxyl radical footprinting. *Methods* **2016**, *103*, 49–56. [CrossRef]
105. Mortimer, S.A.; Weeks, K.M. Time-resolved RNA SHAPE chemistry. *J. Am. Chem. Soc.* **2008**, *130*, 16178–16180. [CrossRef]
106. Roy, R.; Hohng, S.; Ha, T. A practical guide to single-molecule FRET. *Nat. Methods* **2008**, *5*, 507–516. [CrossRef]
107. Feng, X.A.; Poyton, M.F.; Ha, T. Multicolor single-molecule FRET for DNA and RNA processes. *Curr. Opin. Struct. Biol.* **2021**, *70*, 26–33. [CrossRef]
108. Lerner, E.; Cordes, T.; Ingargiola, A.; Alhadid, Y.; Chung, S.; Michalet, X.; Weiss, S. Toward dynamic structural biology: Two decades of single-molecule Forster resonance energy transfer. *Science* **2018**, *359*, eaan1133. [CrossRef]
109. Ha, T.; Zhuang, X.; Kim, H.D.; Orr, J.W.; Williamson, J.R.; Chu, S. Ligand-induced conformational changes observed in single RNA molecules. *Proc. Natl. Acad. Sci. USA* **1999**, *96*, 9077–9082. [CrossRef]
110. Abeysirigunawardena, S.C.; Kim, H.; Lai, J.; Ragunathan, K.; Rappe, M.C.; Luthey-Schulten, Z.; Ha, T.; Woodson, S.A. Evolution of protein-coupled RNA dynamics during hierarchical assembly of ribosomal complexes. *Nat. Commun.* **2017**, *8*, 492. [CrossRef]
111. Lai, D.; Proctor, J.R.; Meyer, I.M. On the importance of cotranscriptional RNA structure formation. *RNA* **2013**, *19*, 1461–1473. [CrossRef]
112. Zhang, J.; Landick, R. A Two-Way Street: Regulatory Interplay between RNA Polymerase and Nascent RNA Structure. *Trends. Biochem. Sci.* **2016**, *41*, 293–310. [CrossRef] [PubMed]
113. Uhm, H.; Kang, W.; Ha, K.S.; Kang, C.; Hohng, S. Single-molecule FRET studies on the cotranscriptional folding of a thiamine pyrophosphate riboswitch. *Proc. Natl. Acad. Sci. USA* **2018**, *115*, 331–336. [CrossRef] [PubMed]
114. Chauvier, A.; St-Pierre, P.; Nadon, J.F.; Hien, E.D.M.; Perez-Gonzalez, C.; Eschbach, S.H.; Lamontagne, A.M.; Penedo, J.C.; Lafontaine, D.A. Monitoring RNA dynamics in native transcriptional complexes. *Proc. Natl. Acad. Sci. USA* **2021**, *118*, e2106564118. [CrossRef] [PubMed]
115. Hua, B.; Panja, S.; Wang, Y.; Woodson, S.A.; Ha, T. Mimicking Co-Transcriptional RNA Folding Using a Superhelicase. *J. Am. Chem. Soc.* **2018**, *140*, 10067–10070. [CrossRef]
116. Wang, M.D.; Schnitzer, M.J.; Yin, H.; Landick, R.; Gelles, J.; Block, S.M. Force and velocity measured for single molecules of RNA polymerase. *Science* **1998**, *282*, 902–907. [CrossRef]
117. Abbondanzieri, E.A.; Greenleaf, W.J.; Shaevitz, J.W.; Landick, R.; Block, S.M. Direct observation of base-pair stepping by RNA polymerase. *Nature* **2005**, *438*, 460–465. [CrossRef]
118. Frieda, K.L.; Block, S.M. Direct observation of cotranscriptional folding in an adenine riboswitch. *Science* **2012**, *338*, 397–400. [CrossRef]
119. Mangeol, P.; Bizebard, T.; Chiaruttini, C.; Dreyfus, M.; Springer, M.; Bockelmann, U. Probing ribosomal protein-RNA interactions with an external force. *Proc. Natl. Acad. Sci. USA* **2011**, *108*, 18272–18276. [CrossRef]
120. Srivastava, A.K.; Schlessinger, D. Coregulation of processing and translation: Mature 5′ termini of *Escherichia coli* 23S ribosomal RNA form in polysomes. *Proc. Natl. Acad. Sci. USA* **1988**, *85*, 7144–7148. [CrossRef]
121. Mangiarotti, G.; Turco, E.; Ponzetto, A.; Altruda, F. Precursor 16S RNA in active 30S ribosomes. *Nature* **1974**, *247*, 147–148. [CrossRef] [PubMed]
122. Qin, B.; Lauer, S.M.; Balke, A.; Vieira-Vieira, C.H.; Burger, J.; Mielke, T.; Selbach, M.; Scheerer, P.; Spahn, C.M.T.; Nikolay, R. Cryo-EM captures early ribosome assembly in action. *Nat. Commun.* **2023**, *14*, 898. [CrossRef] [PubMed]
123. Desai, V.P.; Frank, F.; Lee, A.; Righini, M.; Lancaster, L.; Noller, H.F.; Tinoco, I., Jr.; Bustamante, C. Co-temporal Force and Fluorescence Measurements Reveal a Ribosomal Gear Shift Mechanism of Translation Regulation by Structured mRNAs. *Mol. Cell* **2019**, *75*, 1007–1019.e1005. [CrossRef] [PubMed]

124. C-TRAP® Optical Tweezers Fluorescence & Label-Free Microscopy. Available online: https://lumicks.com/products/c-trap-optical-tweezers-fluorescence-label-free-microscopy/ (accessed on 6 April 2023).
125. Young, G.; Hundt, N.; Cole, D.; Fineberg, A.; Andrecka, J.; Tyler, A.; Olerinyova, A.; Ansari, A.; Marklund, E.G.; Collier, M.P.; et al. Quantitative mass imaging of single biological macromolecules. *Science* **2018**, *360*, 423–427. [CrossRef] [PubMed]
126. Wang, Y.; Zhao, Y.; Bollas, A.; Wang, Y.; Au, K.F. Nanopore sequencing technology, bioinformatics and applications. *Nat. Biotechnol.* **2021**, *39*, 1348–1365. [CrossRef]
127. Smith, A.M.; Jain, M.; Mulroney, L.; Garalde, D.R.; Akeson, M. Reading canonical and modified nucleobases in 16S ribosomal RNA using nanopore native RNA sequencing. *PLoS ONE* **2019**, *14*, e0216709. [CrossRef]
128. Leger, A.; Amaral, P.P.; Pandolfini, L.; Capitanchik, C.; Capraro, F.; Miano, V.; Migliori, V.; Toolan-Kerr, P.; Sideri, T.; Enright, A.J.; et al. RNA modifications detection by comparative Nanopore direct RNA sequencing. *Nat. Commun.* **2021**, *12*, 7198. [CrossRef]
129. Begik, O.; Mattick, J.S.; Novoa, E.M. Exploring the epitranscriptome by native RNA sequencing. *RNA* **2022**, *28*, 1430–1439. [CrossRef]
130. Begik, O.; Lucas, M.C.; Pryszcz, L.P.; Ramirez, J.M.; Medina, R.; Milenkovic, I.; Cruciani, S.; Liu, H.; Vieira, H.G.S.; Sas-Chen, A.; et al. Quantitative profiling of pseudouridylation dynamics in native RNAs with nanopore sequencing. *Nat. Biotechnol.* **2021**, *39*, 1278–1291. [CrossRef]
131. Hendra, C.; Pratanwanich, P.N.; Wan, Y.K.; Goh, W.S.S.; Thiery, A.; Goke, J. Detection of m6A from direct RNA sequencing using a multiple instance learning framework. *Nat. Methods* **2022**, *19*, 1590–1598. [CrossRef]
132. Mateos, P.A.; Sethi, A.J.; Ravindran, A.; Guarnacci, M.; Srivastava, A.; Xu, J.; Woodward, K.; Yuen, Z.W.S.; Mahmud, S.; Kanchi, M.; et al. Simultaneous identification of m6A and m5C reveals coordinated RNA modification at single-molecule resolution. *bioRxiv* **2023**. [CrossRef]
133. Aw, J.G.A.; Lim, S.W.; Wang, J.X.; Lambert, F.R.P.; Tan, W.T.; Shen, Y.; Zhang, Y.; Kaewsapsak, P.; Li, C.; Ng, S.B.; et al. Determination of isoform-specific RNA structure with nanopore long reads. *Nat. Biotechnol.* **2021**, *39*, 336–346. [CrossRef]
134. Wrzesinski, J.; Bakin, A.; Nurse, K.; Lane, B.G.; Ofengand, J. Purification, cloning, and properties of the 16S RNA pseudouridine 516 synthase from *Escherichia coli*. *Biochemistry* **1995**, *34*, 8904–8913. [CrossRef]
135. O'Reilly, F.J.; Xue, L.; Graziadei, A.; Sinn, L.; Lenz, S.; Tegunov, D.; Blotz, C.; Singh, N.; Hagen, W.J.H.; Cramer, P.; et al. In-cell architecture of an actively transcribing-translating expressome. *Science* **2020**, *369*, 554–557. [CrossRef]
136. Ladouceur, A.M.; Parmar, B.S.; Biedzinski, S.; Wall, J.; Tope, S.G.; Cohn, D.; Kim, A.; Soubry, N.; Reyes-Lamothe, R.; Weber, S.C. Clusters of bacterial RNA polymerase are biomolecular condensates that assemble through liquid-liquid phase separation. *Proc. Natl. Acad. Sci. USA* **2020**, *117*, 18540–18549. [CrossRef] [PubMed]
137. Lelek, M.; Gyparaki, M.T.; Beliu, G.; Schueder, F.; Griffie, J.; Manley, S.; Jungmann, R.; Sauer, M.; Lakadamyali, M.; Zimmer, C. Single-molecule localization microscopy. *Nat. Rev. Methods Primers* **2021**, *1*, 39. [CrossRef] [PubMed]
138. Wang, S.; Moffitt, J.R.; Dempsey, G.T.; Xie, X.S.; Zhuang, X. Characterization and development of photoactivatable fluorescent proteins for single-molecule-based superresolution imaging. *Proc. Natl. Acad. Sci USA* **2014**, *111*, 8452–8457. [CrossRef] [PubMed]
139. Weng, X.; Bohrer, C.H.; Bettridge, K.; Lagda, A.C.; Cagliero, C.; Jin, D.J.; Xiao, J. Spatial organization of RNA polymerase and its relationship with transcription in *Escherichia coli*. *Proc. Natl. Acad. Sci. USA* **2019**, *116*, 20115–20123. [CrossRef]
140. Lafontaine, D.L.J. Birth of Nucleolar Compartments: Phase Separation-Driven Ribosomal RNA Sorting and Processing. *Mol. Cell* **2019**, *76*, 694–696. [CrossRef]
141. Ruland, J.A.; Kruger, A.M.; Dorner, K.; Bhatia, R.; Wirths, S.; Poetes, D.; Kutay, U.; Siebrasse, J.P.; Kubitscheck, U. Nuclear export of the pre-60S ribosomal subunit through single nuclear pores observed in real time. *Nat. Commun.* **2021**, *12*, 6211. [CrossRef]
142. Balzarotti, F.; Eilers, Y.; Gwosch, K.C.; Gynna, A.H.; Westphal, V.; Stefani, F.D.; Elf, J.; Hell, S.W. Nanometer resolution imaging and tracking of fluorescent molecules with minimal photon fluxes. *Science* **2017**, *355*, 606–612. [CrossRef] [PubMed]
143. Deguchi, T.; Iwanski, M.K.; Schentarra, E.M.; Heidebrecht, C.; Schmidt, L.; Heck, J.; Weihs, T.; Schnorrenberg, S.; Hoess, P.; Liu, S.; et al. Direct observation of motor protein stepping in living cells using MINFLUX. *Science* **2023**, *379*, 1010–1015. [CrossRef]
144. Wolff, J.O.; Scheiderer, L.; Engelhardt, T.; Engelhardt, J.; Matthias, J.; Hell, S.W. MINFLUX dissects the unimpeded walking of kinesin-1. *Science* **2023**, *379*, 1004–1010. [CrossRef] [PubMed]
145. Turk, M.; Baumeister, W. The promise and the challenges of cryo-electron tomography. *FEBS Lett.* **2020**, *594*, 3243–3261. [CrossRef] [PubMed]
146. Erdmann, P.S.; Hou, Z.; Klumpe, S.; Khavnekar, S.; Beck, F.; Wilfling, F.; Plitzko, J.M.; Baumeister, W. In situ cryo-electron tomography reveals gradient organization of ribosome biogenesis in intact nucleoli. *Nat. Commun.* **2021**, *12*, 5364. [CrossRef]
147. Earnest, T.M.; Lai, J.; Chen, K.; Hallock, M.J.; Williamson, J.R.; Luthey-Schulten, Z. Toward a Whole-Cell Model of Ribosome Biogenesis: Kinetic Modeling of SSU Assembly. *Biophys. J.* **2015**, *109*, 1117–1135. [CrossRef]

Disclaimer/Publisher's Note: The statements, opinions and data contained in all publications are solely those of the individual author(s) and contributor(s) and not of MDPI and/or the editor(s). MDPI and/or the editor(s) disclaim responsibility for any injury to people or property resulting from any ideas, methods, instructions or products referred to in the content.

Article

Loss of the DYRK1A Protein Kinase Results in the Reduction in Ribosomal Protein Gene Expression, Ribosome Mass and Reduced Translation

Chiara Di Vona [1,2,*,†], Laura Barba [1,2,†], Roberto Ferrari [3] and Susana de la Luna [1,2,4,5,*]

1. Centre for Genomic Regulation (CRG), The Barcelona Institute of Science and Technology (BIST), Dr Aiguader 88, 08003 Barcelona, Spain
2. Centro de Investigación Biomédica en Red en Enfermedades Raras (CIBERER), 28029 Madrid, Spain
3. Department of Chemistry, Life Sciences and Environmental Sustainability, University of Parma, Viale delle Scienze 23/A, 43124 Parma, Italy; roberto.ferrari1@unipr.it
4. Department of Medicine and Life Sciences, Universitat Pompeu Fabra (UPF), Dr Aiguader 88, 08003 Barcelona, Spain
5. Institució Catalana de Recerca i Estudis Avançats (ICREA), Passeig Lluís Companys 23, 08010 Barcelona, Spain
* Correspondence: chiara.divona@crg.eu (C.D.V.); susana.luna@crg.eu (S.d.l.L.)
† These authors contributed equally to this work.

Citation: Di Vona, C.; Barba, L.; Ferrari, R.; de la Luna, S. Loss of the DYRK1A Protein Kinase Results in the Reduction in Ribosomal Protein Gene Expression, Ribosome Mass and Reduced Translation. *Biomolecules* **2024**, *14*, 31. https://doi.org/10.3390/biom14010031

Academic Editors: Brigitte Pertschy and Ingrid Zierler

Received: 17 November 2023
Revised: 19 December 2023
Accepted: 21 December 2023
Published: 25 December 2023

Copyright: © 2023 by the authors. Licensee MDPI, Basel, Switzerland. This article is an open access article distributed under the terms and conditions of the Creative Commons Attribution (CC BY) license (https://creativecommons.org/licenses/by/4.0/).

Abstract: Ribosomal proteins (RPs) are evolutionary conserved proteins that are essential for protein translation. RP expression must be tightly regulated to ensure the appropriate assembly of ribosomes and to respond to the growth demands of cells. The elements regulating the transcription of RP genes (RPGs) have been characterized in yeast and *Drosophila*, yet how cells regulate the production of RPs in mammals is less well understood. Here, we show that a subset of RPG promoters is characterized by the presence of the palindromic TCTCGCGAGA motif and marked by the recruitment of the protein kinase DYRK1A. The presence of DYRK1A at these promoters is associated with the enhanced binding of the TATA-binding protein, TBP, and it is negatively correlated with the binding of the GABP transcription factor, establishing at least two clusters of RPGs that could be coordinately regulated. However, DYRK1A silencing leads to a global reduction in RPGs mRNAs, pointing at DYRK1A activities beyond those dependent on its chromatin association. Significantly, cells in which DYRK1A is depleted have reduced RP levels, fewer ribosomes, reduced global protein synthesis and a smaller size. We therefore propose a novel role for DYRK1A in coordinating the expression of genes encoding RPs, thereby controlling cell growth in mammals.

Keywords: ribosomal proteins; TCTCGCGAGA; translation; DYRK1A; transcription

1. Introduction

Ribosomes are cellular machines that translate mRNA into protein, and in mammals, they are formed by the large 60S subunit and the small 40S subunit. The 60S subunit is comprised of the 5S, 5.8S and 28S rRNAs associated with 52 ribosomal proteins (RPs), and the smaller 40S subunit is made up of the 18S rRNA plus 35 RPs. Ribosome biogenesis is a complex process that involves more than 200 different factors: rRNAs, small nucleolar RNAs and canonical and auxiliary RPs [1]. The three RNA polymerases (Pol) participate in the transcription of the ribosomal components, with Pol I responsible for transcribing the 28S, 18S and 5.8S rRNAs, Pol III transcribing the 5S rRNAs and Pol II responsible for the transcription of all the protein coding genes involved in ribosome biogenesis, including the RP genes (RPGs). Therefore, the coordinated expression of these components is required to ensure the correct assembly and proper functioning of ribosomes [2]. Indeed, the dysregulation of ribosome biogenesis is associated with a group of human diseases that are collectively known as ribosomopathies [3]; moreover, alterations to RP expression contribute to cancer cell growth [4].

The coding sequences of RPGs have been highly conserved over evolution, unlike the features of their promoters and the machinery involved in their transcriptional regulation. As such, RPGs are organized into operons in prokaryotes [5], whereas the situation is much more complex in the case of eukaryotes, with multiple genes widely scattered across the genome [6]. The main elements involved in the transcriptional regulation of RPGs have been characterized thoroughly in *Saccharomyces cerevisiae* [7], in which the repressor activator protein 1 (Rap1p) and the Fhl1p forkhead transcription factor (TF) are constitutively bound to the RPG promoters, coordinating RPG expression [8]. In higher eukaryotes, most studies have focused on the differential enrichment of TF binding motifs within RPG promoters [9–12]. In particular, several DNA sequences are found over-represented in human RPG promoters. The polypyrimidine TCT motif is found close to the transcription start site (TSS) of RPGs, and it is thought to play a dual role in the initiation of both transcription and translation [13]. This motif is recognized by the TATA-box-binding protein (TBP)-related factor 2 (TRF2) in *Drosophila* [14,15], yet it remains unclear whether there is functional conservation with its human TBP-like 1 (TBPL1) homolog. Around 35% of the RPG promoters contain a TATA box in the -25 region and an additional 25% contain A/T-rich sequences in this region [16]. Other motifs frequently found are those for SP1, the GA-binding protein (GABP) and the yin yang 1 (YY1) TFs [16]. In addition, the E-box TF MYC is a key regulator of ribosomal biogenesis, enhancing the expression of RPGs [17]. Finally, a de novo motif (M4 motif) was found enriched in human and mouse RPG promoters [9]. RPG mRNA expression displays tissue- and development-specific patterns, both in human and mouse [6,18,19]. Hence, RPG expression could be regulated by specific combinations of TFs in different organisms and/or physiological conditions.

The M4 motif matches the palindromic sequence that is bound by the dual-specificity tyrosine-regulated kinase 1A (DYRK1A) protein kinase [20]. DYRK1A fulfills many diverse functions by phosphorylating a wide range of substrates [21–23], and it is a kinase with exquisite gene-dosage dependency. On the one hand, DYRK1A overexpression in individuals with trisomy 21 has been associated to several of the pathological symptoms associated with Down syndrome (DS) [24]. On the other hand, de novo mutations in one *DYRK1A* allele cause a rare clinical syndrome known as DYRK1A haploinsufficiency syndrome (OMIM#614104) [25–27]. DYRK1A has also been proposed as a pharmacological target for neurodegenerative disorders, diabetes and cancer [22,23,28,29]. We have shown that DYRK1A is a transcriptional activator when recruited to proximal promoter regions of a subset of genes that are enriched for the palindromic motif TCTCGCGAGA [20]. DYRK1A phosphorylates serine residues 2, 5 and 7 within the C-terminal domain (CTD) of the catalytic subunit of Pol II [20]. This activity takes over that of the general TF p-TEFb at gene loci involved in myogenic differentiation [30]. The interaction of DYRK1A with the CTD depends on a run of histidine residues in its noncatalytic C-terminus, which also promotes the nucleation of a phase-separated compartment that is functionally associated with transcriptional elongation [31]. Here, we have analyzed the occupancy of RPG promoters by DYRK1A in depth, performing a comprehensive analysis of the promoter occupancy by other factors whose binding motifs are differentially enriched in human RPG promoter regions. Our results indicate that most of these factors are found at almost all RPG promoters, irrespective of the presence of their cognate binding sites. By contrast, DYRK1A associates with a subset of human and mouse RPG promoters that contain the TCTCGCGAGA motif. Moreover, physiological levels of DYRK1A are required to maintain RPG transcript levels independently of the binding of DYRK1A to their promoters, and this effect could at least in part contribute to the global reduction in ribosome mass and protein synthesis when DYRK1A is silenced. Therefore, our results expand the functional spectrum of the DYRK1A kinase, indicating that it contributes to the regulation of cell growth in mammalian cells.

2. Materials and Methods

2.1. Cell Culture and Lentivirus-Mediated Transduction

Lentiviral transduction of short hairpin (sh)RNAs was used to downregulate DYRK1A expression, and the generation of the lentiviral stocks and the infection conditions are detailed in the Supplementary Methods. The protocols to determine the cell cycle profile and cell volume are also included in the Supplementary Methods. To analyze global protein synthesis, T98G cells were incubated for 90 min in methionine-free Dulbecco's modified Eagle's medium (DMEM; GIBCO, Waltham, MA, USA) with 10% dialyzed fetal bovine serum (FBS; GIBCO), metabolically labeled for 20 min with ^{35}S-Met (50 µCi ^{35}S-Met, 1175 Ci/mmol, Perkin Elmer) and then lysed in SDS lysis buffer. The protein extracts were resolved by SDS-PAGE and the incorporation of ^{35}S-methionine was detected by the autoradiography of the dried gel using film or a Phosphoimager (Typhoon Trio, GE Healthcare, Chicago, IL, USA).

2.2. Preparation of Polysome and Ribosome-Enriched Fractions

Polysome profiles were obtained from approximately 1×10^7 T98G cells. Protein synthesis was arrested by incubation with cycloheximide (CHX, 100 µg/mL). The cells were washed in phosphate-buffered saline (PBS) containing CHX (100 µg/mL), collected in 1 mL of polysome lysis buffer (10 mM Tris-HCl with a pH of 7.4, 100 mM NaCl, 10 mM MgCl$_2$, 1% Triton X-100, 20 mM dithiothreitol [DTT], 100 µg/mL CHX, 0.25% sodium deoxycholate) and frozen rapidly in liquid nitrogen. Cell debris and nuclei were eliminated by centrifugation (12,000× g, 5 min, 4 °C) and the nucleic acid concentration in the supernatants was assessed by measuring the A$_{260}$ in a NanoDrop™ (Thermo Fisher Inc, Waltham, MA, USA). Samples with A$_{260}$ ≈ 10 were loaded onto a 10–50% linear sucrose gradient prepared in polysome gradient buffer (20 mM Tris-HCl with a pH of 7.4, 100 mM NH$_4$Cl, 10 mM MgCl$_2$, 0.5 mM DTT, 100 µg/mL CHX) and centrifuged in a Beckman SW41Ti rotor (35,000 rpm, 3 h, 4 °C). Profiles were obtained by continuous monitoring of the A$_{254}$ (Econo-UV Monitor and Econo-Recorder model 1327; Bio-Rad Laboratories, Hercules, CA, USA). To calculate the polysome:monosome ratio, the polysome and monosome area under the curve was measured with ImageJ (1.50i) [32].

To isolate the total ribosome fraction, cells were collected in sucrose buffer (250 mM sucrose, 250 mM KCl, 5 mM MgCl$_2$, 50 mM Tris-HCl with a pH of 7.4, 0.7% Nonidet P-40), the cytosol was isolated by centrifugation (750× g, 10 min, 4 °C) and then centrifuged again to obtain a postmitochondrial supernatant (12,500× g, 10 min, 4 °C). The supernatant was adjusted to 0.5 M KCl, and the volume equivalent to OD$_{260}$ = 5 was loaded onto a sucrose cushion (1 M sucrose, 0.5 M KCl, 5 mM MgCl$_2$, 50 mM Tris-HCl with a pH of 7.4) and centrifuged in a Beckman TLA 100.3 rotor (250,000× g, 2 h, 4 °C).

2.3. Mass Spectrometry (MS) Analysis

Proteins in the ribosome-enriched pellets were identified and quantified by free-label MS analysis using an LTQ-Orbitrap Fusion Lumos (Thermo Fisher Inc) mass spectrometer. The sample preparation, chromatography and MS analysis are detailed in the Supplementary Methods. For the peptide identification, a precursor ion mass tolerance of 7 ppm was used for MS1, with trypsin as the chosen enzyme and up to three miscleavages allowed. The fragment ion mass tolerance was set to 0.5 Da for MS2. The oxidation of methionine and N-terminal protein acetylation were used as variable modifications, whereas carbamidomethylation on cysteine was set as a fixed modification. In the analysis of phosphorylated peptides, phosphorylation of serine, threonine and tyrosine were also set as variable modifications. The false discovery rate (FDR) was set to a maximum of 5% in the peptide identification. Protein quantification was retrieved from the protein TOP3 Area node from Proteome Discoverer (v2.3). For normalization, a correction factor was applied: sum TOP3 replicate "n"/average sum TOP3 all replicates. Normalized abundance values were log$_2$-transformed, and the fold change (FC) and p-values were calculated. Two independent experiments, each with three biological replicates, were performed on T98G cells transduced with shControl or shDYRK1A lentiviruses, and only those proteins detected in at least three replicates of any condition were quantified. For the RP stoichiometry, the intensity of each RP was defined relative to the intensities of all RPs. The RP protein/mRNA ratios were obtained using the $\log_2$-transformed normalized protein abundance and the $\log_2$-transformed normalized RNA counts from RNA-Seq experiments.

2.4. Chromatin Immunoprecipitation (ChIP)

Detailed information on sample preparation is provided in the Supplementary Methods. DNA libraries were generated with the Ovation® Ultralow Library System V2 (NuGEN Technologies, San Carlos, CA, USA). Libraries were sequenced with 50 bp single-end reads on an Illumina Hiseq-2500 sequencer at the CRG Genomics Unit. The ChIP-Seq analysis was performed as described [33], with few modifications (see Supplementary Methods). To analyze the RPG promoter occupancy, datasets from The Encyclopedia of DNA Elements (ENCODE) Consortium were used and are listed in Table S2 [34]. Read numbers were normalized to reads per million (RPM) in both this work and the ENCODE datasets.

2.5. RNA-Seq

RNA was isolated with the RNeasy extraction kit (Qiagen, Germantown, MD, USA) and the samples were treated with DNase I (see Supplementary Methods for full details). For T98G cell spike-in normalization, equal numbers of T98G cells for each condition were mixed with a fixed number of *D. melanogaster* Kc167 cells (1:4 ratio). Libraries were prepared with the TruSeq Stranded mRNA Sample Prep Kit v2 (Illumina, Cambridge, UK) and sequenced with Illumina Hiseq-2500 to obtain 125 bp pair-ended reads. Differential gene expression was assessed with the DESeq2 (1.30.1) package in R, filtering genes that had >10 average normalized counts per million [35]. For the spike-in libraries, the size factor of each replicate was calculated according to exogenous *Drosophila* spike-in reads. Expression was considered to be altered when the p-value ≤ 0.05, and the $\log_2$FC was above 0.7 and below -0.7 for up- and downregulated genes, respectively.

2.6. Quantitative PCR (qPCR)

PCR reactions were performed in triplicate in 384-well plates with SYBR Green (Roche, Basel, Switzerland) and specific primers using a Roche LC-480 machine. The crossing point was calculated for each sample with the Lightcycler 480 1.2 software. No PCR products were observed in the absence of template and all primer sets gave narrow single melting point curves. For the ChIP-qPCR, a 1/10 dilution of ChIP DNA was used as the template for the PCR reaction, and a 1/1000 dilution of input DNA was used as the standard for normalization. For the RT-qPCR, a 1/10 dilution of the cDNAs was used and expression of the *D. melanogaster* gene *Act42A* was used for normalization. The sequences of primers are listed in Tables S3 and S4.

2.7. Computational Tools and Statistical Analysis

Full details for the computational tools used to analyze the ChIP-Seq, RNA-Seq and proteomics data are included in the Supplementary Methods. To calculate the statistical significance, the normality of the samples was evaluated with the Shapiro–Wilk normality test (Prism 5, v5.0d), and parametric or nonparametric tests were used accordingly. Statistical significance was calculated with two-tailed Mann–Whitney or Student's tests for unpaired samples or with a Wilcoxon matched-pairs signed-ranks test (Prism 5, v5.0d), and a p-value ≤ 0.05 was considered as significant. All experiments were performed independently at least three times.

3. Results

3.1. DYRK1A Is Recruited to the Proximal Promoter Regions of the Canonical RPGs

Our prior analysis of DYRK1A recruitment to chromatin showed an enrichment in gene-ontology terms related to ribosome biogenesis and translational regulation [20]. In addition, a de novo motif analysis found a sequence similar to the DYRK1A-associated motif to be over-represented in the promoter regions of human RPGs [9]. Accordingly, we examined whether DYRK1A was recruited to human RPGs, considering the genes encoding the 80 canonical RPs plus 10 paralogues [6] (a new naming system for RPs has been proposed [36] and listed in Table S5). In the experiments on different human cell lines, the ChIP-Seq analysis showed DYRK1A bound upstream of the TSS and, in general, within 500 bp of the TSS of a subset of the RPGs (Figure 1A). No ChIP signal was detected at either the gene bodies or the transcriptional termination sites (TTS; Figure S1). A clear reduction in the chromatin-associated DYRK1A was observed at the target RPG promoters in cells where the levels of DYRK1A were depleted by the lentiviral delivery of a shRNA targeting DYRK1A, reflecting the specific recruitment of the kinase (Figures 1B and S1). Finally, the presence of DYRK1A at the RPG promoter regions was further confirmed in independent ChIP-qPCR experiments (Figure 1C).

Around 25% of the promoters of all the RPGs were occupied by DYRK1A in the human cell lines analyzed, with considerable overlap among them (Figure 1D,E). No association with any particular ribosomal subunit was observed, since DYRK1A occupancy was detected similarly at the promoters of RPGs from both the large- and small-ribosome subunits (Figure 1E). Furthermore, the DYRK1A ChIP-Seq data from the mESCs also showed the presence of this kinase at the promoters of a subset of mouse RPGs, which coincided well with the DYRK1A-positive subset in the human cells (Figure 1F,G). Together, these results indicate that DYRK1A is recruited to proximal promoter regions of a subset of RPGs in different human and mouse cell lines, suggesting that the chromatin association of DYRK1A with these RPGs represents a general and conserved function for the kinase.

3.2. The TCTCGCGAGA Motif Marks the Subset of the RPG Promoters Positive for DYRK1A

Around 25% of the human RPG promoters contain a TCTCGCGAGA motif (Figure 2A,B), and DYRK1A bound to these in one, two or in all the human cell lines analyzed, except for *RPL10* and *RPS5*, and the RPG paralogs *RPL10L*, *RPL22L1* and *RPS4Y2* (Table S5). Likewise, DYRK1A was almost exclusively detected at RPG promoters containing the TCTCGCGAGA motif in the mESCs (Table S6), as further evidence of its functional conservation. A central model motif analysis (CMEA) [37] showed the TCTCGCGAGA motif to be positioned precisely around the center of the DYRK1A-associated peaks within the RPG promoters (Figure 2C), suggesting that it serves as a platform for DYRK1A recruitment. The kinase was also found to associate with a small number of RPG promoters without this palindromic sequence (Table S5), although its detection was poorer in these cases (Figure 2D). In conclusion, the TCTCGCGAGA motif appears to be a major determinant for the efficient recruitment of DYRK1A to the RPGs; moreover, given that the variability in DYRK1A occupancy in the different cell lines was mostly restricted to the RPG promoters without the motif, it is possible that DYRK1A associates with promoters containing this motif independently of the cell type, whereas the occupancy of other RPG promoters might be context-specific.

The TCTCGCGAGA motif has been shown to work as a promoter element that drives bidirectional transcription [38]. In around 30% of the RPGs, the TSS lies within a 500 bp window from the TSS of the genes transcribed in the opposite direction; in these cases, the RPG is always transcribed more strongly (Table S7). However, no specific bias for the presence of the TCTCGCGAGA motif could be observed, since some RPGs with bidirectional transcription contained this motif (*RPL12*, *RPL15*, *RPL23A*), while others did not (*RPL34*, *RPL9*, *RPS18*; Table S7). The transcriptional repressor zinc finger and BTB domain containing 33 (ZBTB33, also known as KAISO) binds directly to the TCTCGCGAGA motif in vitro when methylated [39]. Indeed, this palindromic motif is included in the Jaspar database of the curated TF binding profiles as a ZBTB33 motif (http://jaspar.genereg.net/matrix/MA0527.1/ (accessed on 1 January 2021)). The ChIP-Seq experiments in the T98G cells did detect ZBTB33 at the majority of the RPG promoters containing the TCTCGCGAGA motif (Figure S2A; Table S8), with signals overlapping those of DYRK1A

(Figure S2B). Moreover, the TCTCGCGAGA motif is positioned centrally within the ZBTB33-bound regions of these RPGs (Figure S2C). Thus, this promoter motif might not only favor DYRK1A recruitment, but also its interaction with other proteins.

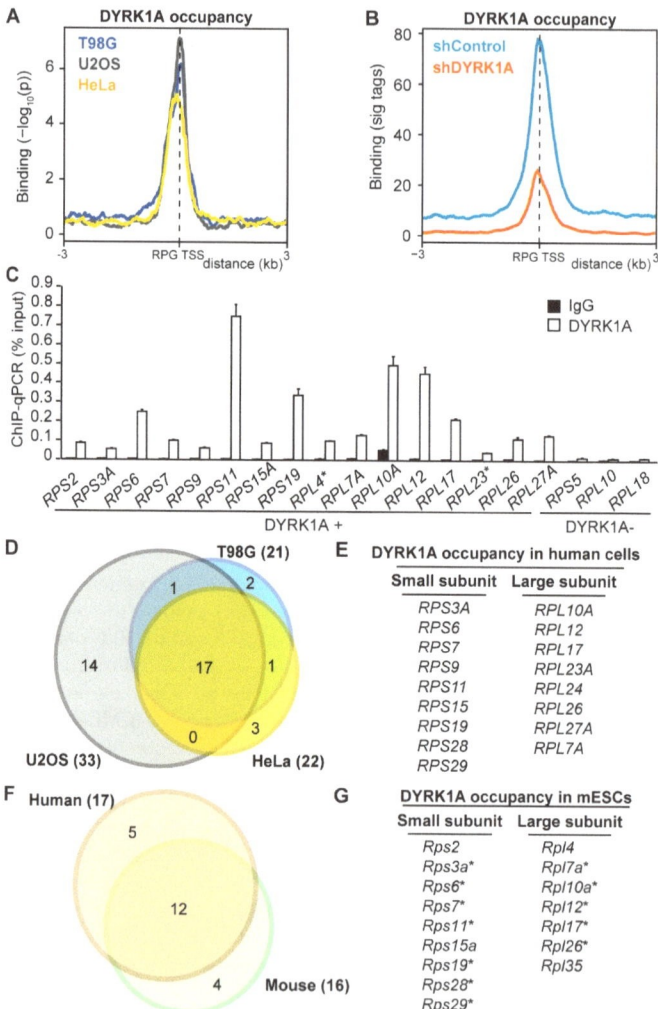

Figure 1. DYRK1A occupies a subset of RPG promoters. (**A**) Distribution of chromatin-bound DYRK1A relative to the TSS of human RPGs in T98G, U2OS and HeLa cells. The y-axis represents the relative protein recruitment ($-\log_{10}$ Poisson p-value) and the offset was set to $+3$ kb from the TSS. (**B**) DYRK1A occupancy relative to the TSS of human RPGs, comparing the shControl and shDYRK1A T98G cells (blue and orange lines, respectively). The y-axis represents the relative protein recruitment quantified as significant (sig) ChIP-Seq tags. The offset was set to ± 3 kb from the TSS (Figure S1B for representative examples). (**C**) Validation of the selected targets by ChIP-qPCR (percentage of input recovery; mean ± SD of three technical replicates). (**D**) Overlap of DYRK1A-associated RPG promoters in T98G, U2OS and HeLa cells. (**E**) List of DYRK1A-positive RPGs common to the three human cell lines. (**F**) Overlap between common DYRK1A-bound RPGs in the human cell lines and mESCs. (**G**) List of RPGs with DYRK1A detected at their promoters in mESCs. The asterisk indicates the coincident occupancy in mESCs with the three human cell lines analyzed.

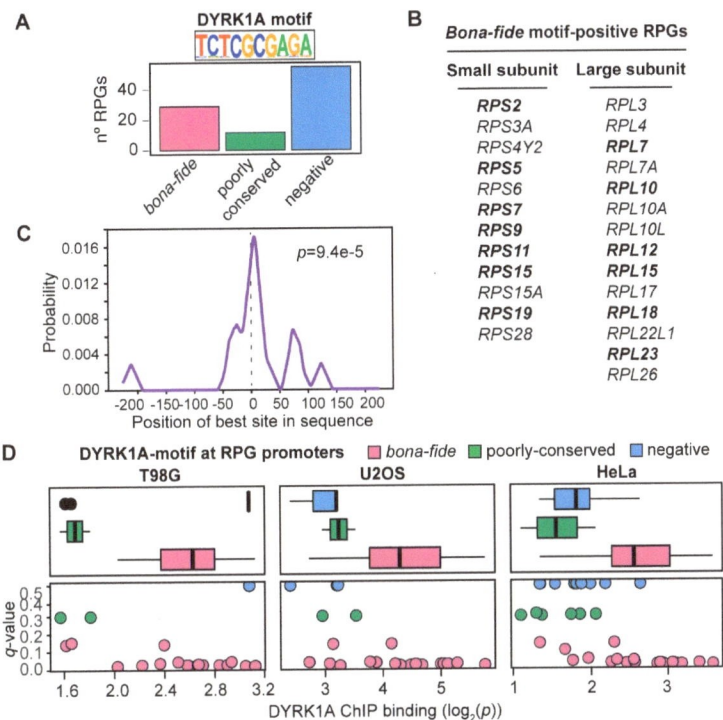

Figure 2. The TCTCGCGAGA motif correlates with significant DYRK1A binding at human RPG promoters. (**A**) Distribution of the TCTCGCGAGA motif (DYRK1A motif) in human RPGs (bona fide: p-value $< 10^{-4}$; poorly conserved: $10^{-4} < p$-value $< 3 \times 10^{-4}$). (**B**) List of human RPGs with the bona fide palindromic motif in their promoters. The RPG promoters containing more than one DYRK1A motif are in bold. (**C**) CentriMo plot showing the distribution of the DYRK1A motif for the RPG promoters that bind DYRK1A. The solid curve shows the positional distribution (averaged over bins of a 10 bp width) of the best site of the DYRK1A motif at each position in the RPG-ChIP-Seq peaks (500 bp). The p-value is for the central enrichment of the motif. (**D**) Box plot and scatter plots (dots represent the individual RPGs) showing the correlation between DYRK1A ChIP binding ($\log_2$ p-value, x-axis) and the conservation of the DYRK1A motif (q-value, y-axis) in T98G, U2OS and HeLa cells. The colors represent the motif-classification indicated.

3.3. Low Functional Conservation of the RPG Core Promoter Elements between Drosophila and Humans

Next, we wondered whether the RPG promoters that bind DYRK1A were characterized by any other feature. Most studies on the regulation of RPG expression in higher eukaryotes have used *Drosophila* as a model system, identifying several sequence elements and TFs that regulate RPG transcription [13–15]. One of them is the TCT motif that is a specific promoter element for the expression of RPGs [13]. Using high-confidence human TSS data, we scanned the human RPG promoters and found the TCT consensus sequence (YC + 1TYTYY; Figure 3A) in 77 of the 86 promoters analyzed (Table S9). However, we did not find any correlation between the TCT motif and the presence of DYRK1A at these promoters (Figure 3B).

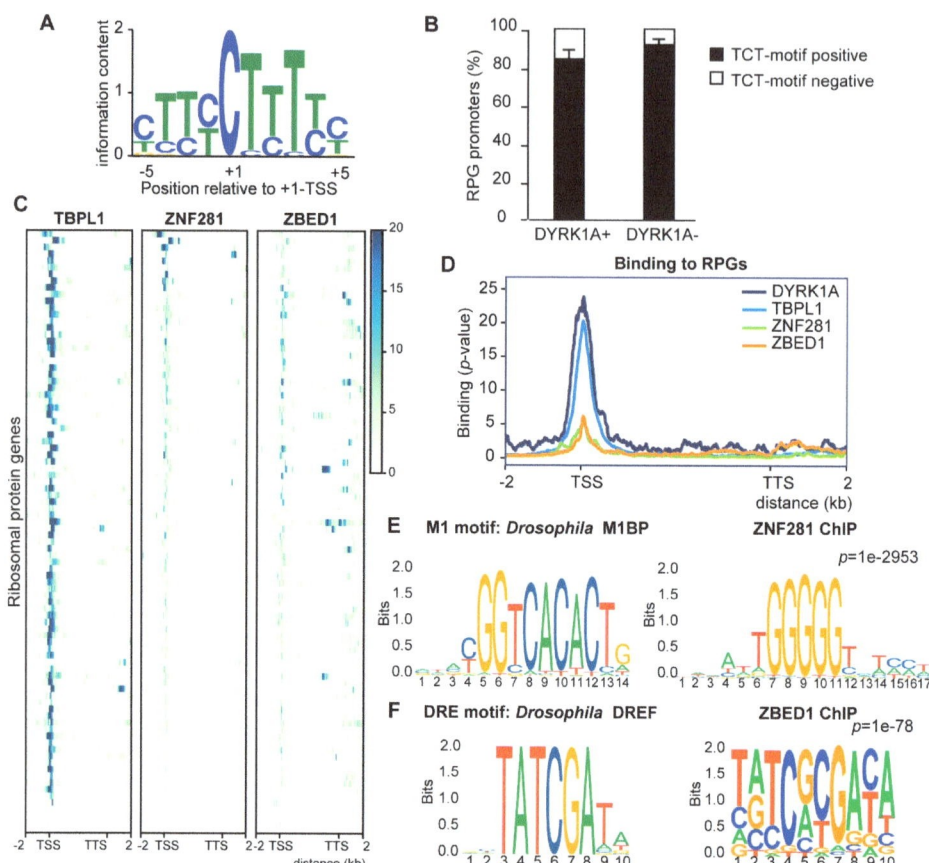

Figure 3. Analysis of the homologs of *Drosophila* RPG expression regulators in mammalian cells. (**A**) Sequence logo of the TCT motif found in human RPGs. (**B**) Percentage of TCT-positive or negative human RPG promoters distributed according to the presence of DYRK1A (average ± stdev of DYRK1A occupancies in T98G, HeLa and U2OS cells). (**C**) Occupancy of human RPGs by TBPL1 (K562, ENCSR783EPA), ZNF281 (HepG2, ENCSR403MJY) and ZBED1 (K562, ENCSR286PCG). The genomic region considered is shown on the *x*-axis. (**D**) Occupancy of the transcription factors shown in panel **C** together with DYRK1A. (**E**) Sequence of the *Drosophila* M1 motif at Jaspar and of the DNA motif enriched in ZNF281-bound regions in human cells. (**F**) Sequence of the DRE motif in *Drosophila* at Jaspar and of the motif enriched in all ZBED1-bound regions in human cells.

In *Drosophila*, the TCT element drives the recruitment of Pol II and, consequently, the transcription of RPGs through the coordinated action of TRF2, but not of TBP, and the TF motif 1 binding protein (M1BP) and DNA-replication-related element (DRE) factor (DREF) [14,15]. Thus, we asked whether the transcriptional regulators of the RPGs were functionally conserved between *Drosophila* and humans, and, if so, how they might be related to the DYRK1A promoter association. We analyzed the presence of the homologs of the fly TFs at the human RPG promoters: TBPL1, the homolog of TRF2; zinc finger protein 281 (ZNF281), the homolog of M1BP; and zinc finger BED-type containing 1 (ZBED1), the homolog of DREF. This analysis showed that TBPL1 binds to the promoter of most RPGs (68 out of 90 in K562 cells; Figure 3C,D; Table S9), similar to its behavior in *Drosophila* [14]; no particular enrichment was detected based on the presence or absence of a TATA box (Figure S3A). The presence of TBPL1 was detected at DYRK1A-positive and negative promoters (Figure S3B), and the results from an unbiased clustering analysis did not allow for the detection of the differential binding of TBPL1 at RPG promoters positive for DYRK1A (Figure S3C). ZNF281 was only detected at seven RPG promoters (Figure 3C; Table S9); in this regard, the sequence motif enriched within the ZNF281 ChIP-Seq dataset differed from the M1BP binding motif (http://jaspar.genereg.net/matrix/MA1459.1/ (accessed on 1 January 2021); Figure 3E) [10]; thus, we cannot assume that M1BP and ZNF281 are fully functional homologs. ZBED1 binds to several human RPG promoters and gene bodies (Figure 3C; Table S9), and the motif enriched in the ChIP-Seq dataset partially overlaps with the *Drosophila* DREF motif (http://jaspar.genereg.net/matrix/MA1456.1/ (accessed on 1 January 2021); Figure 3E). However, no enrichment for this motif was found within the human RPG promoters. ZBED1 might recognize the TCTCGCGAGA motif [40], yet the CMEA analysis did not find a unimodal and centered distribution of the TCTCGCGAGA motif within the ZBED1 ChIP peaks (Figure S3D). Hence, ZBED1 did not appear to bind directly to the TCTCGCGAGA motif, which is consistent with data suggesting that the TCTCGCGAGA motif and the human DRE are distinct regulatory elements [38]. No information on a general transcriptional effect of ZBED1 on RPGs is available other than that its depletion reduced the transcription of *RPS6*, *RPL10A* and *RPL12* genes in human foreskin fibroblasts [40]. Finally, ZBED1 binding to human RPGs showed no particular correlation with the presence of DYRK1A (Figure S3E). Together, these data suggest that there is little functional conservation of the core promoter elements of RPGs between *Drosophila* and humans, either in cis or trans. Furthermore, the association to chromatin of the TF homologs to those identified in *Drosophila* does not appear to depend on the presence of the TCTCGCGAGA motif or the binding of DYRK1A to human RPG promoters.

3.4. GABP and DYRK1A Are Differentially Distributed at the RPG Promoters

We then asked whether DYRK1A recruitment was associated with the presence of TFs whose binding sites are known to be over-represented at human RPG promoters, such as TBP, MYC, SP1, GABP and YY1. Instead of using motif occurrence, as in previous studies analyzing RPG promoter architecture [11,16], we took advantage of genome-wide mapping data for each of the TFs. Thus, though TBP was predicted to be differentially enriched at the human RPG promoters based on the presence of the TATA box [16], the analysis of TBP occupancy found TBP bound to nearly all human RPG promoters (93%) in the different cell lines (Figure 4A,B; Table S9). Therefore, TBP appears to be a general component of the human RPG transcriptional machinery, irrespective of the presence of a TATA consensus, a TATA-like sequence or the complete absence of such motifs (Figure 4C). Consistent with a role for TBP in the assembly of the preinitiation complex (PIC) at the RPG promoters, the presence of the TFIID subunit TBP-associated factor 1 (TAF1) was strongly correlated with that of TBP (Figure 4A; Table S9). Notably, we observed stronger TBP binding at the DYRK1A-enriched RPG promoters than at the RPG promoters devoid of DYRK1A (Figures 4D,E and S4A), suggesting that these two factors might cross-talk.

The presence of YY1, SP1 and MYC was also detected in almost all RPG promoters irrespective of the presence of cognate binding sites (Figure S4B; Table S9), and an unbiased clustering analysis showed no differential distribution of any of these factors based on the presence of DYRK1A (Figure S4C). By contrast, DYRK1A and GABP were distributed into one cluster that included those RPG promoters with a high DYRK1A occupancy and a low GABP presence (Figures 4F and S4D, cluster 1), and another in which promoters were depleted of DYRK1A with a strong GABP occupation (Figures 4F and S4D, cluster 2). Indeed, while the DYRK1A-associated TCTCGCGAGA motif was mostly over-represented in cluster 1, the GABP motif is a hallmark of cluster 2 (Figure 4G). These results suggest that the presence of DYRK1A labels a specific subset of RPGs that might respond distinctly to those labeled by GABP.

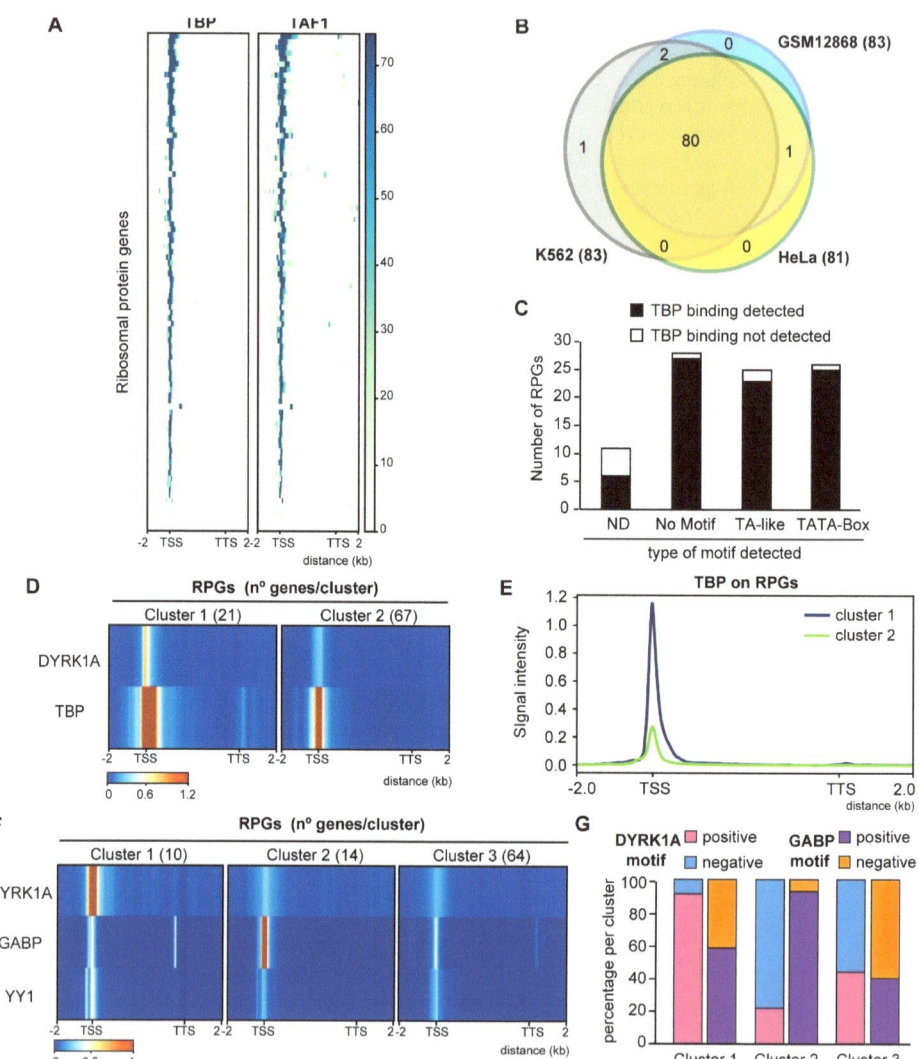

Figure 4. Analysis of TFs whose binding sites are over-represented at human RPG promoters. (**A**) TBP and TAF1 occupancy in human RPGs, showing the genomic region considered on the x-axis (K562 cells, GSE31477 and ENCSR000BKS). (**B**) Overlap of TBP occupancy at RPGs in different human cell lines (GSE31477). (**C**) Relationship of the presence of a TATA box or TA-like sequences in human RPG promoters according to Perry et al. (2005) [16] and TBP binding in HeLa cells (ND, not determined). (**D**) Unbiased k-mean clustering of the average binding of DYRK1A (HeLa, this work) and TBP (HeLa, GSE31477) on human RPGs. The color scale bar indicates the binding score and the genomic region considered is shown on the x-axis. (**E**) Metagene plot showing TBP occupancy relative to human RPGs in HeLa cells according to the clusters shown in Figure 4D. The y-axis represents the relative protein recruitment quantified as ChIP-Seq reads. (**F**) Unbiased k-mean clustering of the average binding of DYRK1A (T98G, this work), GABP and YY1 (SK-N-SH, GSE32465) to the RPGs. The color scale bar indicates the binding score and the genomic region considered is shown on the x-axis. (**G**) Percentage of RPG promoters containing a DYRK1A or a GABP motif in each of the clusters shown in Figure 4F.

3.5. The Expression of RPGs Is Sensitive to DYRK1A Depletion

Based on the ability of DYRK1A to regulate transcription when recruited to chromatin [20], we wondered whether the transcription of the RPGs might be modulated by DYRK1A. Indeed, the TCTCGCGAGA motif is a cis element for the regulation of the expression of several RPGs, such as *RPL7A* [41], *RPS6*, *RPL10A*, *RPL12* [40] and *RPS11* [20]. Furthermore, we demonstrated that the *RPS11* promoter responds to DYRK1A in a kinase- and motif-dependent manner [20]. As described in *Drosophila* [14], the transcripts of most human RPGs were in the group of the top 5% most strongly transcribed genes in the cell lines analyzed (Figure S5A–F). Globally, the expression of genes that have promoters occupied by DYRK1A is stronger than that of the genes in which DYRK1A is absent (Figure S5G). The same tendency towards a stronger expression of DYRK1A-bound targets was observed when assessing the RPGs, although the differences were not statistically significant in any of the cell lines analyzed (Figures 5A and S5H,I). The ChIP-Seq data of Pol II for RPGs revealed a profile corresponding to actively transcribed genes, with Pol II occupancy detected all along the gene bodies, with the exception of those RPGs not expressed in the cell line analyzed (Figure 5B). No differences in the distribution of Pol II were found between the DYRK1A-positive and negative RPGs (Figure 5B).

We assessed next whether the absence of DYRK1A affected the expression of its target genes by comparing the RNA expression of cells infected with a control lentivirus or a lentivirus expressing a shRNA to DYRK1A, and we used *Drosophila* RNA for the spike-in normalization. The majority of the differentially expressed genes were downregulated in response to DYRK1A depletion (Figure S6A). Indeed, these downregulated genes were enriched in the subset of genes with DYRK1A at their promoters (Figure S6B), supporting a role for DYRK1A in transcriptional activation. A general reduction in RPG transcripts was observed in DYRK1A-silenced cells (Figure 5C), and the analysis indicated that RPGs whose promoters were occupied by DYRK1A and those without DYRK1A were affected to a similar extent (Figure 5C,D), suggesting that the effect of DYRK1A on RPG expression goes beyond its direct binding to promoters. The reduction in the transcript levels of the selected DYRK1A target and nontarget RPGs was validated by RT-qPCR using two different shRNAs directed against DYRK1A (Figure S7A,C). The analysis of the occupancy of Pol II at the RPGs showed a general decrease in Pol II along the RPG gene bodies upon DYRK1A depletion at both the DYRK1A-positive and DYRK1A-negative RPGs (Figure 5E), pointing to a transcriptional effect as responsible for the reduction in the RPG transcript steady-state levels. All these results depict a complex scenario in which the physiological levels of DYRK1A are important for maintaining RPG mRNA levels through at least two different, although not necessarily exclusive, mechanisms: on the one hand, through its recruitment to the proximal promoter regions of a subset of RPGs; on the other hand, through a not-yet-identified mechanism that impacts the levels of transcribing Pol II at the RPGs.

3.6. DYRK1A Depletion Causes a Reduction in the Ribosome Content

We next wondered whether the downregulation of the RPG transcript levels induced by DYRK1A silencing would be reflected at the protein level. As such, we used MS to quantify the ribosome composition in the control cells and in DYRK1A-depleted cells using a cytosolic fraction enriched in ribosomes from T98G cells (see Materials and Methods for details). Our dataset had a strong overlap with other studies defining the human riboproteome [42,43] (Figure S8A) with the biological functions enriched related to protein synthesis (Figure S8B; Table S10). Other functions, such as oxidative phosphorylation, RNA transport/processing or endocytosis, probably reflect cosedimenting complexes; the enrichment in nuclear proteins related with splicing has also been described [44]. Nevertheless, core RPs represent more than 75% of the protein mass in the fraction analyzed (Figure 6A).

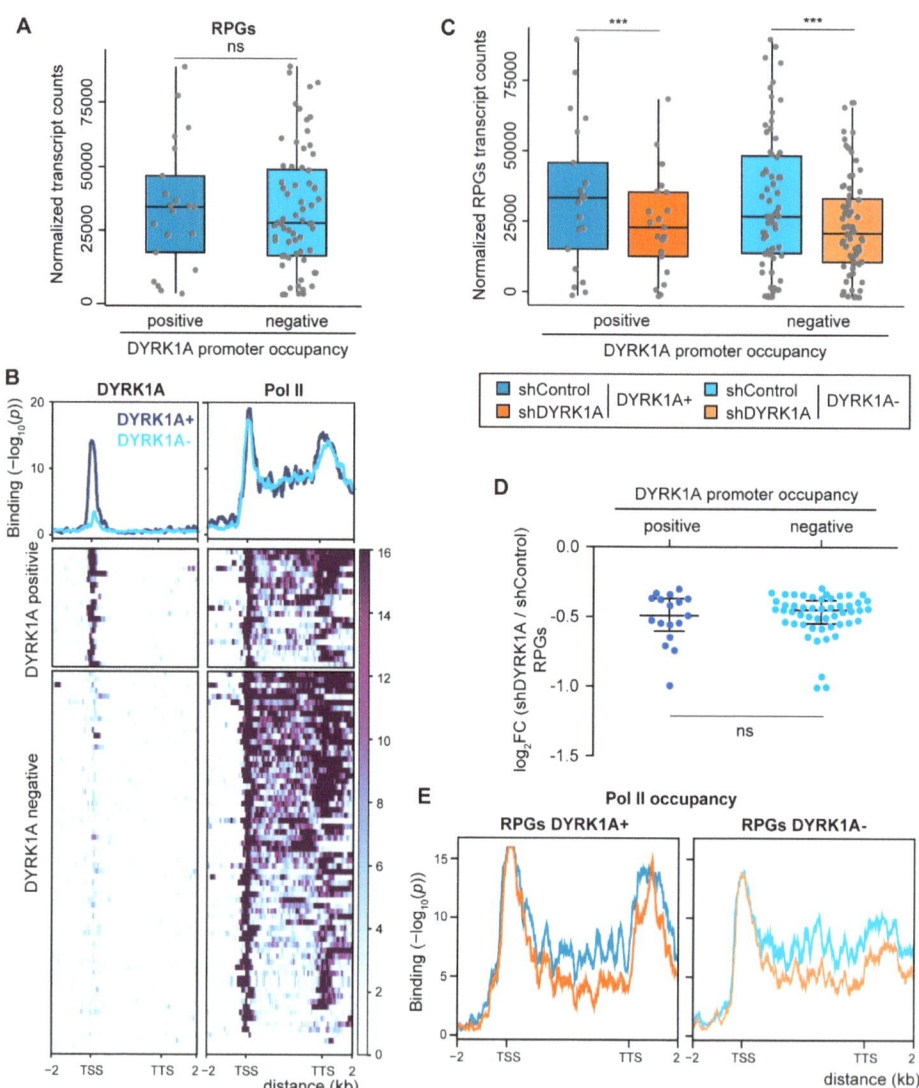

Figure 5. Depletion of DYRK1A causes a general reduction in RPG expression. (**A**) RNA levels of human RPGs (normalized counts) separated into two clusters according to the DYRK1A presence at their promoters (unpaired two-tailed Mann–Whitney test, ns = not significant). (**B**) Bottom panel, DYRK1A and Pol II chromatin occupancy of human RPGs depicted as metagenes and separated into two clusters according to the presence/absence of DYRK1A at the promoters. The binding score ($-\log_{10}$ Poisson p-value) is indicated by the color scale bar. Top panel, density plot corresponding to the mean value of the heatmap (blue and light-blue lines correspond to DYRK1A-positive or DYRK1A-negative RPGs, respectively). The binding score ($-\log_{10}$ Poisson p-value) is indicated on the y-axis. (**C**) Expression of RPGs (normalized counts) in T98G cells classified according to the presence (positive) or absence (negative) of DYRK1A at their promoters, and comparing shRNA Control cells

(blue and light-blue, respectively) or shDYRK1A (orange and light-orange, respectively; Wilcoxon matched-pairs signed-ranks test, *** $p < 10^{-5}$). The reduction in DYRK1A is shown in Figure S7B. (**D**) The graph presents the log$_2$FC for the RPG mRNAs (adj p-value < 0.05) in shDYRK1A vs. shControl and clusterized according to DYRK1A presence at their promoters (two-tailed Mann–Whitney test, ns = not significant). (**E**) Density plots corresponding to the mean value of Pol II binding in RPGs clusterized as positive and negative for DYRK1A binding in shControl (blue and light-blue lines) and shDYRK1A conditions (orange and light-orange lines).

The MS data allowed for the detection and quantification of all RPs except for RPL10L and RPL39L, which showed low mRNA levels or which were not expressed at all in T98G cells, and for RPL41 that is usually not detected by MS approaches [42]. In addition, the data indicated a lower relative abundance of the RPS27L, RPL3L, RPL7L1, RPL22L1, RPL26L1 and RPL36AL paralogs than their corresponding pairs (Figure S9A,B; Table S11), suggesting that they are under-represented in the ribosomes from T98G cells. However, we cannot rule out that the variation in stoichiometry could be due to some RPs being loosely bound to the ribosome; thus, their presence may be affected by the method for the cell extract preparation or because the pools of RPs performing extraribosomal functions might not be present in the cell fraction analyzed. We observed variability in the protein/mRNA ratio in the T98G cells, with extreme cases for most of the weakly expressed paralogs, like RPL3L, RPL7L1 or RPL22L1 (Figure S9C). Although we cannot rule out that some RPs are under-represented in the fraction analyzed, the results are consistent with published data showing that the amounts of RPs correlate poorly with their corresponding mRNA levels in other cellular contexts [45]. In agreement with the RNA data, our analysis did not reveal significant differences in the relative abundance of RPs encoded by DYRK1A-positive or DYRK1A-negative genes (Figure 6B).

We next focused on the alterations induced by silencing DYRK1A. The datasets revealed significant differences in proteins associated with specific functional categories in the ribosomal-enriched fractions from the control and DYRK1A-depleted cells. Proteins in the oxidative phosphorylation category were enriched in the shDYRK1A cell fraction (Figure 6C), including components of the mitochondrial electron transport chain (COX6B1, COX7A2L, SDHB) or mitochondrial ATP synthases (Table S10). Conversely, the ribosome category was enriched in the protein dataset reduced in the fraction from shDYRK1A cells (Figure 6C; Table S10). Indeed, DYRK1A depletion significantly reduced the total ribosome mass when calculated relative to the total amount of protein in the fraction analyzed by MS (Figure 6A). In accordance with the general effect of DYRK1A on the RPG transcript levels, less RP amounts were found for the RPGs containing or lacking DYRK1A at their promoters (Figures 6B and S9A,B). Notably, the rRNA content tended to fall (Figure S9D), supporting the decrease in the ribosome amounts. The protein/mRNA ratios correlated strongly between the control and DYRK1A-silenced cells (Figure 6D), suggesting that changes in the protein abundance were largely due to the changes in the RNA abundance upon DYRK1A depletion. Finally, no big differences in the RP stoichiometry were observed between the shControl and shDYRK1A cells (Figure 6E; RPL22L1, p-value = 0.022; RPL27, p-value = 0.0411; two-tailed unpaired Mann–Whitney). All these results indicate that cells respond to DYRK1A depletion by reducing the steady-state levels of the RPs, and that this effect occurs, at least in part, at the transcript level.

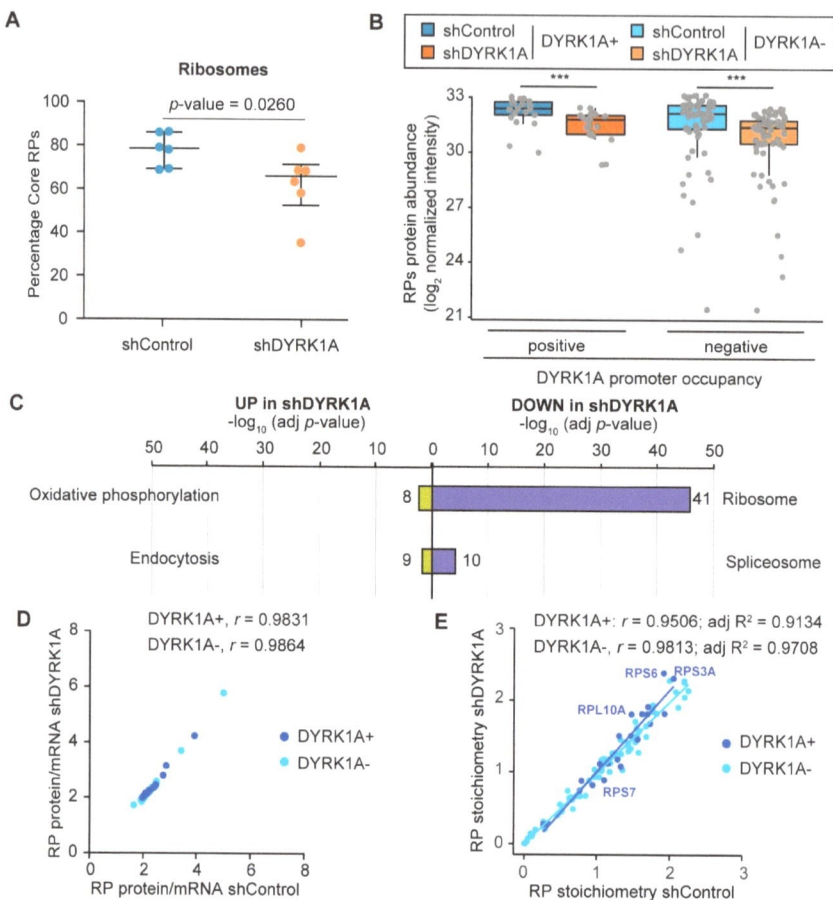

Figure 6. DYRK1A depletion causes a reduction in the ribosome content. (**A**) The relative RP values (the sum of all RP TOP3 values/the sum all TOP3 values) for each sample of shControl (blue) and shDYRK1A (brown) T98G cells are represented, also showing the median and IQRs (n = 6, two-tailed Mann–Whitney test). (**B**) Levels of the RPs ($\log_2$ of normalized peptide intensities) separated into two clusters according to the presence of DYRK1A at the promoters of their corresponding genes, both in shDYRK1A and shControl T98G cells (Wilcoxon matched-pairs signed-rank test, *** $p < 10^{-4}$). The reduction in DYRK1A is shown in Figure S9E. (**C**) Functional categorization of the proteins found more abundant in the shDYRK1A cells (UP) or in the shControl cells (DOWN) (proteins with a p-value < 0.05 in the comparisons or present in only one of the conditions were used). The number of proteins identified in each category is shown; see also Table S10. (**D,E**) Correlation analysis of the ratio of RP protein and mRNA abundances (**D**) and RP stoichiometry (**E**) of the shControl and shDYRK1A T98G cells. RP stoichiometry is defined as the intensity of each RP relative to the intensity of all RPs. A color code was used to indicate the presence (+)/absence (−) of DYRK1A at the RPG promoter regions. The Spearman's correlation coefficient is shown for each subset. In (**E**), the value for the adjusted R2 for each set is also included.

3.7. DYRK1A Plays a Role in Cell-Size Control by Regulating Protein Synthesis

Altered ribosome biogenesis can lead to major defects in translation; thus, we assessed the impact of DYRK1A on the translational status of cells. The functional status of ribosomes was first analyzed by polysome profiling on sucrose gradients upon DYRK1A silencing. The downregulation of DYRK1A diminished the polysome fractions, which corresponded to those ribosomes engaged in active translation, with a concomitant increase in the monosome peak (Figure 7A,B). These results indicate that DYRK1A depletion leads to polysome disorganization. Indeed, there was a significant reduction in the translational rates in the DYRK1A-depleted cells when measured through radiolabeled ^{35}S-methionine incorporation (Figure 7C). This effect was specific, as it was clearly observed with two distinct shRNAs targeted to DYRK1A (Figure 7D).

In eukaryotes, cell growth is coupled to cell cycle progression; therefore, the global translation rates change during the cell cycle [46]. Defects in the cell cycle have been associated with alterations in DYRK1A levels in different cellular environments [47]. Indeed, we found that DYRK1A silencing alters the cell cycle balance in T98G cells, augmenting the population of cells in the G1 phase (Figure S10A). Thus, we next checked whether the shift in the cell cycle phases was associated with the reduced translation rates. As such, the T98G cells were arrested in G1 by serum deprivation (Figure S10B), which led to a strong reduction in the rate of translation (Figure 7D). Serum addition for 30 min induced protein synthesis (Figure 7D), with no changes in the cell cycle profiles (Figure S10B). In these conditions, lower rates of translation persisted in the cells with silenced DYRK1A relative to the control cells (Figure 7D), suggesting that the reduction in protein synthesis upon DYRK1A silencing is independent of the alterations in the cell cycle.

As DYRK1A is a highly pleiotropic kinase, we wondered whether the reduction in translation could be due to an effect of DYRK1A on other signaling pathways that regulate protein synthesis. Therefore, we analyzed the effect of DYRK1A on one of the major signaling pathways that regulates protein translation, the mechanistic target of rapamycin (mTOR) pathway [48], assessing Thr389 phosphorylation of the RPS6 kinase beta-1 (RPS6KB1 or p70S6K) that represents a late event in mTOR pathway activation. DYRK1A depletion did not alter Thr389–p70S6K phosphorylation (Figure S10C,D); likewise, the MS data showed no differences in the amount of RPS6 peptides phosphorylated at Ser235, Ser236 and Ser240 (Figure S10E), all targets of p70S6K downstream of mTOR [48]. Accordingly, DYRK1A does not appear to affect the mTOR pathway under regular growth conditions. Other signaling pathways, like the cellular stress and unfolded protein response, can inhibit protein synthesis through the Ser51 phosphorylation of the translation initiation factor eIF2α [49]. However, the levels of Ser51–eIF2α phosphorylation remained unchanged in the absence of the DYRK1A (Figure S10F,G), indicating that the DYRK1A-dependent inhibition of translation is not mediated by eIF2α phosphorylation. Moreover, DYRK1A was not detected in the ribosome-enriched fraction by MS or in immunoblots of polysome-associated fractions (Figure S10H), suggesting that it is not tightly bound to actively translating ribosomes and that it probably does not act directly on polysomes. We would like to point out that our results are based on the behavior of tumor cell lines as experimental systems. Therefore, the functional interaction of DYRK1A with cell growth regulators in other physiological backgrounds cannot be excluded. In fact, a reduced soma size of cortical layer V neurons has been observed in a conditional *Dyrk1a* null mouse model, which was related to mTOR dysregulation [50].

Finally, as reduced protein synthesis might affect cell mass, we checked whether the cell size was affected by the loss of DYRK1A activity. Indeed, DYRK1A-silenced HeLa and T98G cells were both significantly smaller than their controls in terms of cell volume (Figure 7E,F). Hence, the fine-tuning of cellular DYRK1A levels is important to assure the proper size of human cells is maintained.

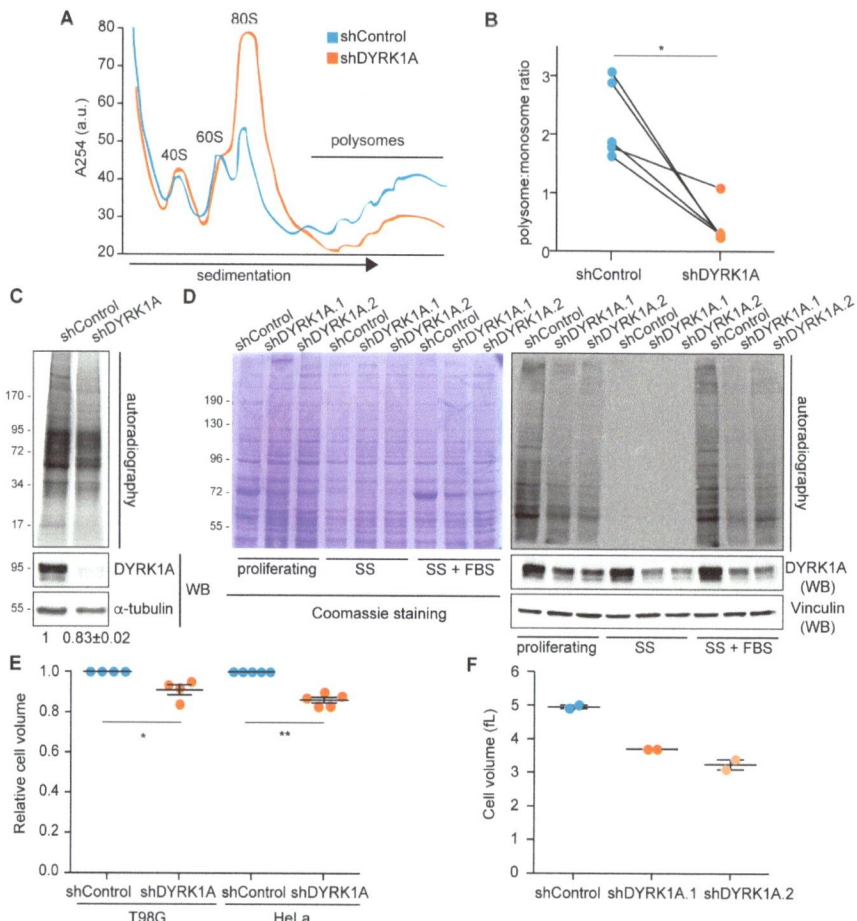

Figure 7. DYRK1A-dependent regulation of protein synthesis might impact cell size. (**A**) Polysome profile of T98G cells in shDYRK1A (orange line) and shControl (blue line) conditions. The position of the 40S, 60S, 80S and polysome peaks is indicated. The *y*-axis shows absorbance at 254 nm in arbitrary units and the *x*-axis corresponds to the different fractions. (**B**) The area under the curve for polysomes and monosomes was measured from the polysome profiles of paired shControl and shDYRK1A experiments, and the polysome:monosome ratio was calculated (the values for each condition in each biological replicate are represented with a color-coded dot and connected with lines; n = 5, shDYRK1A.1 [3] and shDYRK1A.2 [2]; Wilcoxon matched-pairs signed-ranks test, * $p < 0.05$). (**C**) Protein synthesis assays were performed by metabolic ^{35}S-methionine labeling in shDYRK1A or shControl T98G cells. DYRK1A levels were analyzed by WB (Supplementary Materials). Quantification of the average radioactive intensity of independent experiments is shown at the bottom of the images (mean ± SEM, n = 3; Wilcoxon matched-pairs signed-ranks test, $p < 10^{-3}$). (**D**) Autoradiography of protein extracts prepared from proliferating, serum-starved cells for 48 h (SS) or serum-starved cells reincubated with FBS for 30 min (SS + FBS) and pulse-labeled with ^{35}S-Met for 20 min. Equal numbers of T98G cells infected with the indicated shRNA lentivectors were used. The reduction in DYRK1A was assessed in WB (Supplementary Materials) and a Coomassie-stained gel as a loading control is also shown. Cell cycle profiles are shown in Figure S10B. (**E**) Cell volume represented in arbitrary units with the control cells set as 1 (mean ± SEM; Mann–Whitney test, * $p < 0.05$, ** $p < 0.01$). (**F**) Cell volume of T98G cells infected with lentivirus expressing two independent shRNAs against DYRK1A or a shRNA Control (n = 2).

4. Discussion

In this study, we show that there is a subset of RPGs in mammals whose promoters are marked by the palindromic TCTCGCGAGA motif and the presence of the protein kinase DYRK1A, as shown by the chromatin recruitment analysis in different human and mouse cell lines. A motif enrichment analysis did not find the TCTCGCGAGA motif within the RPG promoters in yeast, basal metazoa or plants [11], although a similar motif (CGCGGCGAGACC) was found within the proximal promoter regions of 28 RPGs in *Caenorhabditis elegans* [12]. In *Drosophila*, no similar motifs were enriched in the RPG promoters [10], though the DRE sequence has been proposed to mirror such a motif [40]. By contrast, the TCTCGCGAGA motif is conserved in vertebrates and it is generally found in DNAse I-accessible regions [38,51], leading to the proposal that it serves as a core promoter element in TATA-less promoters associated with CpG islands [38]. These findings would suggest that the TCTCGCGAGA motif is a cis-regulatory element that arose later in evolution and that it might be linked to coevolution with regulatory factors acting in trans. Besides DYRK1A, the transcriptional repressor ZBTB33, the T-cell factors Tcf7l2 and Tcf1 and the BTG3-associated nuclear protein (BANP) have been shown to use the TCTCGCGAGA motif as a chromatin recruitment platform [39,51–53], and we confirm here the presence of ZBTB33 at the RPG promoters that contain this motif. Whether the TCTCGCGAGA motif interacting proteins compete or collaborate in the regulation of common target genes, including RPGs, is an issue that merits further exploration.

In *Drosophila*, RPG expression is regulated by a combination of two TFs, M1BP and DREF, which are associated with distinct subsets of RPGs through binding to their corresponding DNA-sequence motifs. These two proteins are responsible for recruiting TRF2, which substitutes for TBP in the assembly of the PIC [13–15]. Our analysis shows no evidence of such a regulatory network in humans. On the one hand, we observe TBP binding to almost all RPG promoters, regardless of whether or not they contain a TATA sequence, which indicates that TBP would be responsible for PIC assembly in human RPGs, as supported by the presence of the largest subunit of TFIID, TAF1. It is worth noting that the presence of DYRK1A at RPG promoters is correlated with more TBP binding, opening the question on the existence of a functional cross-talk between DYRK1A and TBP. TBPL1, a TRF2 homolog, also binds to almost all RPGs, consistent with the finding that TBPL1 is recruited to the PIC, not as a substitute for TBP but rather along with it, in mouse testis [54]. In *Drosophila*, the depletion of TBP or TBPL1 individually does not affect the expression of the TCT-promoter-bearing RPGs [55], suggesting that these two core TFs might function redundantly. Whether this is the case in mammalian cells is yet unknown. No correlation between DYRK1A presence at RPG promoters and more TBPL1 binding has been detected, though, in this case, we were unable to use datasets from the same cell type and we cannot exclude the impact of cell-type specificity. By contrast, we do not detect significant enrichment of the human M1BP and DREF homologs ZNF281 and ZBED1 at the human RPG promoters. Therefore, a completely different set of regulators must exist in mammals to interpret and to respond dynamically to growth and stress cues.

With the exception of the TCT motif, most of the sequences implicated in regulating the expression of RPGs in humans are only present in a subset of RPGs [13], including the binding motifs for the SP1, GABP, MYC and YY1 TFs. We have extended these results by analyzing the presence of these TFs at RPG promoter regions, revealing a more general distribution than that inferred through the presence of their binding sites. Therefore, no specific bias was found for the recruitment of TFs, like SP1, YY1 or MYC, to promoters containing the TCTCGCGAGA motif or positive for DYRK1A recruitment. By contrast, the RPGs that associate with DYRK1A are characterized by reduced GABP binding. The distinct distribution of DYRK1A and GABP could allow for the differential regulation of subsets of RPGs in response to a variety of stimuli. Notably, both *DYRK1A* and *GABP* are located on human chromosome 21 and, when in trisomy, their overexpression might contribute to the general increase in RPG mRNA transcripts in the brain of individuals with DS (Table S12). We are aware that a limitation of our study is that the comparative analysis of DYRK1A and TFs' occupancies has been performed using data from different sources. When possible, datasets from the same cell line have been used (for instance, HeLa for TBP) or from a similar origin (for instance, neural cell lines T98G and SK-N-SH for GABP and YY1).

The depletion of the DYRK1A results in the downregulation of RPG transcripts, affecting both RPGs with DYRK1A bound at their promoters and RPGs devoid of DYRK1A. These results resemble those found with the manipulation of MYC levels, resulting in changes in RPG transcripts not directly associated to the presence of MYC at the promoter regions of those genes [56,57]. In the case of RPGs bound by DYRK1A within their promoter, this reduction could be a direct effect of the loss of DYRK1A at their proximal promoters and the subsequent reduction in Pol II CTD phosphorylation, as shown for *RPS11* [20], with the reduction in transcript levels being a combination of alterations at both the initiation and elongation steps. However, additional mechanisms must exist to explain that the decrease in the RPG mRNA levels occurs in a general manner. The reduction in Pol II occupancy at the RPGs upon DYRK1A depletion would suggest that transcription might be a target. Even so, we cannot exclude that the reduction in DYRK1A levels may induce post-transcriptional effects acting on RPG mRNA steady-state levels, or may alter the complex interactions between RPG mRNA synthesis/degradation and ribosome biogenesis that have been described in yeast directed to ensure fidelity in ribosome assembly [58–60]. Finally, the DYRK1A-dependent effect on increasing the population of cells in G1 could also be at play.

Our results demonstrate that the production of RPs and, ultimately, the number of ribosomes in proliferating cells depends on the physiological levels of DYRK1A. This could be a direct consequence of the alterations in the RPG transcript levels, since the protein/mRNA ratios are not significantly affected upon DYRK1A silencing. Additionally, compensating mechanisms at the level of ribosome biogenesis might also operate [1]. In either case, the shortage of ribosomes upon a loss of DYRK1A provokes translational dysfunction, with the cell size reduction as one of the possible phenotypic outputs. As mentioned above for the DYRK1A-dependent impact on the mRNA steady-state levels, the increase in the G1 population could also be a contributing factor, or, alternatively, a consequence of delayed growth; however, a DYRK1A-dependent reduction in cell size in postmitotic neurons has been described [50]. The reduction in ribosomes could either globally affect translation or may result in transcript-specific translational control. In this context, the pool of mRNAs associated with polysomes that respond differentially to DYRK1A downregulation still needs to be characterized. Together with the identification of cis-regulatory elements in these transcripts, this information will surely help to discriminate between the two possibilities and establish a mechanistic framework. Nonetheless, our findings do not rule out the existence of other DYRK1A-dependent effectors that contribute to altered ribosome biogenesis and/or translation: such a multilayer regulatory effect is not uncommon in growth regulators, as it is the case of the mTOR protein kinase [48].

The physiological context for the activity of DYRK1A on translational control remains a matter of speculation at this stage. For instance, DYRK1A has been associated with the regulation of cell proliferation, and given that cell growth and proliferation are intimately linked [46,61], this kinase could couple the cell cycle with protein synthesis by maintaining the amounts of RPs. In addition, DYRK1A plays essential roles in central nervous system development, not only influencing cell numbers but also their differentiation [21]. Indeed, a reduction in the size of the soma of cortical layer V neurons has been shown in conditional *Dyrk1a*-null mice [50]. The authors showed reduced mTOR-dependent signaling, but the RP accumulation dysregulation, as we have observed in tumor cell lines, might also represent a contributing factor. It is also possible that the effects of DYRK1A on cell mass may affect other tissues, particularly since heterozygous mouse models exhibit a global reduction in body size [62]. In this regard, DYRK1A has been shown to phosphorylate Pol II at gene loci involved in myogenic differentiation [30], a process that requires increased protein synthetic rates [46]. In a different context, defects in protein production are closely related to cancer, since enhanced translation is required to boost cell proliferation [3,4] and DYRK1A has both positive and negative effects on cell proliferation, depending on the tumor context [22,23]. Finally, mutations in specific RPGs produce very unique phenotypes [3], including craniofacial anomalies and urogenital malformations [63]. In addition, *RPL10* mutations have been linked to neurodevelopmental conditions, including autism spectrum disorders and microcephaly [64], and, indeed, translation is a process targeted in autism-associated disorders [65]. All of these features are hallmarks of *DYRK1A* haploinsufficiency syndrome in humans or in animal models with *Dyrk1a* dysregulation [26,66].

5. Conclusions

In summary, our findings have uncovered the protein kinase DYRK1A as a novel player in the regulation of translation and cell growth in mammals. This discovery adds complexity to our understanding of DYRK1A's role in cellular biology. The study also raises questions about the mechanisms by which mammalian cells regulate the transcription of ribosomal protein genes, prompting comparisons with existing knowledge from studies in yeast or flies. Furthermore, the findings create opportunities for future investigations to explore the connections between the mechanistic aspects of DYRK1A activity and the pathological consequences that arise from its dysregulation.

Supplementary Materials: The following supporting information can be downloaded at: https://www.mdpi.com/article/10.3390/biom14010031/s1, Supplementary Material and Methods; Figure S1: Supporting data for Figure 1; Figure S2: The transcription factor ZBTB33 is recruited to RPG-positive promoters for DYRK1A and the TCTCGCGAGA DNA motif; Figure S3: Recruitment of TBPL1 and ZBED1 to human RPG promoters; Figure S4: Recruitment of TBP, MYC, SP1 and YY1 to RPG promoters in comparison with that of DYRK1A; Figure S5: RPG mRNA expression in different human cell lines; Figure S6: Global alterations in gene expression upon DYRK1A depletion; Figure S7: Supporting data for the alterations in RPG transcript levels in DYRK1A-depleted cells; Figure S8: Riboproteome analysis; Figure S9: RP expression analysis based on MS quantitative data in T98G comparing shControl and shDYRK1A. Figure S10: Supporting data for Figure 7. Table S1: Antibodies used in this research; Table S2: ENCODE datasets used in the analysis of chromatin occupancy; Table S3: Primers for ChIP-qPCR; Table S4: Primers for RT-qPCR; Table S5: DYRK1A binding to RPGs in *Homo sapiens*; Table S6: DYRK1A binding to RPGs in *Mus musculus*; Table S7: RPGs within bidirectional transcription pairs; Table S8: ZBTB33 binding to RPGs in *Homo sapiens*; Table S9: TF binding to RPGs in *Homo sapiens*; Table S10: KEGG data from Figures 6C and S8B; Table S11: RP MS data; Table S12: Differential RPG expression in Down syndrome individuals. References [13,20,33,35,37,67–80] are cited in Supplementary Materials.

Author Contributions: C.D.V.: conceptualization, formal analysis (high-throughput sequencing data), investigation, visualization and writing. L.B.: investigation, methodology and visualization. R.F.: formal analysis. S.d.l.L.: conceptualization, formal analysis, funding acquisition, investigation, supervision, visualization and writing. All authors have read and agreed to the published version of the manuscript.

Funding: This research was supported by the Spanish Ministry of Science and Innovation (BFU2016-76141-P, PID2019-107185GB-I00, AEI/FEDER) and Secretaria d'Universitats i Recerca del Departament d'Empresa i Coneixement de la Generalitat de Catalunya (grant number 2021SGR01229) to S.d.l.L. and the Departments of Excellence' program of the Italian Ministry for University and Research (MIUR, 2018-2022 and MUR, 2023-2027) to R.F. L.B. was a FPU predoctoral fellow (FPU13/02400). We acknowledge the support of the Spanish Ministry of Science and Innovation through the Centro de Excelencia Severo Ochoa (CEX2020-001049-S, MCIN/AEI/10.13039/501100011033) and the Generalitat de Catalunya through the CERCA program. We are grateful to the CRG/UPF Proteomics Unit, which is part of the Spanish Infrastructure for Omics Technologies (ICTS OmicsTech) and a member of the PRB3-ProteoRed (PT17/0019, Instituto de Salud Carlos III and ERDF).

Institutional Review Board Statement: Not applicable.

Informed Consent Statement: Not applicable.

Data Availability Statement: All the raw and processed sequencing data generated in this study have been submitted to the NCBI GEO repository under accession number GSE155809. The raw proteomics data have been submitted to the Proteomics Identifications Database (PRIDE) under accession number PXD022966.

Acknowledgments: We thank all the members of Susana de la Luna's laboratory for their helpful discussions, Alicia Raya for technical assistance, the members of Juana Díez's laboratory (Molecular Virology Lab, UPF) for assistance with the polysome profiling, Enrique Blanco (Epigenetics Events in Cancer Lab, CRG) for helpful discussions regarding the RNA-Seq analysis, Sarah Bonnin (CRG Bioinformatics Unit) for advice on the bioinformatics analysis, Aránzazu Rosado (Genome architecture Lab, CRG) for kindly providing the Kc167 *Drosophila* cells, Giorgio Dieci (Dept. of Chemistry, Life Sciences and Environmental Sustainability, University of Parma) for the critical reading of the manuscript and Mark Sefton for English-language editing. We acknowledge the assistance of Eduard Sabidó and Eva Borras from the CRG/UPF Proteomics Unit and support from the CRG Genomics Facility.

Conflicts of Interest: The authors declare no conflicts of interests.

References

1. Pena, C.; Hurt, E.; Panse, V.G. Eukaryotic ribosome assembly, transport and quality control. *Nat. Struct. Mol. Biol.* **2017**, *24*, 689–699. [CrossRef]
2. Petibon, C.; Malik Ghulam, M.; Catala, M.; Abou Elela, S. Regulation of ribosomal protein genes: An ordered anarchy. *Wiley Interdiscip. Rev. RNA* **2021**, *12*, e1632. [CrossRef]
3. Aspesi, A.; Ellis, S.R. Rare ribosomopathies: Insights into mechanisms of cancer. *Nat. Rev. Cancer* **2019**, *19*, 228–238. [CrossRef]
4. Bustelo, X.R.; Dosil, M. Ribosome biogenesis and cancer: Basic and translational challenges. *Curr. Opin. Genet. Dev.* **2018**, *48*, 22–29. [CrossRef]
5. McGary, K.; Nudler, E. RNA polymerase and the ribosome: The close relationship. *Curr. Opin. Microbiol.* **2013**, *16*, 112–117. [CrossRef]
6. Gupta, V.; Warner, J.R. Ribosome-omics of the human ribosome. *RNA* **2014**, *20*, 1004–1013. [CrossRef]
7. Bosio, M.C.; Fermi, B.; Dieci, G. Transcriptional control of yeast ribosome biogenesis: A multifaceted role for general regulatory factors. *Transcription* **2017**, *8*, 254–260. [CrossRef]
8. Knight, B.; Kubik, S.; Ghosh, B.; Bruzzone, M.J.; Geertz, M.; Martin, V.; Dénervaud, N.; Jacquet, P.; Ozkan, B.; Rougemont, J.; et al. Two distinct promoter architectures centered on dynamic nucleosomes control ribosomal protein gene transcription. *Genes. Dev.* **2014**, *28*, 1695–1709. [CrossRef]
9. Roepcke, S.; Zhi, D.; Vingron, M.; Arndt, P.F. Identification of highly specific localized sequence motifs in human ribosomal protein gene promoters. *Gene* **2006**, *365*, 48–56. [CrossRef]
10. Ma, X.; Zhang, K.; Li, X. Evolution of *Drosophila* ribosomal protein gene core promoters. *Gene* **2009**, *432*, 54–59. [CrossRef]
11. Perina, D.; Korolija, M.; Roller, M.; Harcet, M.; Jelicic, B.; Mikoc, A.; Cetkovic, H. Over-represented localized sequence motifs in ribosomal protein gene promoters of basal metazoans. *Genomics* **2011**, *98*, 56–63. [CrossRef] [PubMed]
12. Sleumer, M.C.; Wei, G.; Wang, Y.; Chang, H.; Xu, T.; Chen, R.; Zhang, M.Q. Regulatory elements of *Caenorhabditis elegans* ribosomal protein genes. *BMC Genom.* **2012**, *13*, 433. [CrossRef]
13. Parry, T.J.; Theisen, J.W.; Hsu, J.Y.; Wang, Y.L.; Corcoran, D.L.; Eustice, M.; Ohler, U.; Kadonaga, J.T. The TCT motif, a key component of an RNA polymerase II transcription system for the translational machinery. *Genes. Dev.* **2010**, *24*, 2013–2018. [CrossRef] [PubMed]
14. Baumann, D.G.; Gilmour, D.S. A sequence-specific core promoter-binding transcription factor recruits TRF2 to coordinately transcribe ribosomal protein genes. *Nucleic Acids Res.* **2017**, *45*, 10481–10491. [CrossRef] [PubMed]
15. Wang, Y.L.; Duttke, S.H.; Chen, K.; Johnston, J.; Kassavetis, G.A.; Zeitlinger, J.; Kadonaga, J.T. TRF2, but not TBP, mediates the transcription of ribosomal protein genes. *Genes. Dev.* **2014**, *28*, 1550–1555. [CrossRef]
16. Perry, R.P. The architecture of mammalian ribosomal protein promoters. *BMC Evol. Biol.* **2005**, *5*, 15. [CrossRef]
17. van Riggelen, J.; Yetil, A.; Felsher, D.W. MYC as a regulator of ribosome biogenesis and protein synthesis. *Nat. Rev. Cancer* **2010**, *10*, 301–309. [CrossRef]
18. Kondrashov, N.; Pusic, A.; Stumpf, C.R.; Shimizu, K.; Hsieh, A.C.; Xue, S.; Ishijima, J.; Shiroishi, T.; Barna, M. Ribosome-mediated specificity in Hox mRNA translation and vertebrate tissue patterning. *Cell* **2011**, *145*, 383–397. [CrossRef]
19. Panda, A.; Yadav, A.; Yeerna, H.; Singh, A.; Biehl, M.; Lux, M.; Schulz, A.; Klecha, T.; Doniach, S.; Khiabanian, H.; et al. Tissue- and development-stage-specific mRNA and heterogeneous CNV signatures of human ribosomal proteins in normal and cancer samples. *Nucleic Acids Res.* **2020**, *48*, 7079–7098. [CrossRef]
20. Di Vona, C.; Bezdan, D.; Islam, A.B.; Salichs, E.; Lopez-Bigas, N.; Ossowski, S.; de la Luna, S. Chromatin-wide profiling of DYRK1A reveals a role as a gene-specific RNA polymerase II CTD kinase. *Mol. Cell* **2015**, *57*, 506–520. [CrossRef]
21. Arbones, M.L.; Thomazeau, A.; Nakano-Kobayashi, A.; Hagiwara, M.; Delabar, J.M. DYRK1A and cognition: A lifelong relationship. *Pharmacol. Ther.* **2019**, *194*, 199–221. [CrossRef] [PubMed]

22. Boni, J.; Rubio-Perez, C.; Lopez-Bigas, N.; Fillat, C.; de la Luna, S. The DYRK family of kinases in cancer: Molecular functions and therapeutic opportunities. *Cancers* **2020**, *12*, 2106. [CrossRef] [PubMed]
23. Rammohan, M.; Harris, E.; Bhansali, R.S.; Zhao, E.; Li, L.S.; Crispino, J.D. The chromosome 21 kinase DYRK1A: Emerging roles in cancer biology and potential as a therapeutic target. *Oncogene* **2022**, *41*, 2003–2011. [CrossRef] [PubMed]
24. Stringer, M.; Goodlett, C.R.; Roper, R.J. Targeting trisomic treatments: Optimizing Dyrk1a inhibition to improve Down syndrome deficits. *Mol. Genet. Genom. Med.* **2017**, *5*, 451–465. [CrossRef] [PubMed]
25. van Bon, B.W.; Coe, B.P.; Bernier, R.; Green, C.; Gerdts, J.; Witherspoon, K.; Kleefstra, T.; Willemsen, M.H.; Kumar, R.; Bosco, P.; et al. Disruptive de novo mutations of *DYRK1A* lead to a syndromic form of autism and ID. *Mol. Psychiatry* **2016**, *21*, 126–132. [CrossRef]
26. Arranz, J.; Balducci, E.; Arato, K.; Sanchez-Elexpuru, G.; Najas, S.; Parras, A.; Rebollo, E.; Pijuan, I.; Erb, I.; Verde, G.; et al. Impaired development of neocortical circuits contributes to the neurological alterations in DYRK1A haploinsufficiency syndrome. *Neurobiol. Dis.* **2019**, *127*, 210–222. [CrossRef]
27. Kurtz-Nelson, E.C.; Rea, H.M.; Petriceks, A.C.; Hudac, C.M.; Wang, T.; Earl, R.K.; Bernier, R.A.; Eichler, E.E.; Neuhaus, E. Characterizing the autism spectrum phenotype in DYRK1A-related syndrome. *Autism Res.* **2023**, *16*, 1488–1500. [CrossRef]
28. Branca, C.; Shaw, D.M.; Belfiore, R.; Gokhale, V.; Shaw, A.Y.; Foley, C.; Smith, B.; Hulme, C.; Dunckley, T.; Meechoovet, B.; et al. Dyrk1 inhibition improves Alzheimer's disease-like pathology. *Aging Cell* **2017**, *16*, 1146–1154. [CrossRef]
29. Scavuzzo, M.A.; Borowiak, M. Two drugs converged in a pancreatic beta cell. *Sci. Transl. Med.* **2020**, *12*, eaba7359. [CrossRef]
30. Yu, D.; Cattoglio, C.; Xue, Y.; Zhou, Q. A complex between DYRK1A and DCAF7 phosphorylates the C-terminal domain of RNA polymerase II to promote myogenesis. *Nucleic Acids Res.* **2019**, *47*, 4462–4475. [CrossRef]
31. Lu, H.; Yu, D.; Hansen, A.S.; Ganguly, S.; Liu, R.; Heckert, A.; Darzacq, X.; Zhou, Q. Phase-separation mechanism for C-terminal hyperphosphorylation of RNA polymerase II. *Nature* **2018**, *558*, 318–323. [CrossRef] [PubMed]
32. Schneider, C.A.; Rasband, W.S.; Eliceiri, K.W. NIH Image to ImageJ: 25 years of image analysis. *Nat. Methods* **2012**, *9*, 671–675. [CrossRef] [PubMed]
33. Ferrari, R.; Su, T.; Li, B.; Bonora, G.; Oberai, A.; Chan, Y.; Sasidharan, R.; Berk, A.J.; Pellegrini, M.; Kurdistani, S.K. Reorganization of the host epigenome by a viral oncogene. *Genome Res.* **2012**, *22*, 1212–1221. [CrossRef] [PubMed]
34. Consortium, E.P. An integrated encyclopedia of DNA elements in the human genome. *Nature* **2012**, *489*, 57–74. [CrossRef] [PubMed]
35. Love, M.I.; Huber, W.; Anders, S. Moderated estimation of fold change and dispersion for RNA-seq data with DESeq2. *Genome Biol.* **2014**, *15*, 550. [CrossRef] [PubMed]
36. Ban, N.; Beckmann, R.; Cate, J.H.; Dinman, J.D.; Dragon, F.; Ellis, S.R.; Lafontaine, D.L.; Lindahl, L.; Liljas, A.; Lipton, J.M.; et al. A new system for naming ribosomal proteins. *Curr. Opin. Struct. Biol.* **2014**, *24*, 165–169. [CrossRef] [PubMed]
37. Bailey, T.L.; Machanick, P. Inferring direct DNA binding from ChIP-seq. *Nucleic Acids Res.* **2012**, *40*, e128. [CrossRef] [PubMed]
38. Mahpour, A.; Scruggs, B.S.; Smiraglia, D.; Ouchi, T.; Gelman, I.H. A methyl-sensitive element induces bidirectional transcription in TATA-less CpG island-associated promoters. *PLoS ONE* **2018**, *13*, e0205608. [CrossRef]
39. Raghav, S.K.; Waszak, S.M.; Krier, I.; Gubelmann, C.; Isakova, A.; Mikkelsen, T.S.; Deplancke, B. Integrative genomics identifies the corepressor SMRT as a gatekeeper of adipogenesis through the transcription factors C/EBPbeta and KAISO. *Mol. Cell* **2012**, *46*, 335–350. [CrossRef]
40. Yamashita, D.; Sano, Y.; Adachi, Y.; Okamoto, Y.; Osada, H.; Takahashi, T.; Yamaguchi, T.; Osumi, T.; Hirose, F. hDREF regulates cell proliferation and expression of ribosomal protein genes. *Mol. Cell Biol.* **2007**, *27*, 2003–2013. [CrossRef]
41. Colombo, P.; Fried, M. Functional elements of the ribosomal protein L7a (rpL7a) gene promoter region and their conservation between mammals and birds. *Nucleic Acids Res.* **1992**, *20*, 3367–3373. [CrossRef] [PubMed]
42. Reschke, M.; Clohessy, J.G.; Seitzer, N.; Goldstein, D.P.; Breitkopf, S.B.; Schmolze, D.B.; Ala, U.; Asara, J.M.; Beck, A.H.; Pandolfi, P.P. Characterization and analysis of the composition and dynamics of the mammalian riboproteome. *Cell Rep.* **2013**, *4*, 1276–1287. [CrossRef] [PubMed]
43. Imami, K.; Milek, M.; Bogdanow, B.; Yasuda, T.; Kastelic, N.; Zauber, H.; Ishihama, Y.; Landthaler, M.; Selbach, M. Phosphorylation of the ribosomal protein RPL12/uL11 affects translation during mitosis. *Mol. Cell* **2018**, *72*, 84–98.e89. [CrossRef] [PubMed]
44. Aviner, R.; Hofmann, S.; Elman, T.; Shenoy, A.; Geiger, T.; Elkon, R.; Ehrlich, M.; Elroy-Stein, O. Proteomic analysis of polyribosomes identifies splicing factors as potential regulators of translation during mitosis. *Nucleic Acids Res.* **2017**, *45*, 5945–5957. [CrossRef] [PubMed]
45. Franks, A.; Airoldi, E.; Slavov, N. Post-transcriptional regulation across human tissues. *PLoS Comput. Biol.* **2017**, *13*, e1005535. [CrossRef]
46. Lloyd, A.C. The regulation of cell size. *Cell* **2013**, *154*, 1194–1205. [CrossRef]
47. Chen, J.Y.; Lin, J.R.; Tsai, F.C.; Meyer, T. Dosage of Dyrk1a shifts cells within a p21-cyclin D1 signaling map to control the decision to enter the cell cycle. *Mol. Cell* **2013**, *52*, 87–100. [CrossRef]
48. Saxton, R.A.; Sabatini, D.M. mTOR signaling in growth, metabolism, and disease. *Cell* **2017**, *168*, 960–976. [CrossRef]
49. Koromilas, A.E. Roles of the translation initiation factor eIF2alpha serine 51 phosphorylation in cancer formation and treatment. *Biochim. Biophys. Acta* **2015**, *1849*, 871–880. [CrossRef]
50. Levy, J.A.; LaFlamme, C.W.; Tsaprailis, G.; Crynen, G.; Page, D.T. Dyrk1a mutations cause undergrowth of cortical pyramidal neurons via dysregulated growth factor signaling. *Biol. Psychiatry* **2021**, *90*, 295–306. [CrossRef]

51. Grand, R.S.; Burger, L.; Gräwe, C.; Michael, A.K.; Isbel, L.; Hess, D.; Hoerner, L.; Iesmantavicius, V.; Durdu, S.; Pregnolato, M.; et al. BANP opens chromatin and activates CpG-island-regulated genes. *Nature* **2021**, *596*, 133–137. [CrossRef] [PubMed]
52. Zhao, C.; Deng, Y.; Liu, L.; Yu, K.; Zhang, L.; Wang, H.; He, X.; Wang, J.; Lu, C.; Wu, L.N.; et al. Dual regulatory switch through interactions of Tcf7l2/Tcf4 with stage-specific partners propels oligodendroglial maturation. *Nat. Commun.* **2016**, *7*, 10883. [CrossRef] [PubMed]
53. De Jaime-Soguero, A.; Aulicino, F.; Ertaylan, G.; Griego, A.; Cerrato, A.; Tallam, A.; Del Sol, A.; Cosma, M.P.; Lluis, F. Wnt/Tcf1 pathway restricts embryonic stem cell cycle through activation of the *Ink4/Arf* locus. *PLoS Genet.* **2017**, *13*, e1006682. [CrossRef] [PubMed]
54. Martianov, I.; Velt, A.; Davidson, G.; Choukrallah, M.A.; Davidson, I. TRF2 is recruited to the pre-initiation complex as a testis-specific subunit of TFIIA/ALF to promote haploid cell gene expression. *Sci. Rep.* **2016**, *6*, 32069. [CrossRef] [PubMed]
55. Serebreni, L.; Pleyer, L.M.; Haberle, V.; Hendy, O.; Vlasova, A.; Loubiere, V.; Nemčko, F.; Bergauer, K.; Roitinger, E.; Mechtler, K.; et al. Functionally distinct promoter classes initiate transcription via different mechanisms reflected in focused versus dispersed initiation patterns. *EMBO J.* **2023**, *42*, e113519. [CrossRef] [PubMed]
56. Wu, C.H.; Sahoo, D.; Arvanitis, C.; Bradon, N.; Dill, D.L.; Felsher, D.W. Combined analysis of murine and human microarrays and ChIP analysis reveals genes associated with the ability of MYC to maintain tumorigenesis. *PLoS Genet.* **2008**, *4*, e1000090. [CrossRef] [PubMed]
57. Perna, D.; Fagà, G.; Verrecchia, A.; Gorski, M.M.; Barozzi, I.; Narang, V.; Khng, J.; Lim, K.C.; Sung, W.K.; Sanges, R.; et al. Genome-wide mapping of Myc binding and gene regulation in serum-stimulated fibroblasts. *Oncogene* **2012**, *31*, 1695–1709. [CrossRef]
58. Grigull, J.; Mnaimneh, S.; Pootoolal, J.; Robinson, M.D.; Hughes, T.R. Genome-wide analysis of mRNA stability using transcription inhibitors and microarrays reveals posttranscriptional control of ribosome biogenesis factors. *Mol. Cell Biol.* **2004**, *24*, 5534–5547. [CrossRef]
59. Gómez-Herreros, F.; Margaritis, T.; Rodríguez-Galán, O.; Pelechano, V.; Begley, V.; Millán-Zambrano, G.; Morillo-Huesca, M.; Muñoz-Centeno, M.C.; Pérez-Ortín, J.E.; de la Cruz, J.; et al. The ribosome assembly gene network is controlled by the feedback regulation of transcription elongation. *Nucleic Acids Res.* **2017**, *45*, 9302–9318. [CrossRef]
60. Gupta, I.; Villanyi, Z.; Kassem, S.; Hughes, C.; Panasenko, O.O.; Steinmetz, L.M.; Collart, M.A. Translational capacity of a cell Is determined during transcription elongation via the Ccr4-Not complex. *Cell Rep.* **2016**, *15*, 1782–1794. [CrossRef]
61. Roux, P.P.; Topisirovic, I. Signaling pathways involved in the regulation of mRNA translation. *Mol. Cell Biol.* **2018**, *38*, e00070-18. [CrossRef]
62. Fotaki, V.; Dierssen, M.; Alcántara, S.; Martínez, S.; Martí, E.; Casas, C.; Visa, J.; Soriano, E.; Estivill, X.; Arbonés, M.L. *Dyrk1A* haploinsufficiency affects viability and causes developmental delay and abnormal brain morphology in mice. *Mol. Cell Biol.* **2002**, *22*, 6636–6647. [CrossRef]
63. Ross, A.P.; Zarbalis, K.S. The emerging roles of ribosome biogenesis in craniofacial development. *Front. Physiol.* **2014**, *5*, 26. [CrossRef]
64. Brooks, S.S.; Wall, A.L.; Golzio, C.; Reid, D.W.; Kondyles, A.; Willer, J.R.; Botti, C.; Nicchitta, C.V.; Katsanis, N.; Davis, E.E. A novel ribosomopathy caused by dysfunction of *RPL10* disrupts neurodevelopment and causes X-linked microcephaly in humans. *Genetics* **2014**, *198*, 723–733. [CrossRef]
65. Borrie, S.C.; Brems, H.; Legius, E.; Bagni, C. Cognitive dysfunctions in intellectual disabilities: The contributions of the Ras-MAPK and PI3K-AKT-mTOR pathways. *Annu. Rev. Genom. Hum. Genet.* **2017**, *18*, 115–142. [CrossRef]
66. Courraud, J.; Chater-Diehl, E.; Durand, B.; Vincent, M.; del Mar Muniz Moreno, M.; Boujelbene, I.; Drouot, N.; Genschik, L.; Schaefer, E.; Nizon, M.; et al. Integrative approach to interpret *DYRK1A* variants, leading to a frequent neurodevelopmental disorder. *Genet. Med.* **2021**, *23*, 2150–2159. [CrossRef]
67. Chiva, C.; Olivella, R.; Borras, E.; Espadas, G.; Pastor, O.; Sole, A.; Sabido, E. QCloud: A cloud-based quality control system for mass spectrometry-based proteomics laboratories. *PLoS ONE* **2018**, *13*, e0189209. [CrossRef]
68. Perkins, D.N.; Pappin, D.J.; Creasy, D.M.; Cottrell, J.S. Probability-based protein identification by searching sequence databases using mass spectrometry data. *Electrophoresis* **1999**, *20*, 3551–3567. [CrossRef]
69. Langmead, B.; Trapnell, C.; Pop, M.; Salzberg, S.L. Ultrafast and memory-efficient alignment of short DNA sequences to the human genome. *Genome Biol.* **2009**, *10*, R25. [CrossRef]
70. Dobin, A.; Davis, C.A.; Schlesinger, F.; Drenkow, J.; Zaleski, C.; Jha, S.; Batut, P.; Chaisson, M.; Gingeras, T.R. STAR: Ultrafast universal RNA-seq aligner. *Bioinformatics* **2013**, *29*, 15–21. [CrossRef]
71. Bluhm, A.; Viceconte, N.; Li, F.; Rane, G.; Ritz, S.; Wang, S.; Levin, M.; Shi, Y.; Kappei, D.; Butter, F. ZBTB10 binds the telomeric variant repeat TTGGGG and interacts with TRF2. *Nucleic Acids Res.* **2019**, *47*, 1896–1907. [CrossRef]
72. Lockstone, H.E.; Harris, L.W.; Swatton, J.E.; Wayland, M.T.; Holland, A.J.; Bahn, S. Gene expression profiling in the adult Down syndrome brain. *Genomics* **2007**, *90*, 647–660. [CrossRef]
73. Nicol, J.W.; Helt, G.A.; Blanchard, S.G., Jr.; Raja, A.; Loraine, A.E. The Integrated Genome Browser: Free software for distribution and exploration of genome-scale datasets. *Bioinformatics* **2009**, *25*, 2730–2731. [CrossRef]
74. Durinck, S.; Moreau, Y.; Kasprzyk, A.; Davis, S.; De Moor, B.; Brazma, A.; Huber, W. BioMart and Bioconductor: A powerful link between biological databases and microarray data analysis. *Bioinformatics* **2005**, *21*, 3439–3440. [CrossRef]

75. Quinlan, A.R.; Hall, I.M. BEDTools: A flexible suite of utilities for comparing genomic features. *Bioinformatics* **2010**, *26*, 841–842. [CrossRef]
76. Ramirez, F.; Ryan, D.P.; Gruning, B.; Bhardwaj, V.; Kilpert, F.; Richter, A.S.; Heyne, S.; Dundar, F.; Manke, T. deepTools2: A next generation web server for deep-sequencing data analysis. *Nucleic Acids Res.* **2016**, *44*, W160–W165. [CrossRef]
77. Grant, C.E.; Bailey, T.L.; Noble, W.S. FIMO: Scanning for occurrences of a given motif. *Bioinformatics* **2011**, *27*, 1017–1018. [CrossRef]
78. Benjamini, Y.; Hochberg, Y. Controlling the False Discovery Rate: A practical and powerful approach to multiple testing. *J. R. Stat. Soc. Series B* **1995**, *57*, 289–300. [CrossRef]
79. Fornes, O.; Castro-Mondragon, J.A.; Khan, A.; van der Lee, R.; Zhang, X.; Richmond, P.A.; Modi, B.P.; Correard, S.; Gheorghe, M.; Baranašić, D.; et al. JASPAR 2020: Update of the open-access database of transcription factor binding profiles. *Nucleic Acids Res.* **2020**, *48*, D87–D92. [CrossRef]
80. Kuleshov, M.V.; Jones, M.R.; Rouillard, A.D.; Fernandez, N.F.; Duan, Q.; Wang, Z.; Koplev, S.; Jenkins, S.L.; Jagodnik, K.M.; Lachmann, A.; et al. Enrichr: A comprehensive gene set enrichment analysis web server 2016 update. *Nucleic Acids Res.* **2016**, *44*, W90–W97. [CrossRef]

Disclaimer/Publisher's Note: The statements, opinions and data contained in all publications are solely those of the individual author(s) and contributor(s) and not of MDPI and/or the editor(s). MDPI and/or the editor(s) disclaim responsibility for any injury to people or property resulting from any ideas, methods, instructions or products referred to in the content.

Article

Dissecting the Nuclear Import of the Ribosomal Protein Rps2 (uS5)

Andreas Steiner [1,2], Sébastien Favre [3], Maximilian Mack [1,2], Annika Hausharter [1], Benjamin Pillet [3], Jutta Hafner [1,2], Valentin Mitterer [1,2], Dieter Kressler [3], Brigitte Pertschy [1,2,*] and Ingrid Zierler [1,2,*]

[1] Institute of Molecular Biosciences, University of Graz, Humboldtstrasse 50, 8010 Graz, Austria; andreas.steiner23@yahoo.com (A.S.); maximilian.mack@uni-graz.at (M.M.); valentin.mitterer@uni-graz.at (V.M.)
[2] BioTechMed-Graz, Mozartgasse 12/II, 8010 Graz, Austria
[3] Unit of Biochemistry, Department of Biology, University of Fribourg, Chemin du Musée 10, 1700 Fribourg, Switzerland; sebastien.favre@unifr.ch (S.F.); benjamin.pillet@unifr.ch (B.P.); dieter.kressler@unifr.ch (D.K.)
* Correspondence: brigitte.pertschy@uni-graz.at (B.P.); ingrid.zierler@uni-graz.at (I.Z.)

Citation: Steiner, A.; Favre, S.; Mack, M.; Hausharter, A.; Pillet, B.; Hafner, J.; Mitterer, V.; Kressler, D.; Pertschy, B.; Zierler, I. Dissecting the Nuclear Import of the Ribosomal Protein Rps2 (uS5). *Biomolecules* **2023**, *13*, 1127. https://doi.org/10.3390/biom13071127

Academic Editor: Leonard B. Maggi, Jr.

Received: 20 April 2023
Revised: 6 July 2023
Accepted: 12 July 2023
Published: 14 July 2023

Copyright: © 2023 by the authors. Licensee MDPI, Basel, Switzerland. This article is an open access article distributed under the terms and conditions of the Creative Commons Attribution (CC BY) license (https://creativecommons.org/licenses/by/4.0/).

Abstract: The ribosome is assembled in a complex process mainly taking place in the nucleus. Consequently, newly synthesized ribosomal proteins have to travel from the cytoplasm into the nucleus, where they are incorporated into nascent ribosomal subunits. In this study, we set out to investigate the mechanism mediating nuclear import of the small subunit ribosomal protein Rps2. We demonstrate that an internal region in Rps2, ranging from amino acids 76 to 145, is sufficient to target a 3xyEGFP reporter to the nucleus. The importin-β Pse1 interacts with this Rps2 region and is involved in its import, with Rps2 residues arginine 95, arginine 97, and lysine 99 being important determinants for both Pse1 binding and nuclear localization. Moreover, our data reveal a second import mechanism involving the N-terminal region of Rps2, which depends on the presence of basic residues within amino acids 10 to 28. This Rps2 segment overlaps with the binding site of the dedicated chaperone Tsr4; however, the nuclear import of Rps2 via the internal as well as the N-terminal nuclear-targeting element does not depend on Tsr4. Taken together, our study has unveiled hitherto undescribed nuclear import signals, showcasing the versatility of the mechanisms coordinating the nuclear import of ribosomal proteins.

Keywords: ribosomal protein; ribosome assembly; nuclear import; Rps2; uS5; importin; Pse1; Kap121; dedicated chaperone; yeast

1. Introduction

The ribosome is a remarkable and extremely efficient macromolecular RNA–protein machine that synthesizes all proteins in the cytoplasm of the eukaryotic cell and is composed of a small 40S and a large 60S subunit. With the help of several hundred different assembly factors, the two subunits are assembled from ribosomal RNAs (rRNAs) and ribosomal proteins (r-proteins) in a highly conserved and complex maturation pathway called ribosome biogenesis, which is best studied in the yeast *Saccharomyces cerevisiae*. This assembly and maturation process starts in the nucleolus, a non-membrane-enclosed subcompartment of the nucleus, progresses in the nucleoplasm, and ends in the cytoplasm, where the two mature subunits form the translation-competent 80S ribosome [1–4].

R-proteins have to overcome two major obstacles before they are assembled into preribosomal particles in the nucle(ol)us. First, as they contain highly basic regions as well as unstructured N- or C-terminal extensions and internal loops, which engage in interactions with the negatively charged rRNA in the ribosome, r-proteins are prone to aggregation as long as they are not assembled into (pre-)ribosomal subunits [5]. Second, since r-proteins are synthesized in the cytoplasm but most of them are incorporated into pre-ribosomes in the nucle(ol)us, they need to be imported into the nucleus through the nuclear pore complex (NPC) before becoming available for ribosome biogenesis [6].

Two classes of proteins are specialized in helping to overcome these obstacles and to ensure the safe delivery of r-proteins at their site of assembly with the rRNA: (1) Dedicated chaperones of r-proteins: More than ten r-proteins were found to require protection by these specialized chaperones [5,7–10]. Most dedicated chaperones bind r-proteins already co-translationally and protect them from aggregation, presumably through shielding positively charged surfaces [5,7,11]. (2) Importins of the karyopherin superfamily: Besides mediating the nuclear import of many diverse substrate proteins (cargoes) across the hydrophobic channel of the NPC [6], selected importins have also been shown to facilitate nuclear import of r-proteins and to prevent their aggregation [12,13].

Importins usually recognize their cargo proteins by binding to basic nuclear localization sequences (NLSs). All importin-ßs have the ability to interact with the hydrophobic FG-repeat meshwork in the central channel of the NPC. Additionally, most importin-ßs, including Pse1 (also called Kap121), Kap123, and Kap104, can recognize the NLS of their cargoes directly; only the importin-ß Kap95 requires an adaptor, the importin-α Srp1 (also called Kap60), for binding to the NLS-containing cargo protein. Importin-α recognizes classical NLS sequences, which can be either monopartite (consensus K-K/R-X-K/R; where X can be any amino acid) or bipartite (K/R-K/R-X_{10-12} K/$R_{3/5}$; with K/$R_{3/5}$ being five residues containing at least three Ks or Rs). The importin-ß Kap104 recognizes PY-NLSs containing an N-terminal basic (or hydrophobic) motif and a C-terminal R/K/H-X_{2-5}-P-Y/L/F motif. Additionally, Kap104 can bind to RGG regions (RG-rich NLSs). Pse1 binds to IK-NLSs with the consensus K-V/I-X-K-X_{1-2}-K/H/R. Importantly, however, many cargoes of the above-named importins do not harbor sequences following the so-far-described NLS consensus sequences, and, moreover, for many importins, no targeting signals have been defined at all [6].

After nuclear import, the importins release their cargo proteins by binding to the small GTPase Ran in its GTP-bound form (RanGTP), which is highly concentrated in the nucleus, thereby controlling the directionality of transport [6]. Nuclear import of r-proteins is believed to be mainly performed by the non-essential importin-β Kap123, with some redundant contribution of the essential Pse1 [13].

In contrast to that notion, studies in recent years by us and others on the coordination of chaperoning and nuclear import of r-proteins revealed that several r-proteins employ importins other than Kap123. Rps3 is imported into the nucleus as a dimer, with one N-terminal domain protected by its dedicated chaperone Yar1 and the second one bound by the Srp1/Kap95 importin-α/importin-β dimer [14,15]. Rpl5 and Rpl11 are imported in complex with their dedicated chaperone Syo1, which functions as a transport adaptor for Kap104 [16]. Rpl4 contains at least five different NLS sequences and is imported into the nucleus in complex with its dedicated chaperone Acl4 by importin Kap104 [17–19]. Last but not least, Rps26 can be imported into the nucleus by Kap123, Kap104, or Pse1, and is then released from the importin in a RanGTP-independent manner by its dedicated chaperone Tsr2 [20].

We are interested in the nuclear import of r-protein Rps2 (also called uS5 [21]). Rps2 has an evolutionarily conserved dedicated chaperone, Tsr4 (PDCD2 in humans), which binds co-translationally to its unstructured N-terminal extension [7,10,22]. In the absence of Tsr4, Rps2 accumulates in the nucleus, suggesting that Tsr4 is required for the efficient incorporation of Rps2 into pre-ribosomes [7]. How Rps2 is imported into the nucleus has, however, remained elusive.

In this study, we uncovered that an internal fragment of Rps2, Rps2(76–145), interacts with the importin-β Pse1 and is sufficient to target a 3xyEGFP reporter into the nucleus, indicating that it contains a functional NLS. Nuclear import of Rps2(76–145)-3xyEGFP was blocked in a *pse1-1* mutant or upon changing three basic residues in Rps2, arginine 95 (R95), R97, and lysine 99 (K99), to alanines (As), suggesting that these residues are part of the NLS. Surprisingly, when fusing a larger N-terminal Rps2 fragment, Rps2(1–145), to the 3xyEGFP reporter, nuclear targeting was no longer disturbed by the R95A, R97A, and K99A exchanges. Moreover, we identified a sequence in the N-terminal part of Rps2 (amino acids 10–28), whose basic residues are essential for the nuclear targeting of the Rps2(1–145)-3xyEGFP fusion protein bearing the R95A, R97A, and K99A mutations, strongly suggesting the presence of a second NLS in the eukaryote-specific N-terminal extension of Rps2. Our results moreover revealed that import, both via the internal and the N-terminal nuclear targeting region, also occurs in the absence of Tsr4.

2. Methods

2.1. Yeast Strains and Genetic Methods

All *S. cerevisiae* strains used in this study are listed in Supplementary Table S1. Yeast plasmids were constructed using standard recombinant DNA techniques and are listed in Supplementary Table S2. All DNA fragments amplified by PCR were verified by sequencing.

The *KAP104* shuffle strain was transformed with the YCplac22-*KAP104* or the pRS314-*kap104-16* plasmid (*TRP1*), respectively, and transformed cells were streaked on 5-FOA (Thermo Scientific) plates to counter-select against the pRS316-*KAP104* (*URA3*) shuffle plasmid. After plasmid shuffling, cells were grown on plates lacking tryptophan (SDC-trp) and transformed with the *LEU2* plasmid expressing Rps2(76–145)-3xyEGFP.

2.2. Fluorescence Microscopy

Yeast strains were grown at 30 °C in SDC media lacking leucine (SDC-leu) to an OD_{600} of ~0.5 (logarithmic growth phase). Cells were imaged by fluorescence microscopy using a Leica DM6 B microscope, equipped with a DFC 9000 GT camera, using the PLAN APO 100× objective, narrow-band GFP or RHOD ET filters, and LasX software. Full-length Rps2 as well as fragments and variants thereof were expressed with a C-terminal 3xyEGFP tag under the transcriptional control of the *ADH1* promoter from a centromeric *LEU2* plasmid. Plasmids expressing these 3xyEGFP reporter proteins were transformed into a Nop58-yEmCherry expressing strain, the C303 wild-type strain, *rps2Δ* and *tsr4Δ rps2Δ* strains containing a centromeric *URA3-RPS2* plasmid, or the indicated importin mutant strains. Since Rps2 is an essential protein, we investigated the localization of the different Rps2-3xyEGFP fusion proteins in strains harboring the wild-type *RPS2* gene either at the chromosomal locus or on a plasmid.

2.3. Yeast Two-Hybrid (Y2H) Assays

Protein–protein interactions between Rps2 (and fragments/variants thereof) and Pse1/Pse1.302C or Tsr4 were analyzed by yeast two-hybrid (Y2H) assays using the reporter strain PJ69-4A. This Y2H strain allows for the detection of both weak (*HIS3* reporter) and strong interactions (*ADE2* reporter). Two plasmids were co-transformed into PJ69-4A, whereby one plasmid was expressing fusions to the Gal4 DNA-binding domain (G4BD, BD, *TRP1* marker) and the other fusions to the Gal4 transcription activation domain (G4AD, AD, *LEU2* marker). For the Rps2-Tsr4 Y2H interaction assays, Rps2 variants, C-terminally fused to the G4BD, and full-length Tsr4, C-terminally fused to the G4AD, were expressed from centromeric (CEN, low-copy) plasmids (pG4BDC22 and pG4ADC111, respectively). For the Rps2-Pse1 Y2H interaction assays, Rps2 variants, either N- or C-terminally fused to the G4BD, and Pse1 or Pse1.302C, C-terminally fused to the G4AD, were expressed from episomal (2μ, high-copy) plasmids (pG4BDN112, pGAG4BDC112, and pGAG4ADC181, respectively).

After the selection of transformants on plates lacking leucine and tryptophan (SDC-leu-trp, -LT), cells were spotted onto SDC-leu-trp plates as well as onto plates lacking histidine, leucine, and tryptophan (SDC-his-leu-trp, -HLT), and lacking adenine, leucine, and tryptophan (SDC-ade-leu-trp, -ALT), respectively. Plates were incubated for 3 days at 30 °C.

2.4. Tandem Affinity Purification (TAP)

For TAP purifications, plasmids expressing Rps2(76–145) or Rps2(76–145).$R_{95}R_{97}K_{99>A}$, C-terminally fused to the TAP tag, or a plasmid expressing the TAP tag alone were transformed into a haploid W303-derived wild-type strain. Cells were grown in 4 l yeast extract peptone dextrose medium (YPD) to an optical density (OD_{600}) of 2 at 30 °C.

TAP purifications were performed in a buffer containing 50 mM Tris-HCl (pH 7.5), 100 mM NaCl, 1.5 mM $MgCl_2$, 0.1% NP-40, 1 mM dithiothreitol (DTT), and 1x Protease Inhibitor Mix FY (Serva). Cells were lysed by mechanical disruption using glass beads and the cell lysate was incubated with 300 µL IgG Sepharose™ 6 Fast Flow (GE Healthcare, Chicago, IL, USA) for 60 min at 4 °C. After incubation, the IgG Sepharose™ beads were transferred into Mobicol columns (MoBiTec, Göttingen, Germany) and washed with 10 mL buffer. Then, TEV protease was added and elution from the beads was performed under rotation for 90 min at room temperature. After the addition of 2 mM $CaCl_2$, TEV eluates were incubated with 300 µL Calmodulin Sepharose™ 4B beads (GE Healthcare) for 60 min at 4 °C. After washing with 5 mL buffer containing 2 mM $CaCl_2$, proteins were eluted from Calmodulin Sepharose™ with elution buffer consisting of 10 mM Tris-HCl (pH 8.0), 5 mM EGTA, and 50 mM NaCl under rotation for 20 min at room temperature. The protein samples were separated on NuPAGE™ 4–12% Bis-Tris gels (Invitrogen, Carlsbad, CA, USA) followed by Western blotting.

2.5. Western Blotting

Western blot analysis was performed using the following antibodies: α-CBP antibody (1:5000; Merck-Millipore, Burlington, MA, USA, cat. no. 07-482), α-Pse1 antibody (1:500; Matthias Seedorf [22]), secondary α-rabbit horseradish peroxidase-conjugated antibody (1:15,000; Sigma-Aldrich, St. Louis, MO, USA, cat. no. A0545). Protein signals were visualized using the Clarity™ Western ECL Substrate Kit (Bio-Rad, Hercules, CA, USA) and captured by the ChemiDoc™ Touch Imaging System (Bio-Rad).

2.6. TurboID-Based Proximity Labeling

Plasmids expressing C-terminal TurboID-tagged bait proteins under the control of the copper-inducible *CUP1* promoter were transformed into the wild-type strain YDK11-5A. Transformed cells were grown at 30 °C in 100 mL SDC-leu medium, prepared with copper-free yeast nitrogen base (FORMEDIUM), to an OD_{600} between 0.4 and 0.5. Then, copper sulfate, to induce expression from the *CUP1* promoter, and freshly prepared biotin (Sigma-Aldrich, St. Louis, MO, USA) were added to a final concentration of 500 µM, and cells were grown for an additional hour, typically reaching a final OD_{600} between 0.6 and 0.8, and harvested by centrifugation at 4000 rpm for 5 min at 4 °C. Then, cells were washed with 50 mL ice-cold H_2O, resuspended in 1 mL ice-cold lysis buffer (LB: 50 mM Tris-HCl (pH 7.5), 150 mM NaCl, 1.5 mM MgCl2, 0.1% SDS, and 1% Triton X-100) containing 1 mM PMSF, transferred to 2 mL safe-lock tubes, pelleted by centrifugation, frozen in liquid nitrogen, and stored at −80 °C. Extracts were prepared, upon the resuspension of cells in 400 µL lysis buffer containing 0.5% sodium deoxycholate and 1 mM PMSF (LB-P/D), by glass bead lysis with a Precellys 24 homogenizer (Bertin Technologies, Montigny-le-Bretonneux, France) set at 5000 rpm using a 3 × 30 s lysis cycle with 30 s breaks in between at 4 °C. Lysates were transferred to 1.5 mL tubes. For complete extract recovery, 200 µL LB-P/D was added to the glass beads and, after brief vortexing, combined with the already transferred lysate. Cell lysates were clarified by centrifugation for 10 min at 13,500 rpm at 4 °C, transferred to a new 1.5 mL tube. Total protein concentration in the clarified cell extracts was determined with the Pierce™ BCA Protein Assay Kit (Thermo Scientific, Waltham, MA, USA) using a microplate reader (BioTek 800 TS). To reduce non-specific binding, 100 µL of Pierce™ High Capacity Streptavidin Agarose Resin (Thermo Scientific) slurry, corresponding to 50 µL of settled beads, were transferred to a 1.5 mL safe-lock tube, blocked by incubation with 1 mL LB containing 3% BSA for 1 h at RT, and then washed four times with 1 mL LB. For the affinity purification of biotinylated proteins, 2 mg of total protein in an adjusted volume of 800 µL LB-P/D was added to the blocked and washed streptavidin beads, and binding was carried out for 1 h at RT on a rotating wheel. Beads were then washed once for 5 min with 1 mL of wash buffer (50 mM Tris-HCl (pH 7.5), 2% SDS), five times with 1 mL LB, and finally five times with 1 mL ABC buffer (100 mM ammonium bicarbonate (pH 8.2)). Bound proteins were eluted by two consecutive incubations with 30 µL 3× SDS sample buffer, containing 10 mM biotin and 20 mM DTT, for 10 min at 75 °C. The eluates were combined in one 1.5 mL safe-lock tube and stored at –20 °C. Upon reduction with DTT and alkylation with iodoacetamide, samples were separated on NuPAGE 4–12% Bis-Tris gels (Invitrogen, Carlsbad, CA, USA), run in NuPAGE 1× MES SDS running buffer (Novex) at 200 V for a total of 12 min. The gels were incubated with Brilliant Blue G Colloidal Coomassie (Sigma-Aldrich) until the staining of proteins was visible. Each lane was cut, from slot to the migration front, into three gel pieces that were, upon their fragmentation into smaller pieces, transferred into separate 1.5 mL low-binding tubes.

Gel pieces were covered with 100–150 µL of ABC buffer, prepared in HPLC-grade H_2O, and incubated for 10 min at RT in a thermoshaker set to 1000 rpm. Then, gel pieces were covered with 100–150 µL of HPLC-grade absolute EtOH and incubated for 10 min at RT in a thermoshaker set to 1000 rpm. These two wash steps were repeated two more times. For the in-gel digestion of proteins, gel pieces were covered with 120 µL of ABC buffer containing 1 µg sequencing-grade modified trypsin (Promega Madison, WI, USA) and incubated overnight at 37 °C with shaking at 1000 rpm. To stop the digestion and recover the peptides, 50 µL of a 2% trifluoroacetic acid (TFA) solution was added, and, after a 10 min incubation at RT with shaking at 1000 rpm, the supernatant was transferred to a new 1.5 mL low-binding tube. The gel pieces were then incubated, again for 10 min at RT with shaking at 1000 rpm, another two times with 100–150 µL EtOH, and these two supernatants were combined with the first supernatant. Finally, using a SpeedVac, the organic solvents were evaporated and the volume was reduced to around 50 µL. Then, 200 µL of buffer A (0.5% acetic acid) were added, and the samples were applied to C18 StageTips [23], equilibrated with 50 µL of buffer B (80% acetonitrile, 0.3% TFA) and washed twice with 50 µL of buffer A, for desalting and peptide purification. StageTips were washed once with 100 µL of buffer A, and the peptides were eluted with 50 µL of buffer B. The solvents were completely evaporated using a SpeedVac. Peptides were resuspended by first adding 3 µL buffer A* (3% acetonitrile, 0.3% TFA) and then 17 µL buffer A*/A (30% buffer A*/70% buffer A), with each solvent addition being followed by vortexing for 10 s. Samples were stored at –80 °C.

LC-MS/MS measurements were performed on a Q Exactive HF-X (Thermo Scientific) coupled to an EASY-nLC 1200 nanoflow-HPLC (Thermo Scientific). HPLC-column tips (fused silica) with 75 µm inner diameter were self-packed with ReproSil-Pur 120 C18-AQ, 1.9 µm particle size (Dr. Maisch GmbH, Ammerbuch, Germany) to a length of 20 cm. Samples were directly applied onto the column without a pre-column. A gradient of A (0.1% formic acid in H_2O) and B (0.1% formic acid in 80% acetonitrile in H_2O) with increasing organic proportion was used for peptide separation (loading of sample with 0% B; separation ramp: from 5–30% B within 85 min). The flow rate was 250 nL/min and for sample application, it was 600 nL/min. The mass spectrometer was operated in the data-dependent mode and switched automatically between MS (max. of 1×10^6 ions) and MS/MS. Each MS scan was followed by a maximum of ten MS/MS scans using a normalized collision energy of 25% and a target value of 1000. Parent ions with a charge state form $z = 1$ and unassigned charge states were excluded for fragmentation. The mass range for MS was $m/z = 370–1750$. The resolution for MS was set to 70,000 and for MS/MS to 17,500. MS parameters were as follows: spray voltage 2.3 kV, no sheath and auxiliary gas flow, ion-transfer tube temperature 250 °C.

The MS raw data files were analyzed with the MaxQuant software package version 1.6.2.10 [24] for peak detection, generation of peak lists of mass-error-corrected peptides, and database searches. The UniProt *Saccharomyces cerevisiae* database (version March 2016), additionally including common contaminants, trypsin, TurboID, and GFP, was used as reference. Carbamidomethylcysteine was set as fixed modification and protein amino-terminal acetylation, oxidation of methionine, and biotin were set as variable modifications. Four missed cleavages were allowed, enzyme specificity was Trypsin/P, and the MS/MS tolerance was set to 20 ppm. Peptide lists were further used by MaxQuant to identify and relatively quantify proteins using the following parameters: peptide and protein false discovery rates, based on a forward–reverse database, were set to 0.01, minimum peptide length was set to seven, and minimum number of unique peptides for identification and quantification of proteins was set to one. The 'match-between-run' option (0.7 min) was used.

For quantification, missing iBAQ (intensity-based absolute quantification) values in the two control purifications from cells expressing either the GFP-TurboID or the NLS-GFP-TurboID bait were imputed in Perseus [25]. For normalization of intensities in each independent purification, iBAQ values were divided by the median iBAQ value, derived from all nonzero values, of the respective purification. To calculate the enrichment of a given protein compared to its average abundance in the two control purifications, the normalized iBAQ values were log2-transformed and those of the control purifications were subtracted from the ones of each respective bait purification. For graphical presentation, the normalized iBAQ value (log10 scale) of each protein detected in a given bait purification was plotted against its relative abundance (log2-transformed enrichment compared to the control purifications). To visualize the effects of the RRK>A mutations on the proximal protein neighborhood of Rps2, the normalized iBAQ value (log10 scale) of each protein detected in the purification from cells expressing a wild-type Rps2 bait protein (full-length Rps2, Rps2(1–145), or Rps2(76–145)) was plotted against its relative abundance (log2-transformed enrichment) compared to the purification from cells expressing the respective RRK>A mutant protein (with prior imputation of missing iBAQ values).

3. Results

3.1. Rps2 Amino Acids 76–145 Are Sufficient to Target the Protein to the Nucleus

Like most other r-proteins, Rps2 assembles into pre-ribosomal particles in the nucleus [26], necessitating nuclear import of newly synthesized Rps2. Yeast Rps2 is recognized co-translationally by its dedicated chaperone Tsr4 in the cytoplasm [7,10]; however, the mechanism by which Rps2 is imported into the nucleus has so far remained elusive.

We first aimed to narrow down the part of Rps2 that is capable of targeting the protein to the nucleus. Since full-length Rps2 is imported into the nucleus, incorporated into pre-ribosomal particles, and subsequently, as a component of these, exported to the cytoplasm, the majority of all cellular Rps2 is present in cytoplasmic 40S subunits. We reasoned, however, that small sub-fragments of Rps2 would most likely not become incorporated into pre-ribosomes and could hence be visualized in the nucleus in case they carry an NLS. We designed a series of Rps2 fragments with overlapping regions, and constructed plasmids encoding these Rps2 fragments (Figure 1A, Supplementary Figure S1A,B), each fused to a 3xyEGFP tag at the C-terminus. In order to differentiate between different types of nuclear localization, we utilized a strain expressing the nucleolar marker protein Nop58 fused to mCherry (Nop58-yEmCherry) from the genomic locus. A localization of 3xyEGFP reporter fusion proteins exclusively in the nucleolus, the site where ribosome biogenesis starts, would result in a perfect overlap with the Nop58-yEmCherry signal. Conversely, a nucleoplasmic localization of the 3xyEGFP reporter fusions would result in a signal adjacent to the Nop58-yEmCherry nucleolar signal with no overlap. Finally, reporter fusions localizing to both nuclear subcompartments would exhibit a larger oval-shaped GFP signal and partially overlap with the Nop58-yEmCherry marker.

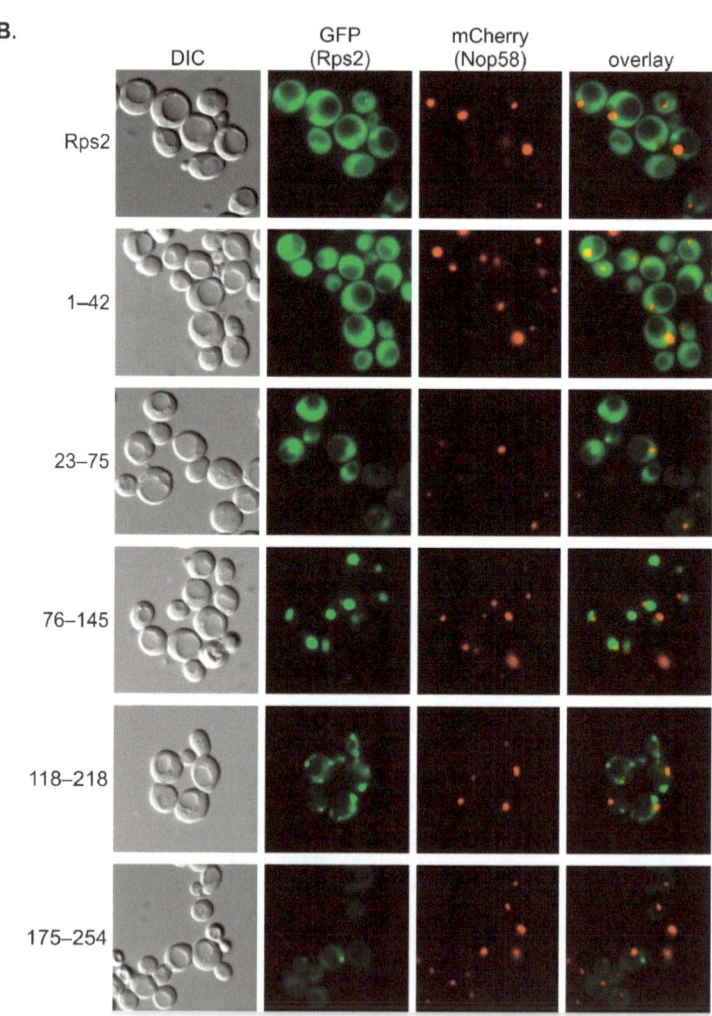

Figure 1. Rps2 residues 76–145 are sufficient to target a 3xyEGFP reporter to the nucleus. (**A**) Schematic representation of Rps2 with secondary structure elements and overview of fragments tested in (**B**). Indicated domains according to the Rps2 structure shown in Supplementary Figure S1B are as follows: N, eukaryote-specific N-terminal extension; 3H, three-helix element; D1, domain one; D2, domain two; C, C-terminal extension. The Tsr4-binding site, as previously determined [7], is indicated on top of the schematic representation. (**B**) Fluorescence microscopy of a strain expressing Nop58-yEmCherry (nucleolar marker) as well as 3xyEGFP fusions of Rps2 or the indicated truncated Rps2 fragments. DIC, differential interference contrast.

We transformed the Nop58-yEmCherry-expressing strain with plasmids encoding the Rps2 fragment 3xyEGFP fusions, and inspected the localization of the reporter proteins by fluorescence microscopy (Figure 1B). As expected, full-length Rps2-3xyEGFP showed a predominantly cytoplasmic signal. Additionally, we occasionally observed, as previously described [7], small dot-like structures that likely correspond to aggregates. The occurrence and size of these dot-like structures was strongly increased for several of the 3xyEGFP-fused Rps2 fragments, particularly for Rps2(118–218) and to a lesser extent also for Rps2(23–75) and Rps2(175–254). Moreover, the localization of Rps2(118–218)-3xyEGFP might correspond to a mitochondrial staining. We speculate that the above Rps2 fragments, as they are no longer embedded in the full-length protein context, are especially prone to misfolding, which may lead to their aberrant localization and/or increase their aggregation. Among these, Rps2(23–75)3xyEGFP localized to the entire nucleus and the cytoplasm, with a stronger signal in the nucleus. Last but not least, we identified two 3xyEGFP-fused Rps2 fragments that did not exhibit, when compared to full-length Rps2, increased aggregate formation: Rps2(1–42)-3xyEGFP localized both to the cytoplasm and the entire nucleus, with a slightly stronger signal in the nucleus. More strikingly, Rps2(76–145)-3xyEGFP localized exclusively to the nucleus, suggesting that this Rps2 fragment is sufficient to target the 3xyEGFP reporter to the nucleus (Figure 1B). The nuclear Rps2(76–145)-3xyEGFP signal appeared weaker in the area overlapping with Nop58-yEmCherry compared to the rest of the nucleus. To conclude, our data suggest that the Rps2(76–145)-3xyEGFP fragment contains a functional NLS, which mediates the targeting of this fragment to the nucleoplasm.

3.2. Rps2 Residues R95, R97, and K99 Are Essential for Nuclear Targeting of Rps2(76–145)-3xyEGFP

As attempts to narrow down the sequence responsible for nuclear targeting by further N- or C-terminal truncation of the Rps2(76–145) fragment resulted in mitochondrial staining or increased aggregate formation, respectively, suggesting misfolding of the resulting 3xyEGFP fusion proteins, we instead used the Rps2(76–145)-3xyEGFP fusion as a starting point to introduce the selected amino acid exchanges into potential NLS segments.

As the sequence of the Rps2(76–145) fragment does not contain any obvious so-far-described NLS, we searched for clusters of basic amino acids that are conserved in eukaryotic Rps2. We considered Rps2 residues 95 to 99 (95-RTRFK-99), notably containing three basic amino acids, as a candidate sequence that might contribute to a non-classical NLS (Figure 2A, Supplementary Figure S1A).

To address whether R95, R97, and K99 indeed contribute to the nuclear import of the Rps2(76–145) fragment, we constructed a plasmid expressing the Rps2(76–145).$R_{95}R_{97}K_{99>A}$-3xyEGFP variant (abbreviated as 76–145 RRK>A in Figures) in which all three basic amino acids were exchanged to As. Next, we compared the localization of Rps2(76–145).$R_{95}R_{97}K_{99>A}$-3xyEGFP with the one of Rps2(76–145)-3xyEGFP in the yeast strain expressing Nop58-yEmCherry by fluorescence microscopy (Figure 2B, Supplementary Figure S2). Indeed, the R95A, R97A, and K99A exchanges (95-ATAFA-99 sequence instead of 95-RTRFK-99) resulted in an almost complete shift of the otherwise nuclear Rps2(76–145)-3xyEGFP fragment to the cytoplasm. We conclude that Rps2 contains a functional NLS within amino acids 76–145, with residues R95, R97, and K99 being essential features of this NLS. Importantly, considering that Tsr4 interacts with the very N-terminal part of Rps2 [7], which is not present in the tested Rps2(76–145) fragment (Figure 2A), import via this sequence has to be independent of Tsr4.

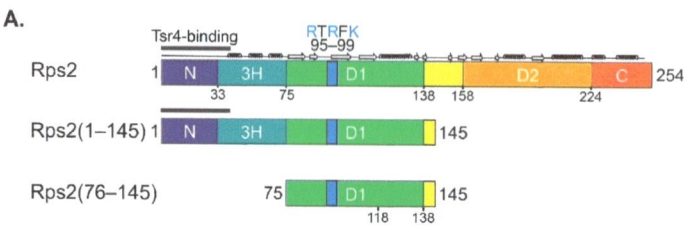

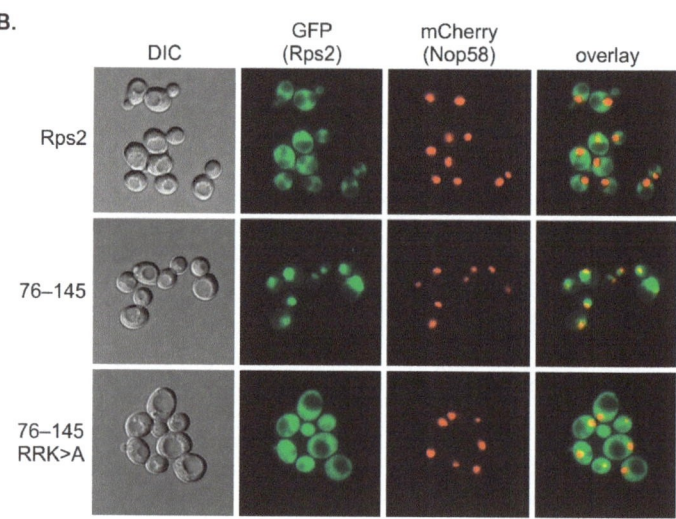

Figure 2. Residues R95, R97, and K99 are essential for nuclear targeting of Rps2(76–145). (**A**) Overview of the main Rps2 fragments tested in this study. (**B**) Fluorescence microscopy of a strain expressing Nop58-yEmCherry as well as Rps2-3xyEGFP or Rps2(76–145)-3xyEGFP with or without exchanges of the three basic residues ($R_{95}R_{97}K_{99>A}$, abbreviated RRK>A), comprised in the 95-RTRFK-99 stretch that are part of the putative NLS. In this experiment, the intensities of the GFP fluorescence signals were adjusted for better comparison. The original, identically processed pictures of the adjusted images are shown in Supplementary Figure S2.

3.3. Import of Rps2(76–145) Is Mediated by Pse1

To gain better insight into the nuclear import of Rps2 mediated by amino acids 76–145, we aimed to identify the importin(s) that recognize the novel Rps2 NLS. To this end, we analyzed the localization of the Rps2(76–145)-3xyEGFP reporter in different importin mutant strains (Figure 3A, Supplementary Figure S3A). The nuclear localization of the Rps2(76–145)-3xyEGFP fusion protein remained largely unaffected in *srp1-31*, *kap95*-ts, and *kap104-16* importin mutants (Supplementary Figure S3A). Rps2(76–145)-3xyEGFP appeared to show a reduced nuclear localization in some cells in the *kap123Δ* strain (Supplementary Figure S3A). To better distinguish whether or not *kap123Δ* cells have a slight Rps2(76–145)-3xyEGFP import defect, we transformed the cells with a plasmid containing the *KAP123* wild-type gene and assessed whether this would result in an increased nuclear signal, which would be an indication for complementation of a potential import defect. As reported before [22], *kap123Δ* cells did not display any growth defects, and as expected, growth was unaltered upon transformation of the *KAP123*-containing plasmid (Supplementary Figure S3B). Moreover, Rps2(76–145)-3xyEGFP displayed a similar localization in *kap123Δ* cells either transformed with *KAP123*-containing or empty plasmid; hence, no complementation of a potential defect was observed (Supplementary Figure S3C). We conclude that *kap123Δ* cells do not show an Rps2(76–145)-3xyEGFP import defect.

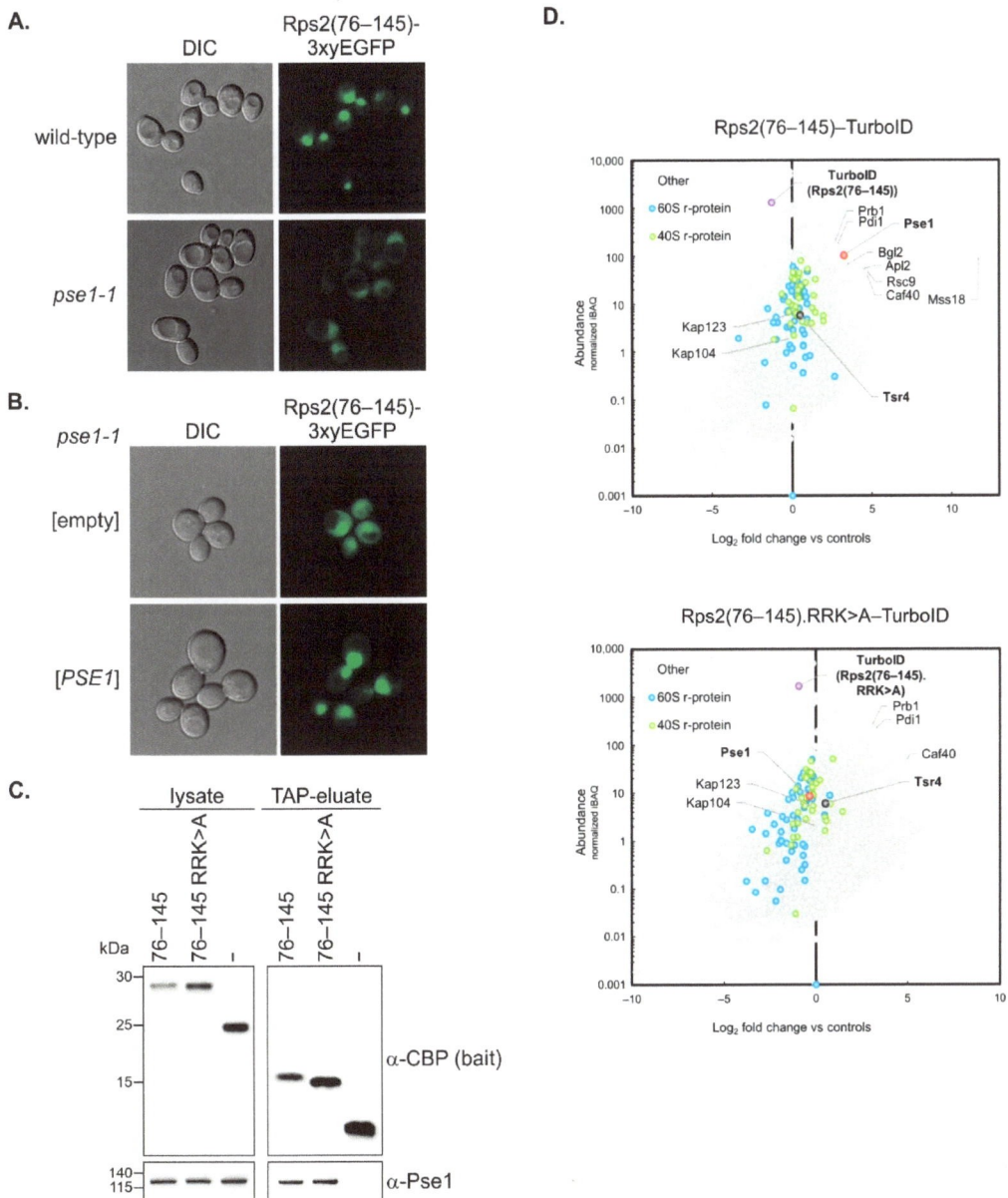

Figure 3. Pse1 mediates nuclear import of Rps2(76–145). (**A**) Localization of Rps2(76–145)-3xyEGFP visualized by fluorescence microscopy in the wild-type strain and the importin mutant strain *pse1-1*. The localization of the fusion protein in additional importin mutant strains is shown in Supplementary Figure S3A. (**B**) Complementation assay. The *pse1-1* strain was transformed with a *PSE1*-harboring *URA3* plasmid or the empty control plasmid, as well as with the Rps2(76–145)-3xyEGFP *LEU2* reporter plasmid, and transformants were inspected by fluorescence microscopy. Growth assays of the same strains are shown in Supplementary Figure S3D. (**C**) Pse1 co-purifies with Rps2 in vivo. Rps2(76–145)-TAP with and without the $R_{95}R_{97}K_{99>A}$ exchanges, as well as the TAP tag alone as negative control (-), were expressed from plasmids in a wild-type strain. After TAP purification, lysates

and TAP eluates were analyzed by Western blotting using α-CBP and α-Pse1 antibodies. (**D**) TurboID-based proximity labeling with Rps2(76–145) and Rps2(76–145).$R_{95}R_{97}K_{99>A}$ as baits. The normalized abundance value (iBAQ, intensity-based absolute quantification; y-axis) of each protein detected in the respective purification is plotted against its relative abundance ($\log_2$-transformed enrichment; x-axis). Relative abundance was calculated compared to the averaged protein abundance in the two control purifications (derived from cells individually expressing the GFP-TurboID and the NLS-GFP-TurboID bait, which accounts for the cytoplasmic and nuclear background, respectively). Proteins that are enriched compared to the negative controls can be found on the right side of the Christmas tree plot. The names of proteins that are particularly enriched, as well as importins Pse1, Kap123, and Kap104 are indicated. The bait proteins and Pse1 are highlighted by bold letters.

Last but not least, a strong reduction in nuclear accumulation of Rps2(76–145)-3xyEGFP was observed in *pse1-1* mutant cells (Figure 3A). Importantly, transformation of the *pse1-1* mutant with a plasmid containing the *PSE1* wild-type gene complemented the growth defect of *pse1-1* mutant cells (Supplementary Figure S3D), as well as their defect in the nuclear import of the Rps2(76–145)-3xyEGFP reporter protein (Figure 3B). Altogether, our data suggest that the NLS within Rps2(76–145) is mainly recognized by the importin-β Pse1.

To further confirm the interaction between Pse1 and Rps2(76–145) and to address whether R95, R97, and K99 are required for this interaction, we performed tandem affinity purification of C-terminally TAP-tagged Rps2(76–145) and Rps2(76–145).$R_{95}R_{97}K_{99>A}$, both expressed from plasmid in a wild-type strain, and compared the extent of Pse1 co-purification (Figure 3C). As expected, Pse1 co-purified with Rps2(76–145)-TAP in a two-step affinity purification, but not with the TAP tag alone (-). Pse1 was also co-purified with Rps2(76–145).$R_{95}R_{97}K_{99>A}$-TAP; however, the enrichment of Pse1 relative to the amounts of the purified bait was clearly less pronounced in the case of the Rps2(76–145).$R_{95}R_{97}K_{99>A}$ protein.

To obtain additional evidence for a preferential binding of Pse1 to wild-type Rps2(76–145) in vivo, we performed TurboID-based proximity labeling to identify the proteins that are in physical proximity of C-terminally TurboID-tagged Rps2(76–145) and Rps2(76–145).$R_{95}R_{97}K_{99>A}$, both expressed from plasmid under the transcriptional control of the copper-inducible *CUP1* promoter (Figure 3D, Supplementary Figure S4 (panels in first row), Supplementary Table S3). Indeed, Pse1 was among the most strongly enriched proteins in the affinity purification of biotinylated proteins from cells expressing wild-type Rps2(76–145), while Pse1 was not enriched when the TurboID experiment was performed with the Rps2(76–145).$R_{95}R_{97}K_{99>A}$ mutant protein. These results suggest that the R95, R97, and K99 exchanges in Rps2(76–145) reduce the binding of Pse1. The reduced binding of Pse1 is most likely the reason for the nuclear import defect observed in *pse1-1* mutant cells. We conclude that Pse1 drives the nuclear import of Rps2(76–145), with R95, R97, and K99 being important determinants for full Pse1 binding.

3.4. Rps2 Contains a Second NLS in Its N-Terminal Extension

Having established that Rps2 contains a functional NLS in an internal region of the r-protein, we went on to test the localization of an Rps2 fragment containing both the NLS and the Tsr4-binding site and constructed a reporter plasmid for the expression of Rps2(1–145)-3xyEGFP (Figure 2A). Fluorescence microscopy of a wild-type strain transformed with the plasmid revealed that Rps2(1–145)-3xyEGFP was, similarly to Rps2(76–145)-3xEGFP, localized in the nucleus (Figure 4A). Next, we assessed the localization of the Rps2(1–145)-3xyEGFP fusion protein additionally carrying the $R_{95}R_{97}K_{99}$>A exchanges in the wild-type strain. Strikingly, in contrast to the strong shift to the cytoplasm of the Rps2(76–145).$R_{95}R_{97}K_{99}$>A-3xEGFP reporter, Rps2(1–145).$R_{95}R_{97}K_{99}$>A-3xEGFP was still mainly found in the nucleus (Figure 4A, Supplementary Figure S5), suggesting that an element within amino acids 1–75 of Rps2 ensures the nuclear targeting of Rps2(1–145), even when the above-identified Pse1-dependent NLS is rendered non-functional by the mutation of residues R95, R97, and K99.

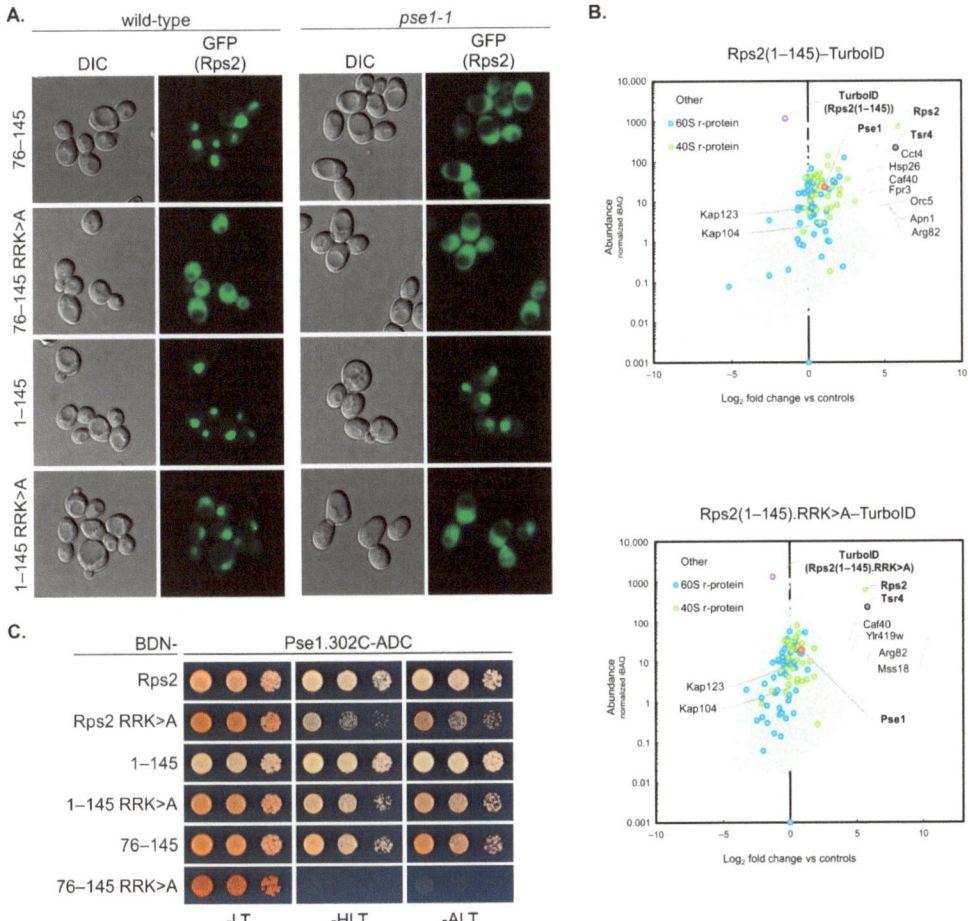

Figure 4. Rps2 contains a second NLS in its N-terminal extension. (**A**) Fluorescence microscopy of a wild-type as well as a *pse1-1* mutant strain expressing 3xyEGFP fusions of the indicated wild-type or mutated Rps2 fragments. In this experiment, the intensities of the GFP fluorescence signals were adjusted for better comparison. The original, identically processed pictures of the adjusted images are shown in Supplementary Figure S5. (**B**) TurboID-based proximity labeling with Rps2(1–145) and Rps2(1–145).$R_{95}R_{97}K_{99}$>A as baits. The normalized abundance value (iBAQ) of each protein detected in the respective purification is plotted against its relative abundance ($\log_2$-transformed enrichment) compared to the averaged abundance in the control purifications (GFP-TurboID and NLS-GFP-TurboID). The names of proteins that are particularly enriched, as well as importins Pse1, Kap123, and Kap104, are indicated. The bait proteins, Tsr4, and Pse1 are highlighted by bold letters. (**C**) Yeast two-hybrid (Y2H) interaction assay between Pse1 lacking the 301 N-terminal amino acids (Pse1.302C), C-terminally fused to the Gal4 activation domain (AD), and Rps2 and the indicated fragments thereof (including, when indicated, the RRK>A exchanges) containing the Gal4 DNA-binding domain (BD) at the N-terminal end. Growth on SDC-his-leu-trp plates (labeled -HLT) indicates a weak interaction; growth on SDC-ade-leu-trp plates (labeled -ALT) indicates a strong Y2H interaction. SDC-leu-trp (labeled -LT) served as growth control. For Y2H assays with the same Rps2 proteins containing the Gal4 DNA-binding domain at the C-terminal end, as well as the Y2H assays between the N- or C-terminally-fused Rps2 variants and full-length Pse1 or the Gal4 activation domain alone (negative control), see Supplementary Figure S6.

We also investigated the localization of the above Rps2-3xyEGFP reporter fusions in the *pse1-1* mutant strain. As described above (Figure 3A), Rps2(76–145)-3xEGFP was shifted to the cytoplasm in the *pse1-1* mutant strain compared to the wild-type strain (Figure 4A). As expected, also the Rps2(76–145).$R_{95}R_{97}K_{99}$>A-3xEGFP reporter showed a predominantly cytoplasmic signal in the *pse1-1* mutant, similar to the wild-type strain. The Rps2(1–145)-3xyEGFP reporter, although still showing the highest signal intensity in the nucleus, was slightly shifted to the cytoplasm, as opposed to the exclusively nuclear signal of the same fragment in the wild-type strain (Figure 4A). An even stronger shift to the cytoplasm was observed for the Rps2(1–145).$R_{95}R_{97}K_{99}$>A-3xEGFP reporter in the *pse1-1* strain (Figure 4A). These results indicate that Pse1 contributes to the nuclear import of the Rps2(1–145) fragment even when R95, R97, and K99 are mutated, suggesting that there has to be a second Pse1-dependent NLS within amino acids 1 to 75 of Rps2. However, none of the fragments was completely shifted to the cytoplasm in the *pse1-1* mutant; therefore, other importins have to contribute to some extent to Rps2(1–145) nuclear import, at least in the absence of Pse1.

Next, we performed TurboID experiments to identify proteins in close proximity to Rps2(1–145) and Rps2(1–145).$R_{95}R_{97}K_{99}$>A (Figure 4B, Supplementary Figure S4, Supplementary Table S3). In both cases, as expected due to the presence of the N-terminal Tsr4-binding region, Tsr4 was strongly enriched in the affinity purification of biotinylated proteins. Pse1 was detected as well, although it was much less enriched than upon expression of TurboID-tagged Rps2(76–145) (Figure 3D). Moreover, the $R_{95}R_{97}K_{99}$>A exchanges only had a minor effect on the extent of Pse1 enrichment in the context of Rps2(1–145) (Figure 4B). TurboID with full-length Rps2 (wild-type or containing the $R_{95}R_{97}K_{99}$>A exchanges) yielded a similar extent of Tsr4 enrichment as observed in the case of Rps2(1–145) wild-type and $R_{95}R_{97}K_{99}$>A mutant protein, while an enrichment of Pse1 could not be observed, potentially due to the short duration of the Rps2–Pse1 interaction compared to interactions of Rps2 in the context of the ribosome (Supplementary Figure S4 and Table S3). To further characterize the effects of the $R_{95}R_{97}K_{99}$>A exchanges on the interaction of the different Rps2 fragments with Pse1, we performed yeast two-hybrid (Y2H) analyses (Figure 4C, Supplementary Figure S6). No interaction of any of the Rps2 variants was detected with full-length Pse1 (Supplementary Figure S6), which was not surprising as importin–cargo interactions are generally only very short-lived in the nucleus due to the fact that the binding of RanGTP to the N-terminal arch of importins mediates cargo release [6]. To prevent cargo dissociation and hence enable productive importin–cargo Y2H interactions in the nucleus, we generated a Pse1 variant with a partial deletion of its N-terminal RanGTP-binding surface (Pse1.302C; starting with amino acid 302) [27]. As anticipated, utilization of the Pse1.302C variant permitted the detection of Y2H interactions between Pse1 and Rps2, Rps2(1–145), and Rps2(76–145) (Figure 4C, Supplementary Figure S6). While the $R_{95}R_{97}K_{99}$>A exchanges almost completely abolished the interaction of Rps2(76–145) with Pse1.302C, they only reduced the interaction of both Rps2 and Rps2(1–145) with Pse1.302C (Figure 4C, Supplementary Figure S6).

Taken together, the above results indicate that the mutant Rps2(1–145) $R_{95}R_{97}K_{99}$>A protein is still capable, albeit less efficiently than the wild-type counterpart, of interacting with Pse1 and can thus still be imported into the nucleus via Pse1. On the other hand, the mutant Rps2(76–145).$R_{95}R_{97}K_{99}$>A protein interacts with Pse1 only poorly and its nuclear import is strongly impaired. We conclude that amino acids 1–75 of Rps2 must harbor a second import signal, which could also be recognized by Pse1.

3.5. Tsr4 Is Not Required for Import Mediated by the N-Terminal Rps2 Region

Considering that Tsr4 binds to the N-terminal region of Rps2 (Figure 2A, Ref. [7]), we speculated that Tsr4 might be involved in this second Rps2 import mechanism. To address this possibility, we examined the localization of Rps2(1–145).$R_{95}R_{97}K_{99}$>A-3xyEGFP in the absence of Tsr4. This analysis is complicated by the fact that Tsr4 is an essential protein; however, in our previous study we found that *tsr4Δ* cells are viable, although displaying a severe slow-growth phenotype, when *RPS2* is provided on a low-copy number plasmid in a *rps2Δ* strain, presumably resulting in an increased *RPS2* copy number compared to the single-copy presence of *RPS2* in a wild-type strain [7]. Building on this knowledge, we transformed the 3xyEGFP reporter plasmids into a *tsr4Δ rps2Δ* strain complemented by a *URA3-RPS2* plasmid and, as a control, into a Tsr4-expressing *rps2Δ URA3-RPS2* strain. As previously observed, the Rps2-3xyEGFP reporter accumulated in the nucleus in cells lacks Tsr4 (*tsr4Δ rps2Δ* [*RPS2*] strain) (Figure 5, Supplementary Figure S7), suggesting that the nuclear import of Rps2 can occur in the absence of Tsr4, and that, moreover, efficient Rps2 incorporation into pre-ribosomes is dependent on Tsr4 [7].

Notably, a strong nuclear accumulation was also observed for Rps2.$R_{95}R_{97}K_{99}$>A-3xyEGFP in the absence of Tsr4, while Rps2(1–145)-3xyEGFP and Rps2(1–145).$R_{95}R_{97}K_{99}$>A-3xyEGFP localized to the nucleus, mostly within intense dot-like structures outside the nucleolus that could correspond to aggregates, both in cells containing or lacking Tsr4 (Figure 5). We conclude that the second import mechanism also utilized by Rps2, involving amino acids 1–75, does not depend on Tsr4.

3.6. The N-Terminal Rps2 NLS Overlaps with the Tsr4-Binding Region

The partial nuclear localization of the Rps2(1–42)-3xyEGFP fusion protein (Figure 1B), together with the occurrence of an RG-rich sequence within the 28 N-terminal amino acids of Rps2 (Supplementary Figures S1A and S8B), which might potentially represent an RG-NLS, prompted us to test whether the very N-terminal region of Rps2 is required for the nuclear import of Rps2(1–145).$R_{95}R_{97}K_{99}$>A-3xyEGFP. Notably, Tsr4 binds approximately to the same region, as suggested by our previous study in which we mapped the Tsr4-binding site to amino acids 1–42 of Rps2 [7]. We reasoned that it might be possible to map the Tsr4-binding site to an even shorter Rps2 fragment by generating further N- and C-terminal truncation variants and testing their capacity to interact with Tsr4 in Y2H assays (Figure 6A, Supplementary Figure S8A). Indeed, Rps2 missing the N-terminal five or ten amino acids still showed full interaction with Tsr4. Moreover, the 33 or 28 N-terminal residues alone were sufficient to confer a robust Y2H interaction with Tsr4, while, as already previously described [7], Rps2(1–22) interacted only weakly with Tsr4. Finally, we combined the above N- and C-terminal truncations which supported full interaction, and found that all tested combinations still interacted equally well with Tsr4, with the shortest tested fragment displaying full interaction being Rps2(10–28) (Figure 6A, Supplementary Figure S8A).

Next, we wanted to address whether mutation of the Tsr4-binding site would hamper the putative N-terminal NLS. To this end, we generated constructs expressing 3xyEGFP fusions of either a variant lacking the 28 N-terminal amino acids of Rps2 (Rps2(29–145)) or a Rps2(1–145) fragment, termed Rps2(1–145).KR_{10-28}>A, having all basic residues within amino acids 10 to 28 (one K and seven R residues; see Supplementary Figure S8B) exchanged to As. Moreover, both variants were generated with and without the $R_{95}R_{97}K_{99}$>A exchanges. The variants containing only the N-terminal manipulations (Rps2(29–145)-3xyEGFP and Rps2(1–145).KR_{10-28}>A-3xyEGFP) still showed a nuclear localization similar to (Rps2(1–145)-3xyEGFP, with a stronger signal in the nucleoplasm than in the nucleolus (Figure 6B, Supplementary Figure S9). In contrast, both variants additionally carrying the exchanges affecting the internal Rps2 NLS (Rps2(29–145).$R_{95}R_{97}K_{99}$>A-3xyEGFP and Rps2(1–145).KR_{10-28}>A/$R_{95}R_{97}K_{99}$>A-3xyEGFP) failed to accumulate in the nucleus. We conclude that besides the NLS within amino acids 76–145, to which R95, R97, and K99 make an essential contribution, Rps2 contains a second NLS in its N-terminal region, which overlaps with the Tsr4-binding site and critically depends on the presence of several basic residues within a short stretch ranging from amino acid 10 to 28. Notably, this sequence stretch is present in the Rps2(1–42)-3xyEGFP reporter fusion, which localizes to both the cytoplasm and nucleus (Figure 1B). We hypothesized that the binding of Tsr4 might affect the nuclear import of this fragment through the N-terminal NLS by potentially modulating the efficiency of importin binding. To test this hypothesis, we again utilized a *tsr4Δ rps2Δ* strain complemented by a *URA3-RPS2* plasmid and, as a control, a *rps2Δ* [*URA3-RPS2*] strain. Both strains were transformed with the Rps2(1–42)-3xyEGFP reporter plasmid. In contrast to the wild-type strain, where the Rps2(1–42)-3xyEGFP signal was stronger in the nucleus compared to the cytoplasm (Figure 1B), the *rps2Δ* [*URA3-RPS2*] strain displayed an even distribution of Rps2(1–42)-3xyEGFP between the cytoplasm and the nucleus (Figure 6C). Interestingly, the *tsr4Δ rps2Δ* [*URA3-RPS2*] strain, lacking Tsr4, exhibited a slight accumulation of the Rps2(1–42)-3xyEGFP reporter fusion in the nucleus, suggesting that nuclear import via the N-terminal NLS of Rps2 might be more efficient in the absence of Tsr4.

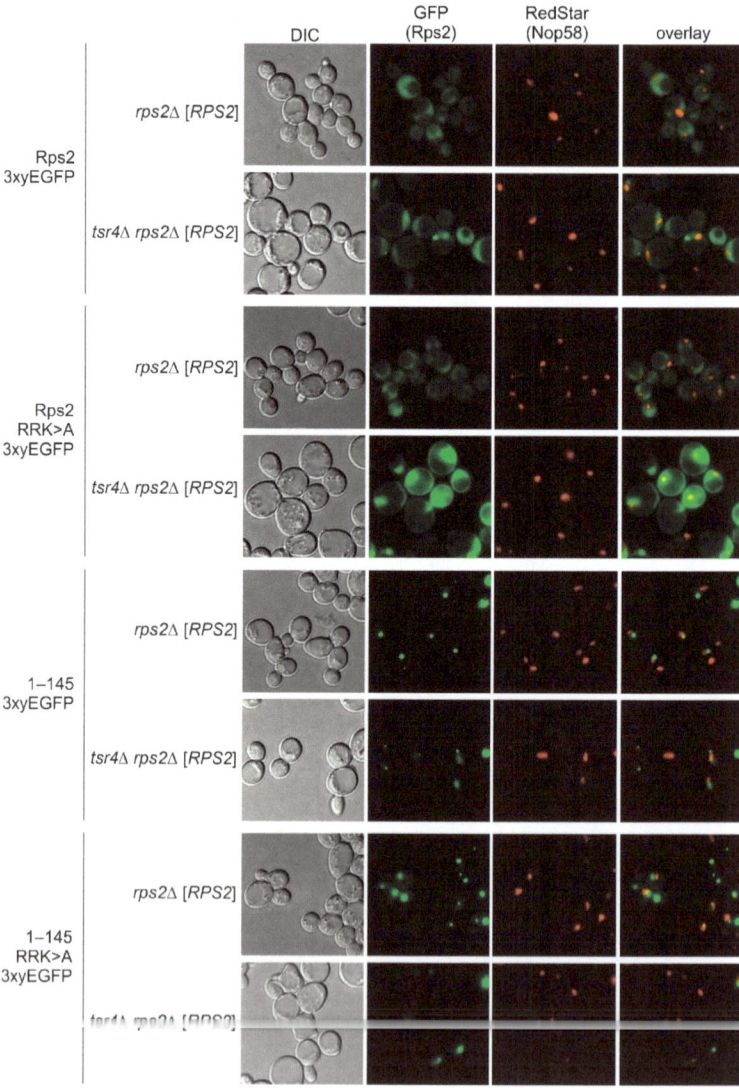

Figure 5. Tsr4 is not required for import mediated by the N-terminal Rps2 region. Fluorescence microscopy of *rps2Δ* and *tsr4Δ rps2Δ* strains, containing a *URA3-RPS2* plasmid, expressing the indicated Rps2-3xyEGFP fusion proteins from *LEU2* plasmids, and a chromosomal C-terminal RedStar2 fusion of Nop58. Each panel was processed individually to make the observed phenotypes more apparent. To allow for the evaluation of the differences in signal intensities, the same panels, but all identically processed, are shown in Supplementary Figure S7.

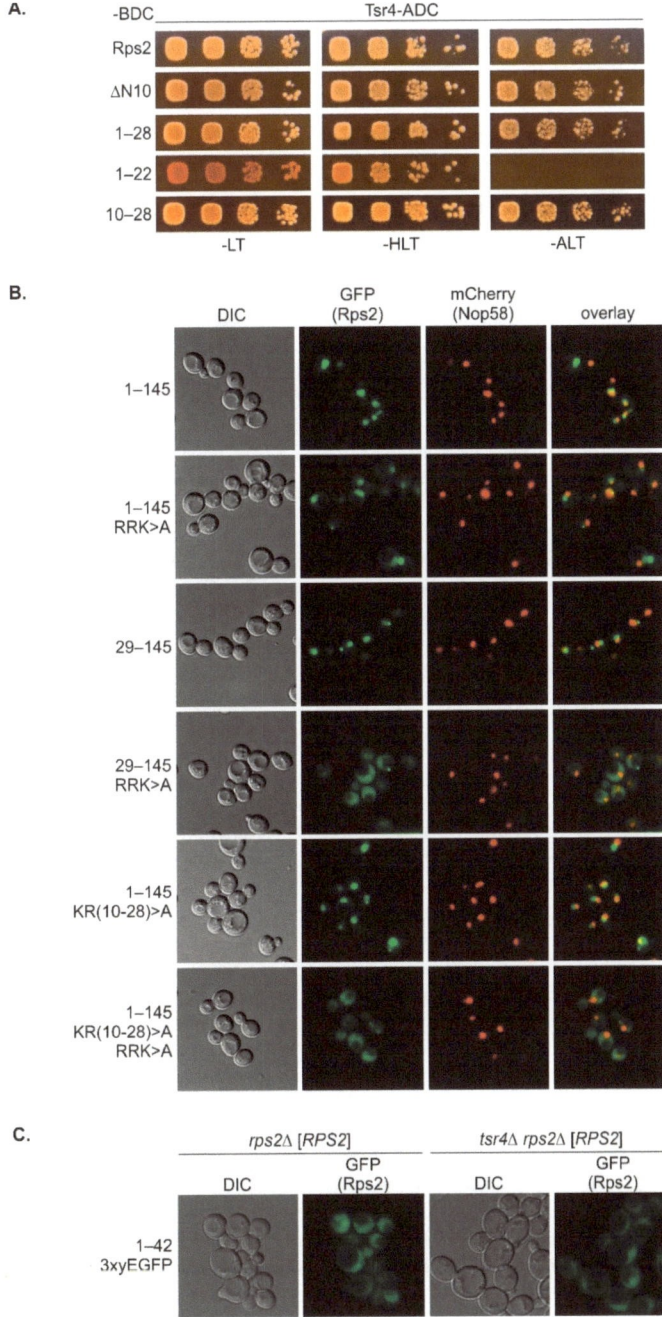

Figure 6. The N-terminal Rps2 NLS overlaps with the Tsr4-binding region. (**A**) Yeast two-hybrid (Y2H) assays between full-length Tsr4, C-terminally fused to the Gal4 activation domain (AD), and Rps2 and fragments thereof, and C-terminally fused to the Gal4 DNA-binding domain (BD). For more details, see the legend of Figure 4C. For results with additional fragments as well as negative controls, see Supplementary Figure S8A. (**B**) Fluorescence microscopy of a strain expressing Nop58-yEmCherry

as well as 3xyEGFP fusions of the indicated wild-type or mutated Rps2 fragments. In this experiment, the intensities of the GFP fluorescence signals were adjusted for better comparison. The original, identically processed pictures of the adjusted images are shown in Supplementary Figure S9. (C) Fluorescence microscopy of $rps2\Delta$ and $tsr4\Delta$ $rps2\Delta$ strains, containing a $URA3$-$RPS2$ plasmid, expressing Rps2(1–42)-3xyEGFP from an $LEU2$ plasmid.

4. Discussion

With this study, we have provided insights into the intricate mechanisms underlying nuclear import of the r-protein Rps2. We found that amino acids 76 to 145 are sufficient to target the protein to the nucleus, with residues R95, R97, and K99 being essential for the nuclear localization of this fragment. The main importin responsible for import via Rps2(76–145) is Pse1. Hence, the preference of Rps2 for Pse1 deviates from the common preference of r-proteins for Kap123, with Pse1 stepping in place mainly in the absence of Kap123 [13]. Our data moreover demonstrate that the mutation of R95, R97, and K99 in the Rps2(76–145) fragment greatly reduces the interaction with Pse1, suggesting that these residues are critical determinants for Pse1 binding. Previous structural analyses of Pse1 in complex with NLS sequences of three different Pse1 cargoes have led to the definition of the IK-NLS with the consensus K-V/I-X-K-X_{1-2}-K/H/R [27,28]. The segment ranging from residues 95 to 99 of Rps2 (RTRFK), however, does not match this consensus. Moreover, it is positioned within a beta-sheet (Supplementary Figure S1B), while IK-NLSs are unstructured [27,28]. Hence, Rps2 likely uses a binding mode that differs from the one reported for the interaction of Pse1 with IK-NLSs, and seems to involve structured elements. In our tandem affinity purification experiment, where we expressed either Rps2(76–145)-TAP or Rps2(76–145).$R_{95}R_{97}K_{99}$>A-TAP from plasmids in a wild-type strain, we observed higher levels of the Rps2 fragment carrying the substitutions in cell lysates compared to the wild-type fragment (Figure 3C). This observation suggests that the substitutions of R95, R97, and K99 to A might induce structural changes that enhance the stability of the Rps2(76–145) fragment. These altered structural features might impede the efficient binding of Pse1, despite promoting protein stability.

The Rps2(1–145) fragment, containing in addition to the above-discussed nuclear-targeting domain also the N-terminal part of Rps2, enters the nucleus as well, even when the residues critical for the nuclear targeting of the Rps2(76–145) fragment are mutated. Moreover, R95A, R97A, and K99A mutation reduces the Y2H interaction of the Rps2(1–145) fragment with Pse1 only slightly, while the interaction of Rps2(76–145) with Pse1 is severely reduced by these exchanges. This suggests that Pse1 may possess an additional binding site within amino acids 1–75 of Rps2. Indeed, Rps2(1–145).$R_{95}R_{97}K_{99}$>A-3xEGFP displayed an increased cytoplasmic signal in *pse1-1* mutant cells compared to wild-type cells, indicating that even if the internal NLS is not available for interaction with Pse1, Pse1 is capable of importing the Rps2(1–145) fragment. It is worth noting that none of the tested Rps2 fragment 3xEGFP fusions showed complete import inhibition in the *pse1-1* mutant, implying that, as also suggested in previous studies (see for example [29–31]), other importins can compensate for the loss of one importin. Nevertheless, the significant defects observed in the *pse1-1* mutant strongly indicate that Pse1 is the primary importin binding to the two Rps2 NLS elements described in this study.

Nuclear import via this N-terminal nuclear-targeting region requires basic residues within the RG-rich, unstructured N-terminal part (amino acids 10–28) of Rps2. It is already known that such RGG regions can function as NLSs for Kap104 [32,33]; however, recognition of RG-rich NLSs by Pse1 has not been reported so far. Although our findings suggest that Rps2(1–145) can still be imported into the nucleus by Pse1 when either the N-terminal or the internal NLS is mutated, it remains unclear whether Pse1 interacts simultaneously with both binding sites in the wild-type scenario, or if it only utilizes one of them at a time. It will be interesting to further define and map the two Pse1-binding regions of Rps2 in the future, which might lead to the definition of novel NLS consensus motifs for Pse1. Notably, while the N-terminal and internal NLSs share some sequence similarities, such as positively charged amino acids with similar spacing (e.g., 95-RTRFK-99 and 17-RNRGR-21), they are embedded in entirely different structural contexts. The N-terminal NLS resides within an unstructured region, whereas the internal NLS lies within a beta-sheet (Supplementary Figure S1B). Consequently, the two NLSs may employ distinct binding modes for Pse1 interaction.

It is important to acknowledge that the basic residues within amino acids 10–28 of Rps2, although being necessary for the nuclear targeting of a Rps2(1–145) fragment with a mutated internal NLS (Figure 6B), are not sufficient for efficient import, as concluded from the fact that a small Rps2 fragment containing these amino acids, Rps2(1–42), does not exclusively localize to the nucleus (Figures 1B and 6C). Hence, additional sequence elements are required for the complete functionality of the N-terminal NLS. Furthermore, it is possible that not all eight positively charged amino acids within Rps2(10–28) are essential for the function of the N-terminal NLS. It is plausible that a few specific residues within this range are crucial for the import via the N-terminal NLS, or that multiple clusters of positively charged amino acids within this sequence can fulfill this function alternatively, as recently reported for the NLS of the viral protein HIV-1 Tat [34]. Future in-depth biochemical and structural studies might provide further insights into the binding modes and interplay between the two Rps2 NLSs.

Importantly, the amino acids required for the function of the N-terminal NLS of Rps2 overlap with the binding site of its dedicated chaperone, Tsr4. Therefore, it was important to investigate whether the presence of Tsr4 affects the nuclear targeting of Rps2 via the N-terminal NLS.

We can exclude the possibility that import mediated by Rps2 amino acids 10–28 occurs via a 'piggyback' mechanism in which Tsr4 binds Rps2 and provides the NLS for the nuclear import of the Rps2-Tsr4 complex, as our data revealed that the presence of Tsr4 is not required for the import involving the N-terminal region of Rps2 (Figure 5). Rps2(1–42)-3xyEGFP even exhibited a stronger nuclear signal in the absence of Tsr4 (Figure 6C), suggesting that its import is less efficient when Tsr4 is present. One potential explanation for this effect is that Tsr4 shields the N-terminal NLS, thereby reducing the efficiency of importin binding. It is yet to be determined whether Tsr4 accompanies Rps2 into the nucleus or dissociates from Rps2 already in the cytoplasm, e.g., upon the binding of importin. Tsr4-GFP does not accumulate in the nucleus upon inhibition of the main exportin Crm1 [10]. However, the human Tsr4 homolog PDCD2 accompanies human RPS2 into the nucleus [35], as does the closely related PDCD2L [36]. Further, our data demonstrate that in the absence of Tsr4, Rps2 accumulates in the nucleolus (Figure 5 and [7]), suggesting that Rps2's efficient incorporation into pre-ribosomal particles is prevented. The simplest explanation for this phenotype would be that Tsr4 functions in promoting Rps2 pre-ribosome incorporation in the nucleus. On the other hand, the more efficient import of the Rps2(1–42) fragment in the absence of Tsr4 suggests that nuclear import might occur after the release of Tsr4. Alternatively, the interaction of Pse1 with the internal NLS of Rps2 may be sufficient to mediate the nuclear targeting of Rps2 bound to Tsr4, even if the N-terminal NLS is not fully accessible to the importin.

The binding of Pse1 to Rps2 could serve a second function beside nuclear import as importins have been implicated in functioning as chaperones for exposed basic domains [12]. The richness in positive charges, together with the flexibility of the Rps2 N-terminal region, might make Rps2 particularly prone to aggregation, which could be the reason why two different, potentially redundant mechanisms for chaperoning of this region have evolved, with the main one relying on a dedicated chaperone and the second one involving an importin.

Interestingly, the N-terminal RG-rich region of Rps2 is absent in bacteria and archaea (Supplementary Figure S1A), suggesting that it serves a eukaryote-specific function, as is the case for a targeting sequence for nuclear import. In contrast, parts of the internal positively charged NLS residues are also found in archaea and bacteria. For instance, *Pyrococcus furiosus* uS5 contains all three of these residues, while *Sulfolobus* and *Bacillus subtilis* have two positively charged amino acids in the corresponding region (Supplementary Figure S1A).

NLS-type motifs have been observed in archaea before, suggesting that NLS sequences may have originated from sequences that originally served other functions [37,38].

Intriguingly, Rps2's unstructured N-terminal region seems to be a hub for the binding of multiple interaction partners (elaborated in detail in a review article by the Bachand group within this Special Issue [39]). Besides the binding partners investigated in this study (Tsr4 and importins), the N-terminal part of Rps2 also likely interacts (at least transiently) with Hmt1, as this enzyme methylates an arginine in the N-terminal region of Rps2 [40,41]. In the human system, RPS2 is bound by PDCD2 or PDCD2L, and is additionally stably bound by the arginine methyl transferase PRMT3, which competes with the zinc finger protein ZNF277 for RPS2 binding [22,36,42,43]. The investigation of the timing and coordination of these manifold interactions will be an interesting subject for future studies.

Supplementary Materials: The following supporting information can be downloaded at https://www.mdpi.com/article/10.3390/biom13071127/s1: Figure S1: Sequence and structure [44] of Rps2; Figure S2: Localization of Rps2(76–145).R95R97K99>A-3xyEGFP; Figure S3: Nuclear import of Rps2(76–145)-3xyEGFP; Figure S4: TurboID-based proximity labeling using Rps2, Rps2(1–145), and Rps2(76–145), all with and without the R95R97K99>A exchanges, as baits; Figure S5: Localization of Rps2-3xyEGFP variants in the wild-type and *pse1-1* mutant strain; Figure S6. Yeast two-hybrid (Y2H) interaction of Rps2 and Pse1; Figure S7: Localization of Rps2-3xyEGFP reporter constructs in the absence of Tsr4; Figure S8. Mapping of the Tsr4-binding region on Rps2; Figure S9: Localization of Rps2-3xyEGFP variants containing the R95R97K99>A exchanges; Table S1: Yeast strains [7,15,18,23,45–49]; Table S2: *S. cerevisiae* plasmids [7,16,43,50,51]; Table S3: TurboID proximity labeling data.

Author Contributions: A.S., S.F., M.M., B.P. (Benjamin Pillet), J.H., D.K., B.P. (Brigitte Pertschy) and I.Z. designed the experiments and interpreted the results. A.S., S.F., M.M., A.H., B.P. (Benjamin Pillet), J.H., D.K. and I.Z. constructed the strains and plasmids. A.S., M.M., A.H., J.H., V.M. and I.Z. performed fluorescence microscopy. S.F. and M.M. performed Y2H experiments. M.M. performed TAP purifications. S.F. performed Turbo-ID. B.P. (Brigitte Pertschy), D.K. and I.Z. wrote the manuscript. All authors commented on the manuscript. All authors have read and agreed to the published version of the manuscript.

Funding: This work was funded by Austrian science fund (FWF) grant P32673, FWF grant P32320, the BioTechMed-Graz Flagship project DYNIMO, and FWF doc.fund 50 'Molecular Metabolism' to B. Pertschy and UFO project PN35 from the 'Land Steiermark' to I.Z. Additionally, B. Pertschy was supported by the 'Land Steiermark' and 'Stadt Graz'. The work carried out in the laboratory of D.K. was funded by the Swiss National Science Foundation (SNSF), project grant 310030_204801.

Data Availability Statement: TurboID results are provided in Supplementary Table S3.

Acknowledgments: We kindly acknowledge Matthias Seedorf for the α-Pse1 antibody and Ed Hurt for the importin mutant strains. We thank Devanarayanan Siva Sankar for measurement of TurboID samples and data analysis in MaxQuant.

Conflicts of Interest: The authors declare no conflict of interest.

References

1. Woolford, J.L., Jr.; Baserga, S.J. Ribosome Biogenesis in the Yeast Saccharomyces cerevisiae. *Genetics* **2013**, *195*, 643–681. [CrossRef] [PubMed]
2. Kressler, D.; Hurt, E.; Bassler, J. Driving ribosome assembly. *Biochim. Biophys. Acta (BBA)-Mol. Cell Res.* **2010**, *1803*, 673–683. [CrossRef]
3. de la Cruz, J.; Karbstein, K.; Woolford, J.L. Functions of ribosomal proteins in assembly of eukaryotic ribosomes in vivo. *Annu. Rev. Biochem.* **2015**, *84*, 93–129. [CrossRef]
4. Klinge, S.; Woolford, J.L. Ribosome assembly coming into focus. *Nat. Rev. Mol. Cell Biol.* **2019**, *20*, 116–131. [CrossRef] [PubMed]
5. Pillet, B.; Mitterer, V.; Kressler, D.; Pertschy, B. Hold on to your friends: Dedicated chaperones of ribosomal proteins: Dedicated chaperones mediate the safe transfer of ribosomal proteins to their site of pre-ribosome incorporation. *Bioessays* **2017**, *39*, 1–12. [CrossRef] [PubMed]
6. Wing, C.E.; Fung, H.Y.J.; Chook, Y.M. Karyopherin-mediated nucleocytoplasmic transport. *Nat. Rev. Mol. Cell Biol.* **2022**, *23*, 307–328. [CrossRef] [PubMed]

7. Rössler, I.; Embacher, J.; Pillet, B.; Murat, G.; Liesinger, L.; Hafner, J.; Unterluggauer, J.J.; Birner-Gruenberger, R.; Kressler, D.; Pertschy, B. Tsr4 and Nap1, two novel members of the ribosomal protein chaperOME. *Nucleic Acids Res.* **2019**, *47*, 6984–7002. [CrossRef]
8. Ting, Y.-H.; Lu, T.-J.; Johnson, A.W.; Shie, J.-T.; Chen, B.-R.; Kumar, S.S.; Lo, K.-Y. Bcp1 Is the Nuclear Chaperone of Rpl23 in Saccharomyces cerevisiae. *J. Biol. Chem.* **2017**, *292*, 585–596. [CrossRef]
9. Liang, K.-J.; Yueh, L.-Y.; Hsu, N.-H.; Lai, J.-S.; Lo, K.-Y. Puf6 and Loc1 Are the Dedicated Chaperones of Ribosomal Protein Rpl43 in Saccharomyces cerevisiae. *Int. J. Mol. Sci.* **2019**, *20*, 5941. [CrossRef]
10. Black, J.J.; Musalgaonkar, S.; Johnson, A.W. Tsr4 Is a Cytoplasmic Chaperone for the Ribosomal Protein Rps2 in Saccharomyces cerevisiae. *Mol. Cell Biol.* **2019**, *39*, e00094-19. [CrossRef]
11. Pausch, P.; Singh, U.; Ahmed, Y.L.; Pillet, B.; Murat, G.; Altegoer, F.; Stier, G.; Thoms, M.; Hurt, E.; Sinning, I.; et al. Co-translational capturing of nascent ribosomal proteins by their dedicated chaperones. *Nat. Commun.* **2015**, *6*, 7494. [CrossRef] [PubMed]
12. Jäkel, S.; Mingot, J.-M.; Schwarzmaier, P.; Hartmann, E.; Görlich, D. Importins fulfil a dual function as nuclear import receptors and cytoplasmic chaperones for exposed basic domains. *EMBO J.* **2002**, *21*, 377–386. [CrossRef] [PubMed]
13. Rout, M.P.; Blobel, G.; Aitchison, J.D. A distinct nuclear import pathway used by ribosomal proteins. *Cell* **1997**, *89*, 715–725. [CrossRef] [PubMed]
14. Mitterer, V.; Gantenbein, N.; Birner-Gruenberger, R.; Murat, G.; Bergler, H.; Kressler, D.; Pertschy, B. Nuclear import of dimerized ribosomal protein Rps3 in complex with its chaperone Yar1. *Sci. Rep.* **2016**, *6*, 36714. [CrossRef]
15. Mitterer, V.; Murat, G.; Réty, S.; Blaud, M.; Delbos, L.; Stanborough, T.; Bergler, H.; Leulliot, N.; Kressler, D.; Pertschy, B. Sequential domain assembly of ribosomal protein S3 drives 40S subunit maturation. *Nat. Commun.* **2016**, *7*, 10336. [CrossRef] [PubMed]
16. Kressler, D.; Bange, G.; Ogawa, Y.; Stjepanovic, G.; Bradatsch, B.; Pratte, D.; Amlacher, S.; Strauß, D.; Yoneda, Y.; Katahira, J.; et al. Synchronizing nuclear import of ribosomal proteins with ribosome assembly. *Science* **2012**, *338*, 666–671. [CrossRef]
17. Huber, F.M.; Hoelz, A. Molecular basis for protection of ribosomal protein L4 from cellular degradation. *Nat. Commun.* **2017**, *8*, 14354. [CrossRef]
18. Pillet, B.; García-Gómez, J.J.; Pausch, P.; Falquet, L.; Bange, G.; de la Cruz, J.; Kressler, D.; Tollervey, D. The Dedicated Chaperone Acl4 Escorts Ribosomal Protein Rpl4 to Its Nuclear Pre-60S Assembly Site. *PLoS Genet.* **2015**, *11*, e1005565. [CrossRef]
19. Stelter, P.; Huber, F.M.; Kunze, R.; Flemming, D.; Hoelz, A.; Hurt, E. Coordinated Ribosomal L4 Protein Assembly into the Pre-Ribosome Is Regulated by Its Eukaryote-Specific Extension. *Mol. Cell* **2015**, *58*, 854–862. [CrossRef]
20. Schütz, S.; Fischer, U.; Altvater, M.; Nerurkar, P.; Peña, C.; Gerber, M.; Chang, Y.; Caesar, S.; Schubert, O.T.; Schlenstedt, G.; et al. A RanGTP-independent mechanism allows ribosomal protein nuclear import for ribosome assembly. *Elife* **2014**, *3*, e03473. [CrossRef]
21. Ban, N.; Beckmann, R.; Cate, J.H.D.; Dinman, J.D.; Dragon, F.; Ellis, S.R.; Lafontaine, D.L.J.; Lindahl, L.; Liljas, A.; Lipton, J.M.; et al. A new system for naming ribosomal proteins. *Curr. Opin. Struct. Biol.* **2014**, *24*, 165–169. [CrossRef] [PubMed]
22. Seedorf, M.; Silver, P.A. Importin/karyopherin protein family members required for mRNA export from the nucleus. *Proc. Natl. Acad. Sci. USA* **1997**, *94*, 8590–8595. [CrossRef] [PubMed]
23. Rappsilber, J.; Mann, M.; Ishihama, Y. Protocol for micro-purification, enrichment, pre-fractionation and storage of peptides for proteomics using StageTips. *Nat. Protoc.* **2007**, *2*, 1896–1906. [CrossRef]
24. Tyanova, S.; Temu, T.; Cox, J. The MaxQuant computational platform for mass spectrometry-based shotgun proteomics. *Nat. Protoc.* **2016**, *11*, 2301–2319. [CrossRef]
25. Tyanova, S.; Temu, T.; Sinitcyn, P.; Carlson, A.; Hein, M.Y.; Geiger, T.; Mann, M.; Cox, J. The Perseus computational platform for comprehensive analysis of (prote)omics data. *Nat. Methods* **2016**, *13*, 731–740. [CrossRef] [PubMed]
26. Schäfer, T.; Strauss, D.; Petfalski, E.; Tollervey, D.; Hurt, E. The path from nucleolar 90S to cytoplasmic 40S pre-ribosomes. *EMBO J.* **2003**, *22*, 1370–1380. [CrossRef] [PubMed]
27. Kobayashi, J.; Matsuura, Y. Structural basis for cell-cycle-dependent nuclear import mediated by the karyopherin Kap121p. *J. Mol. Biol.* **2013**, *425*, 1852–1868. [CrossRef]
28. Kobayashi, J.; Hirano, H.; Matsuura, Y. Crystal structure of the karyopherin Kap121p bound to the extreme C-terminus of the protein phosphatase Cdc14p. *Biochem. Biophys. Res. Commun.* **2015**, *463*, 309–314. [CrossRef]
29. Kimura, M.; Morinaka, Y.; Imai, K.; Kose, S.; Horton, P.; Imamoto, N. Extensive cargo identification reveals distinct biological roles of the 12 importin pathways. *Elife* **2017**, *6*, e21184. [CrossRef]
30. Baade, I.; Spillner, C.; Schmitt, K.; Valerius, O.; Kehlenbach, R.H. Extensive Identification and In-depth Validation of Importin 13 Cargoes. *Mol. Cell. Proteom.* **2018**, *17*, 1337–1353. [CrossRef]
31. Sydorskyy, Y.; Dilworth, D.J.; Yi, E.C.; Goodlett, D.R.; Wozniak, R.W.; Aitchison, J.D. Intersection of the Kap123p-mediated nuclear import and ribosome export pathways. *Mol. Cell Biol.* **2003**, *23*, 2042–2054. [CrossRef] [PubMed]
32. Bourgeois, B.; Hutten, S.; Gottschalk, B.; Hofweber, M.; Richter, G.; Sternat, J.; Abou-Ajram, C.; Göbl, C.; Leitinger, G.; Graier, W.F.; et al. Nonclassical nuclear localization signals mediate nuclear import of CIRBP. *Proc. Natl. Acad. Sci. USA* **2020**, *117*, 8503–8514. [CrossRef] [PubMed]
33. Gonzalez, A.; Mannen, T.; Çağatay, T.; Fujiwara, A.; Matsumura, H.; Niesman, A.B.; Brautigam, C.A.; Chook, Y.M.; Yoshizawa, T. Mechanism of karyopherin-β2 binding and nuclear import of ALS variants FUS(P525L) and FUS(R495X). *Sci. Rep.* **2021**, *11*, 3754. [CrossRef] [PubMed]

34. Kurnaeva, M.A.; Zalevsky, A.O.; Arifulin, E.A.; Lisitsyna, O.M.; Tvorogova, A.V.; Shubina, M.Y.; Bourenkov, G.P.; Tikhomirova, M.A.; Potashnikova, D.M.; Kachalova, A.I.; et al. Molecular Coevolution of Nuclear and Nucleolar Localization Signals inside the Basic Domain of HIV-1 Tat. *J. Virol.* **2022**, *96*, e0150521. [CrossRef] [PubMed]
35. Landry-Voyer, A.-M.; Bergeron, D.; Yague-Sanz, C.; Baker, B.; Bachand, F. PDCD2 functions as an evolutionarily conserved chaperone dedicated for the 40S ribosomal protein uS5 (RPS2). *Nucleic Acids Res.* **2020**, *48*, 12900–12916. [CrossRef] [PubMed]
36. Landry-Voyer, A.-M.; Bilodeau, S.; Bergeron, D.; Dionne, K.L.; Port, S.A.; Rouleau, C.; Boisvert, F.-M.; Kehlenbach, R.H.; Bachand, F. Human PDCD2L Is an Export Substrate of CRM1 That Associates with 40S Ribosomal Subunit Precursors. *Mol. Cell Biol.* **2016**, *36*, 3019–3032. [CrossRef]
37. Melnikov, S.; Ben-Shem, A.; Yusupova, G.; Yusupov, M. Insights into the origin of the nuclear localization signals in conserved ribosomal proteins. *Nat. Commun.* **2015**, *6*, 7382. [CrossRef]
38. Melnikov, S.; Kwok, H.-S.; Manakongtreecheep, K.; van den Elzen, A.; Thoreen, C.C.; Söll, D. Archaeal ribosomal proteins possess nuclear localization signal-type motifs: Implications for the origin of the cell nucleus. *Mol. Biol. Evol.* **2020**, *37*, 124–133. [CrossRef]
39. Landry-Voyer, A.-M.; Mir Hassani, Z.; Avino, M.; Bachand, F. Ribosomal Protein uS5 and Friends: Protein-Protein Interactions Involved in Ribosome Assembly and Beyond. *Biomolecules* **2023**, *13*, 853. [CrossRef]
40. Lipson, R.S.; Webb, K.J.; Clarke, S.G. Rmt1 catalyzes zinc-finger independent arginine methylation of ribosomal protein Rps2 in Saccharomyces cerevisiae. *Biochem. Biophys. Res. Commun.* **2010**, *391*, 1658–1662. [CrossRef]
41. Young, B.D.; Weiss, D.I.; Zurita-Lopez, C.I.; Webb, K.J.; Clarke, S.G.; McBride, A.E. Identification of Methylated Proteins in the Yeast Small Ribosomal Subunit: A Role for SPOUT Methyltransferases in Protein Arginine Methylation. *Biochemistry* **2012**, *51*, 5091–5104. [CrossRef] [PubMed]
42. Dionne, K.L.; Bergeron, D.; Landry-Voyer, A.-M.; Bachand, F. The 40S ribosomal protein uS5 (RPS2) assembles into an extra-ribosomal complex with human ZNF277 that competes with the PRMT3-uS5 interaction. *J. Biol. Chem.* **2019**, *294*, 1944–1955. [CrossRef] [PubMed]
43. Swiercz, R.; Person, M.D.; Bedford, M.T. Ribosomal protein S2 is a substrate for mammalian PRMT3 (protein arginine methyltransferase 3). *Biochem. J.* **2005**, *386*, 85–91. [CrossRef] [PubMed]
44. Ben-Shem, A.; Garreau de Loubresse, N.; Melnikov, S.; Jenner, L.; Yusupova, G.; Yusupov, M. The structure of the eukaryotic ribosome at 3.0 Å resolution. *Science* **2011**, *334*, 1524–1529. [CrossRef]
45. Thomas, B.J.; Rothstein, R. Elevated recombination rates in transcriptionally active DNA. *Cell* **1989**, *56*, 619–630. [CrossRef]
46. Kressler, D.; Doère, M.; Rojo, M.; Linder, P. Synthetic lethality with conditional *dbp6* alleles identifies Rsa1p, a nucleoplasmic protein involved in the assembly of 60S ribosomal subunits. *Mol. Cell Biol.* **1999**, *19*, 8633–8645. [CrossRef]
47. James, P.; Halladay, J.; Craig, E.A. Genomic libraries and a host strain designed for highly efficient two-hybrid selection in yeast. *Genetics* **1996**, *144*, 1425–1436. [CrossRef]
48. Loeb, J.D.; Schlenstedt, G.; Pellman, D.; Kornitzer, D.; Silver, P.A.; Fink, G.R. The yeast nuclear import receptor is required for mitosis. *Proc. Natl. Acad. Sci. USA* **1995**, *92*, 7647–7651. [CrossRef]
49. Goffin, L.; Vodala, S.; Fraser, C.; Ryan, J.; Timms, M.; Meusburger, S.; Catimel, B.; Nice, E.C.; Silver, P.A.; Xiao, C.-Y.; et al. The unfolded protein response transducer Ire1p contains a nuclear localization sequence recognized by multiple beta importins. *Mol. Biol. Cell* **2006**, *17*, 5309–5323. [CrossRef]
50. Bhutada, P.; Favre, S.; Jaafar, M.; Hafner, J.; Liesinger, L.; Unterweger, S.; Bischof, K.; Darnhofer, B.; Siva Sankar, D.; Rechberger, G.; et al. Rbp95 binds to 25S rRNA helix H95 and cooperates with the Npa1 complex during early pre-60S particle maturation. *Nucleic Acids Res.* **2022**, *50*, 10053–10077. [CrossRef]
51. Sikorski, R.S.; Hieter, P. A system of shuttle vectors and yeast host strains designed for efficient manipulation of DNA in Saccharomyces cerevisiae. *Genetics* **1989**, *122*, 19–27. [CrossRef] [PubMed]

Disclaimer/Publisher's Note: The statements, opinions and data contained in all publications are solely those of the individual author(s) and contributor(s) and not of MDPI and/or the editor(s). MDPI and/or the editor(s) disclaim responsibility for any injury to people or property resulting from any ideas, methods, instructions or products referred to in the content.

Review

Ribosomal Protein uS5 and Friends: Protein–Protein Interactions Involved in Ribosome Assembly and Beyond

Anne-Marie Landry-Voyer †, Zabih Mir Hassani †, Mariano Avino and François Bachand *

Dept of Biochemistry & Functional Genomics, Université de Sherbrooke, Sherbrooke, QC J1E 4K8, Canada; anne-marie.landry-voyer@usherbrooke.ca (A.-M.L.-V.); zabihullah.mir.hassani@usherbrooke.ca (Z.M.H.); mariano.avino@usherbrooke.ca (M.A.)
* Correspondence: f.bachand@usherbrooke.ca
† These authors contributed equally to this work.

Abstract: Ribosomal proteins are fundamental components of the ribosomes in all living cells. The ribosomal protein uS5 (Rps2) is a stable component of the small ribosomal subunit within all three domains of life. In addition to its interactions with proximal ribosomal proteins and rRNA inside the ribosome, uS5 has a surprisingly complex network of evolutionarily conserved non-ribosome-associated proteins. In this review, we focus on a set of four conserved uS5-associated proteins: the protein arginine methyltransferase 3 (PRMT3), the programmed cell death 2 (PDCD2) and its PDCD2-like (PDCD2L) paralog, and the zinc finger protein, ZNF277. We discuss recent work that presents PDCD2 and homologs as a dedicated uS5 chaperone and PDCD2L as a potential adaptor protein for the nuclear export of pre-40S subunits. Although the functional significance of the PRMT3–uS5 and ZNF277–uS5 interactions remain elusive, we reflect on the potential roles of uS5 arginine methylation by PRMT3 and on data indicating that ZNF277 and PRMT3 compete for uS5 binding. Together, these discussions highlight the complex and conserved regulatory network responsible for monitoring the availability and the folding of uS5 for the formation of 40S ribosomal subunits and/or the role of uS5 in potential extra-ribosomal functions.

Keywords: uS5; dedicated chaperone; PDCD2; PDCD2L; PRMT3; ZNF277; ribosome biogenesis

Citation: Landry-Voyer, A.-M.; Mir Hassani, Z.; Avino, M.; Bachand, F. Ribosomal Protein uS5 and Friends: Protein–Protein Interactions Involved in Ribosome Assembly and Beyond. *Biomolecules* **2023**, *13*, 853. https:// doi.org/10.3390/biom13050853

Academic Editors: Brigitte Pertschy and Ingrid Rössler

Received: 26 April 2023
Revised: 15 May 2023
Accepted: 16 May 2023
Published: 18 May 2023

Copyright: © 2023 by the authors. Licensee MDPI, Basel, Switzerland. This article is an open access article distributed under the terms and conditions of the Creative Commons Attribution (CC BY) license (https:// creativecommons.org/licenses/by/ 4.0/).

1. Introduction

Despite the expanding number of roles played by noncoding RNAs, proteins remain key actors involved in nearly every operation required for cellular life, from proliferation to differentiation, internal organization, intercellular communication, and cell death. In order to set in motion the synthesis of new proteins, the information encoded by genes as messenger RNAs (mRNAs) is decoded into polymers of amino acids by a highly complex cellular machine, the ribosome, in a process known as translation. The ribosome is one of the most important ribonucleoprotein complexes in the cell, as demonstrated by its essential role in protein synthesis, its highly conserved nature, and its dominating abundance in most cell types. In fact, the fundamental structure and function of the ribosome were highly conserved throughout the evolution from bacteria to humans [1,2]. Since the topic of this review will focus on eukaryotes, the following paragraphs will refer to the eukaryotic ribosome unless otherwise indicated.

The 80S ribosome is a large RNA–protein complex with a molecular mass of 4.3 megadalton in humans [3] and is composed of two independent subunits: the 40S (or small) and 60S (or large) ribosomal subunits. The 40S ribosomal subunit consists of 33 different ribosomal proteins (RPs) and a single ribosomal RNA (rRNA), the 1869-nt-long 18S rRNA, whereas the 60S ribosomal subunit contains 47 RPs and three different RNA molecules: the 5S (121-nt), 5.8S (157-nt), and 28S (5070-nt) rRNAs [3]. While the 40S subunit contains the decoding center that monitors the complementarity of mRNA and tRNA during translation, the peptidyl-transferase center and the exit tunnel, in which the nascent polypeptide emerges out the ribosome, are at the heart of the 60S ribosomal subunit [4].

The synthesis of new ribosomes is one of the most energy demanding and complex processes occurring in eukaryotic cells. In addition to the four rRNAs and 80 RPs, ribosome biogenesis involves the coordinated action of the three cellular RNA polymerases, several hundred ribosome biogenesis factors (RBFs), as well as about 200 small nucleolar RNAs (snoRNAs) [5]. Ribosome synthesis begins in the nucleolus, where nascent rRNA is transcribed by RNA polymerase I (RNAPI) and assembled co-transcriptionally into a 90S pre-ribosomal particle via the spatio-temporal recruitment of several RPs, RBFs, and snoRNPs. Following endonucleolytic cleavage of the primary transcript between 18S and 5.8S rRNAs sequences, pre-40S and pre-60S particles will subsequently follow distinct maturation pathways. Whereas this endonucleolytic cleavage step mainly occurs co-transcriptionally in budding yeast [6], the extent to which this internal cleavage step is co-transcriptional in mammalian cells remains unclear. RNAPIII transcribes the fourth rRNA, the 5S rRNA, which joins the pre-60S particle in the nucleolus as part of the 5S RNP complex [7]. After transiting through the nucleoplasm, pre-40S and pre-60S particles are independently exported to the cytoplasm where they will be further processed to ultimately become competent for translation [5,7,8].

While the main role of the 80 RPs is to assist in the folding of the four rRNA molecules into a three-dimensional structure required for the precise interaction of mRNA codons with tRNA anticodons, the coordinated incorporation of RPs into their corresponding pre-ribosomal particle is critical for the stepwise assembly of mature ribosomal subunits. Specifically, functional studies in yeast and human cells show that deficiency of most RPs affects specific steps of pre-rRNA processing associated with pre-90S, pre-40S, and/or pre-60S maturation, which usually coincides with the timing of RP incorporation into pre-ribosomes [9–12]. Accordingly, most genes that code for RPs are essential for cellular proliferation and viability as well as for embryonic development in multi-cellular organisms [4].

The eukaryotic ribosomal protein uS5 (also referred as RPS2), which is homologous to the prokaryotic 30S ribosomal protein S5 (RPS5/rpsE), is one of the largest RPs of the 40S ribosomal subunit. Interestingly, the past 20 years has seen the identification of several evolutionarily conserved uS5-associated proteins. However, the biological significance of the interaction of uS5 with many of these proteins remains to be defined. In this review, we begin by revising the knowledge on the functional role of uS5 in the late stages of pre-40S maturation as well as evidence supporting that uS5 contributes to translation fidelity. We next outline the considerable list of conserved uS5-associated proteins and discuss their functions in ribosome biogenesis and beyond.

2. Structural Features of Eukaryotic uS5 and Role in Translation

Human uS5 is a 293-amino-acids-long protein with a molecular mass of approximately 31 kDa that shows cytoplasmic expression in most tissues [13]. Analyses of actively translating ribosomes by cryo-electron microscopy (cryo-EM) [14] reveal that uS5 is located on the solvent-exposed side of the 40S ribosomal subunit (Figure 1A). Specifically, in the context of the mature 40S ribosomal subunit, uS5 is physically connected with ribosomal proteins eS21, uS2, uS8, uS4, and uS3. Residues of uS5 also interact with the 18S rRNA via a double-stranded RNA-binding-like domain (Figure 1B, magenta; PROSITE entry PS50881) and the conserved S5 C-terminal domain (Figure 1B, orange; PROSITE entry PS00585). Like many other ribosomal proteins, uS5 adopts a globular structure that is associated with disordered N- and C-terminal extensions [4,14] (Figure 1B). More precisely, the first 56 and last 14 amino acids of uS5 were not modeled from the cryo-EM analysis of active ribosomes [14] and show very low structural confidence scores as predicted by AlphaFold [15], consistent with disordered regions (Figure 1B). Notably, both N- and C-terminal extensions are unique to eukaryotic uS5 and absent in the *E. coli* homolog (Figure 1C). As shown in Figure 1D, the eukaryotic-specific N-terminal extension of uS5 is rich in arginine and glycine residues. Arginine-glycine (RG)-rich motifs have been associated with mediating interactions with RNA and protein as well as contributing to nuclear localization [16]. Interestingly, several arginine residues in the N-terminal RG-rich extension of uS5 are targeted by asymmetric dimethylation (see section on PRMT3), a uS5 post-translational modification that appears to be evolutionarily conserved [17–19]. Finally, human uS5 would include an unconventional nuclear localization signal (NLS), between lysine-159 and threonine-232, which would allow uS5 to be transported to the nucleus by various import receptors [20].

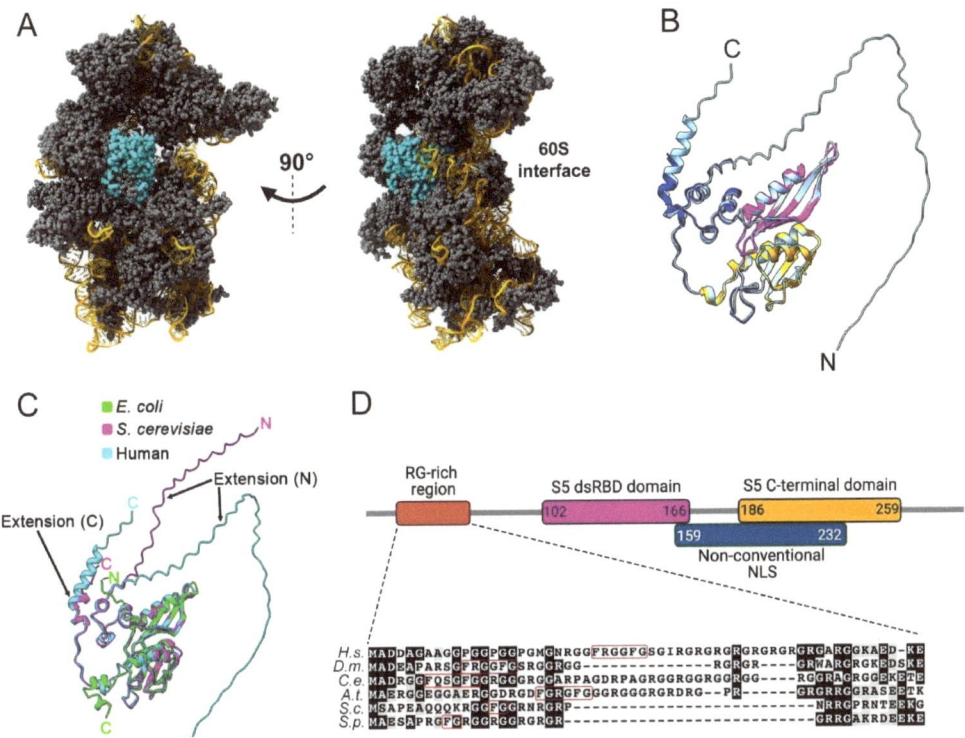

Figure 1. Structural characteristics of the 40S ribosomal protein uS5. (**A**) Cryo-EM structure of the actively translating 40 ribosomal subunit (PDB entry 5AJ0), left, and rotated 90 degrees, right. uS5 is shown in cyan, while the other 40S ribosomal proteins are colored in grey. The 18S rRNA is shown in orange. (**B**) Superposition of the tertiary structures of uS5 extracted from the active 40S ribosomal subunit (PDB entry 5AJ0; dark blue, magenta, and orange colors) and predicted by AlphaFold (pale blue). Note that the uS5 structure from the 40S subunit (PDB 5AJ0) represents only amino acids D57 to T278. The double-stranded RNA-binding-like domain and the conserved S5 C-terminal domain are shown in pink and orange, respectively, while the N- and C-terminal extensions are only seen in the AlphaFold model (pale blue). (**C**) Superposition of the AlphaFold tertiary structures of uS5 from *E. coli* (lime, P0A7W1), *S. cerevisiae* (magenta, P25443), and human (cyan, P15880) showing eukaryotic-specific N- and C-terminal extensions. (**D**) Motifs and functional domains of uS5 are shown. Numbers indicate the amino acid positions of each domain. Alignment and shading were generated using ClustalW and Boxshade software. Sequences are from *Homo sapiens* (H.s.), *Drosophila melanogaster* (D.m.), *Caenorhabditis elegans* (C.e.). *Arabidopsis thaliana* (A.t.), *Saccharomyces cerevisiae* (S.c.), and *Schizosaccharomyces pombe* (S.p.). The FXXXFG and FG motifs are boxed in red.

uS5 has been shown to be important for translation fidelity in *E. coli*, especially for a conserved glycine at position 28 [21–24]. *E. coli* uS5, together with uS3 and uS4, form part of the tunnel through which mRNA enters the small subunit of the ribosome to reach the interface between the two subunits [25]. While uS3 and uS4 act as RNA helicases, uS5 orients the incoming mRNA for proper codon reading in the ribosome A site [26]. In eukaryotes, based on cryo-EM structures of the yeast pre-initiation complex following AUG recognition [27], recent findings using *Saccharomyces cerevisiae* also support a role for eukaryotic uS5 in translation fidelity, especially at the level of translational initiation [28]. Accordingly, substitutions of uS5 residues identified as proximal to mRNA nucleotides 8 to 13 downstream of the AUG start codon [27] were shown to enhance translation initiation at suboptimal start codons [28], suggesting that uS5–mRNA contacts may contribute to the stability and thermodynamics of the eukaryotic preinitiation complex. Recent studies in mammals also support the role of uS5 in translation fidelity, as a substitution of alanine-226 for a tyrosine in uS5 leads to increased mistranslation in human cells [29] and muscle atrophy in mice [30].

3. uS5 Is an Essential Protein Required for 40s Ribosomal Subunit Production

Whereas most yeast RPs are encoded by duplicated paralogous genes, uS5 is one of the few RPs encoded by a single gene in both budding and fission yeast. *uS5* is an essential gene in budding yeast as its deletion in *S. cerevisiae* yields inviable spores [31]. Accordingly, uS5 expression is required for ribosome biogenesis. A conditional mutant strain of uS5 in *S. cerevisiae* results in the accumulation of 20S rRNA precursors; yet, it also shows a reduction in newly made 20S pre-rRNA molecules in the cytoplasm, suggesting a role for uS5 in the export of pre-40S particles [10]. As for budding yeast, uS5 is also essential for cell viability in fission yeast, and knockdown of uS5 results in the complete inhibition of 40S ribosomal subunit production [32]. Notably, *Schizosaccharomyces pombe* cells depleted of uS5 showed only a small fraction of pre-rRNA matured into 20S precursors, suggesting that a large fraction of pre-40S is actively turned-over in the absence of uS5 [32]. In *Drosophila*, *uS5* was identified as the allele associated with the "string of pearl (sop)" recessive female sterile mutants [33]. The *sop* allele is associated with reduced *uS5* mRNA levels, oogenesis and early development defects, larval lethality, and a Minute-like phenotype [33]. The Minute syndrome in *Drosophila*—which includes delayed development, low fertility and viability, and decreased body size—is thought to arise as a consequence of suboptimal protein synthesis that results from reduced levels of cellular ribosomes [34]. In mammals, most of our knowledge about uS5 comes from studies performed on immortalized cell lines. Consistent with findings in yeast and *Drosophila*, *uS5* is an essential gene in most tested human cancer cell lines [35], thereby making uS5 a potentially interesting target for cancer vulnerabilities [36]. Biochemical and structural data obtained from human cells indicate that uS5 is incorporated at late stages of pre-40S particle assembly prior to nuclear export [11,37]. Accordingly, knockdown of uS5 in human cell lines results in the accumulation of 21S pre-rRNA, suggesting the uncoupling of cleavage at sites A0–1 in the 5' external transcribed spacer sequence [11,38], as well as increase detection of 18S-E precursors in the nucleus, consistent with delayed nuclear export of pre-40S particles [11].

Collectively, the current data support a conserved role for uS5 in the late stages of 40S ribosomal subunits assembly. Consistent with this conclusion, recent cryo-EM analyses of pre-40S intermediates isolated prior to nuclear export suggest that uS5 is incorporated into nucleoplasmic pre-40S particles [37]. Interestingly, although data generally support that the ribosome assembly process is largely conserved between yeast and human cells [7], recent results suggest that uS5 may incorporate pre-40S particles at different time points between yeast and humans [37]. Specifically, *S. cerevisiae* uS5 was detected in pre-40S particles before the incorporation of uS2 and eS21, whereas, in human cells, the timing of uS5 incorporation coincided with the insertion of uS2 and eS21, suggesting that uS2–uS5–eS21 are incorporated as a cluster in humans [37].

A Multifaceted Network of uS5-Associated Proteins

The identification of evolutionarily conserved uS5-associated proteins has been the focus of several studies in the past two decades. Such studies have provided new insights into the processes and mechanisms that promote uS5 expression and incorporation into ribosomes, as well as possible yet-to-be-defined extra-ribosomal functions. The next sections will focus on the best-characterized and -conserved uS5-associated proteins: PDCD2, PDCD2L, PRMT3, and ZNF277.

4. PDCD2 and PDCD2L: uS5-Associated Paralogs That Take Part in Human Ribosome Assembly

During the process of establishing that the uS5–PRMT3 complex, which was originally identified in fission yeast [17], is conserved in humans, a set of novel and highly specific PRMT3 interactors were identified in addition to uS5, including strong enrichments of the PDCD2 and PDCD2-like (PDCD2L) proteins [39]. Biochemical assays further demonstrated that uS5 bridges the association between PRMT3 and PDCD2/PDCD2L, as depletion of uS5 totally prevented the copurification of PRMT3 and PDCD2/PDCD2L [39]. *PDCD2* and *PDCD2L* are paralogous genes conserved through evolution, with homologs from bacteria to animals but not in archaebacteria. Based on sequence analysis, *PDCD2* is thought to have arisen from the duplication of the *PDCD2L* gene prior to the divergence of animals, fungi, and plants from a common ancestor [40]. Homologs of human PDCD2 and PDCD2L paralogs are also found in mice (Pdcd2 and Pdcd2l), in *Drosophila* (Zfrp8 and Trus), and in fission yeast (Trs401 and Trs402); however, a single homolog is found in budding yeast (Tsr4). As shown in Figure 2A, PDCD2 and PDCD2L (34% identical; 52% similar) belong to a family of proteins containing N- and C-terminal TYPP (Tsr4, YwqG, PDCD2L, PDCD2) domains [40], each consisting of GGxP and $Cx_{1-2}C$-like motifs as well as a highly conserved glutamine (Q) residue (see Figure 2A). In PDCD2, the N- and C-terminal TYPP motifs are interrupted by the insertion of a MYND-type zinc finger, which was shown to be involved in transcriptional repression via protein–protein interactions [41,42]. On the other hand, PDCD2L lacks the MYND zinc finger but contains a leucine-rich nuclear export sequence (NES) that enables PDCD2L to exit the nucleus in a CRM1-dependent manner [39] (Figure 2A). While the MYND domain is conserved in *Drosophila* Zfrp8 (Figure 2B), it is not found in the *S. cerevisiae* homolog of PDCD2 (Tsr4). The C-terminal TYPP domain also appears to be degenerated in yeast Tsr4 (Figure 2B, note lack of $Cx_{1-2}C$ motif), resulting in a predicted structure that is markedly different from other PDCD2 homologs (Figure 2C). The functional role of the TYPP domain has not been well studied, though it is thought to facilitate chaperoning activity and protein–protein interactions [40]. Indeed, substitutions that modify key residues conserved in the TYPP domain of PDCD2 completely abolish the association between PDCD2 and uS5 in human cells [38].

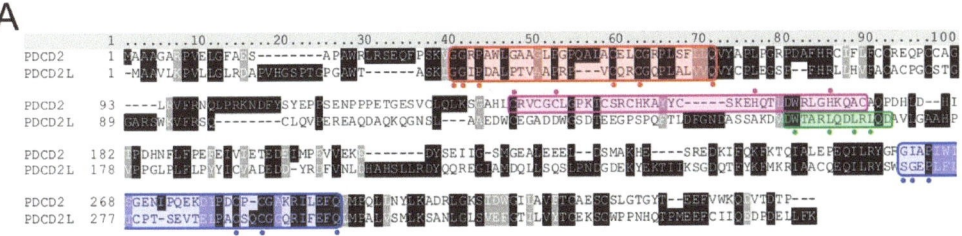

Figure 2. *Cont.*

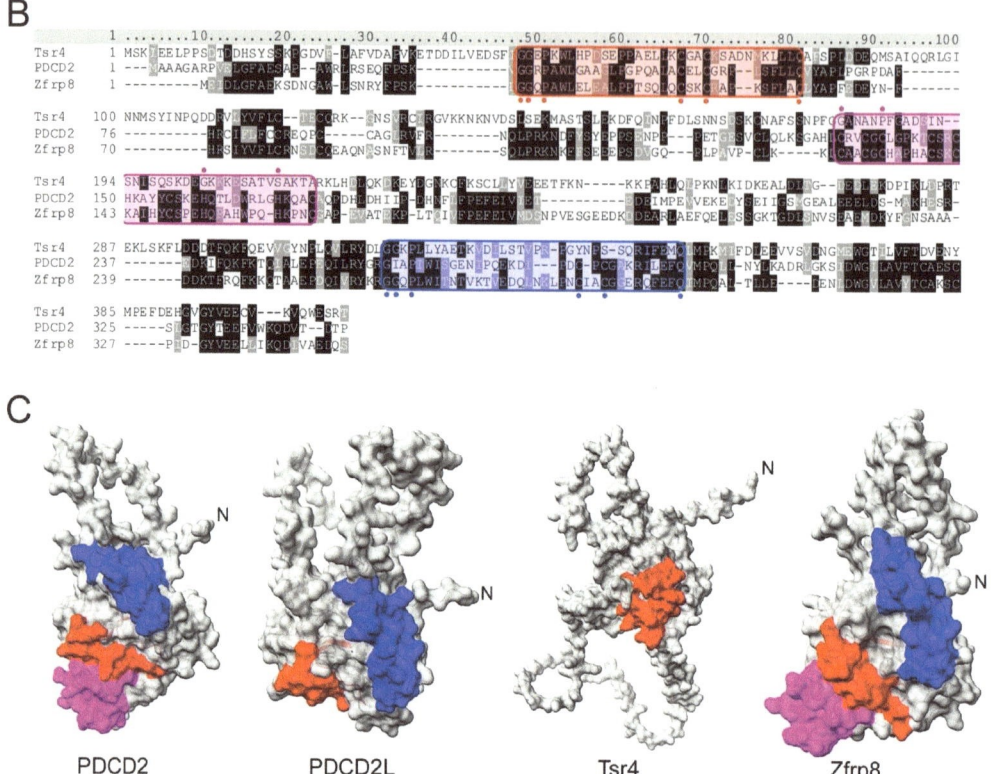

Figure 2. Sequence and structural analysis of PDCD2 and PDCD2L paralogs. (**A**) Amino acid sequence alignment of human PDCD2 and PDCD2L. Both proteins harbor N- and C-terminal TYPP domains (highlighted in red and blue, respectively) with conserved GxxP, $Cx_{1-2}C$, and Q residues highlighted with circles. Whereas PDCD2 contains a MYND zinc finger domain (in magenta with critical cysteine and histidine residues indicated by circles marked above), PDCD2L harbors a leucine-rich NES consensus sequence (green), $\Phi x_{2-3}\Phi x_{2-3}\Phi x\Phi$, where Φ represents large hydrophobic residues (indicated by green circles marked underneath). (**B**) *S. cerevisiae* Tsr4 lacks the MYND zinc finger domain and its C-terminal TYPP domain is degenerated. (**C**) AlphaFold structures for human PDCD2 (Q16342), human PDCD2L (Q9BRP1), yeast Tsr4 (P25040), and *Drosophila* Zfrp8 (Q9W1A3). Red: N-terminal TYPP domain; Blue: C-terminal TYPP domain; Magenta: MYND domain.

5. PDCD2 Is a Conserved Dedicated Chaperone for uS5

The *PDCD2* gene was originally identified in a screen for mRNAs upregulated upon apoptosis in rat cells [43]. However, subsequent experiments failed to support a correlation between *PDCD2* mRNA expression and apoptosis [44,45]. Since then, PDCD2 has been associated with the pathogenesis of several disorders, including cancer [44,46–52], Parkinson's disease [53], and fragile X syndrome [54]. Along with its potential role in diseases, PDCD2 is also implicated in development. In mice, PDCD2 is essential for stem cell viability and proliferation, and its absence leads to early embryonic lethality [55]. Although the aforementioned studies establish a clear role for PDCD2 in the development of human disorders as well as during embryonic development, the molecular function of PDCD2 had remained largely elusive until recently. Indeed, a set of elegant studies in budding yeast and human cell lines both support the conclusion that Tsr4/PDCD2 functions as an evolutionarily conserved chaperone dedicated for uS5 [38,56,57].

Previous work had already suggested the involvement of the yeast homolog of PDCD2/PDCD2L in ribosome biogenesis. Specifically, a screen for candidate genes involved in ribosome biogenesis identified *TSR4* (Twenty S rRNA accumulation 4) as a gene required for 40S ribosomal subunit production [58]. A few years later, it was reported that Zfrp8 (Tsr4/PDCD2/PDCD2L homolog in *Drosophila*) depletion results in reduced cytoplasmic level of three RPs, including uS5 [59]. Consistent with these observations, PDCD2 copurifies with uS5 in both yeast and human cells and show binding via two-hybrid assays [38,56,57], suggesting a direct interaction between PDCD2 and uS5 that is evolutionarily conserved. Although the structure of the uS5–PDCD2 complex remains to be determined experimentally, we used AlphaFold-Multimer [60] to generate models of the human uS5–PDCD2 complex. Figure 3A shows the best confident relaxed structure with the highest predicted Local Distance Difference Test (pLDDT). Alternative predicted models showed highly similar pLDDT values, indicating uniformity among the predicted structures. Whereas the overall globular structure of uS5 remained largely unchanged in the context of the uS5–PDCD2 heterodimer relative to the uS5 monomer, residues 20–50 in the N-terminal disordered region of uS5 exhibited an increased confidence score and a considerably reduced predicted position error in the uS5–PDCD2 complex compared to the same region in the uS5 monomer (Figure 3B). In contrast, the C-terminal region of uS5 (aa 273–293) appears to be more disordered in the context of the uS5–PDCD2 complex relative to the uS5 monomer (Figure 3B). Interestingly, the disordered N-terminal extension of uS5 (see Figure 1B) is predicted to fold into a hydrophobic pocket located in the N-terminal half of human PDCD2 (Figure 3C). Notably, two phenylalanine residues of human uS5 (Phe25 and Phe29) are buried inside hydrophobic core regions of PDCD2 (Figure 3D). Consistent with this model, an FXXXFG motif can be found in the N-terminal extension of uS5 from humans, fruit flies, nematodes, and plants, whereas a single phenylalanine-glycine (FG motif) is found in uS5 from budding and fission yeasts (see Figure 1D). Although this remains a predicted model, the rearrangement of the uS5 unstructured N-terminal extension into a relatively stable structure in the context of the uS5–PDCD2 heterodimer is consistent with data in yeast showing that the first 42 amino acids of uS5 appear sufficient for interaction with *S. cerevisiae* Tsr4 [56,57]. The minimal PDCD2 interaction domain of uS5 in metazoans remains to be determined. The AlphaFold-Multimer prediction of the human uS5–PDCD2 complex also suggests the insertion of the uS5 dsRBD into a C-shaped opening formed by amino acids 204 to 239 of PDCD2 (Figure 3A,E), which is likely to stabilize the heterodimer.

Studies in both yeast and human cells indicate that Tsr4/PDCD2 recognizes uS5 co-translationally and that Tsr4/PDCD2 is required for the accumulation of newly synthesized uS5 [38,56,57]. Consistent with the view that Tsr4/PDCD2 recognizes nascent uS5 is the fact that the N-terminal disordered region of uS5 is required for the formation of a stable Tsr4–uS5 complex in yeast [56,57], which is also suggested by the prediction of the human PDCD2–uS5 complex shown in Figure 3. The underlying mechanism of the specific co-translational recruitment remains unclear, however. It is possible that PDCD2/Tsr4 might have some degree of affinity for the *uS5* mRNA, and thus, that the recruitment is initiated prior to *uS5* translation initiation. The loss of function of PDCD2/Tsr4 phenocopies that of uS5 deficiency: reduced 40S production; 20S and 21S pre-rRNA accumulation in yeast and humans, respectively; and reduced incorporation of uS5 into pre-40S particles [38,56,57]. These findings, the co-translational binding of PDCD2 to nascent uS5, and the lack of identification of ribosome assembly factors in the interaction network PDCD2 support a conserved role of PDCD2 as a dedicated chaperone to uS5 [38,56,57].

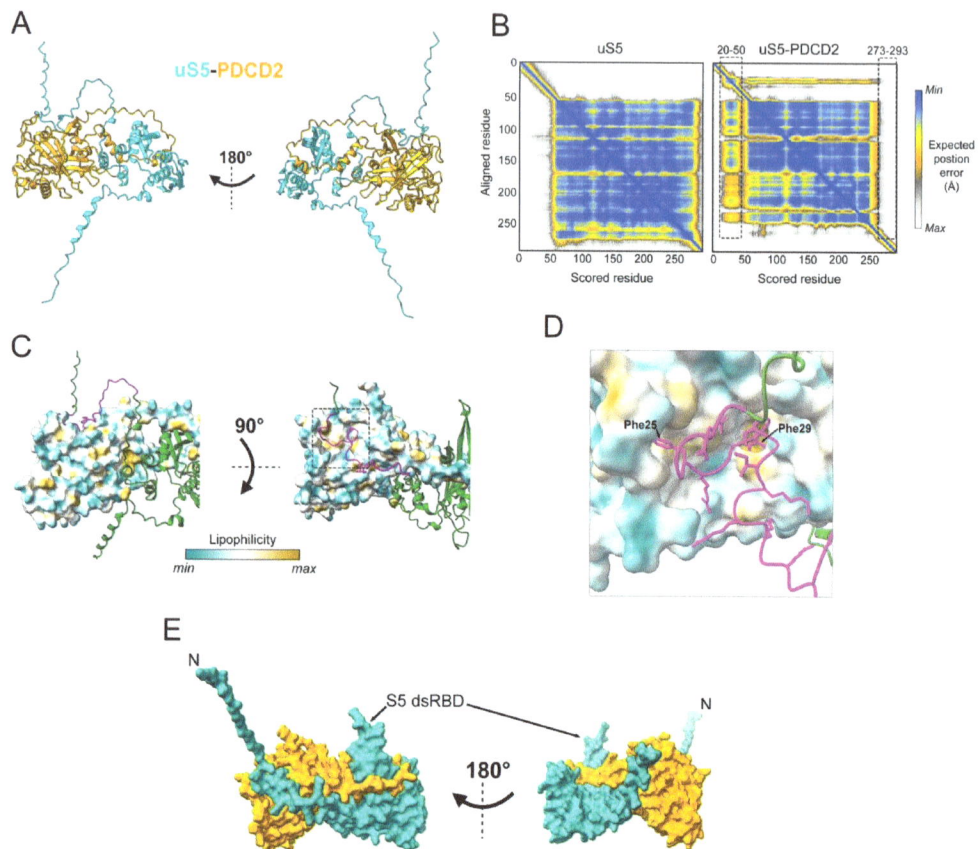

Figure 3. Predicted model of the human uS5–PDCD2 complex. (**A**) AlphaFold-Multimer [60] prediction of the human uS5 (cyan)–PDCD2 (orange) complex, left, and rotated 180 degrees, right. (**B**) AlphaFold-predicted aligned error plot for the uS5 monomer (left) and uS5–PDCD2 complex (right), highlighting residues 20–50 of uS5 confidently predicted to interact with PDCD2 and residues 273–293 that show reduced predicted position error. (**C**) Surface representation of PDCD2 lipophilicity with ribbon-like structure of uS5 (green), left, and rotated 90 degrees, right. Residues 20–50 of uS5 are colored in magenta. (**D**) Phe25 and Phe29 residues of human uS5 are predicted to be embedded in hydrophobic core regions of PDCD2. (**E**) Surface representation of the uS5 (cyan)–PDCD2 (orange) complex, left, and rotated 180 degrees, right. A C-shaped region of PDCD2 (aa 204–239) wraps around the S5 dsRBD of uS5.

How does PDCD2 promote the incorporation of uS5 into the pre-40S particle? In human cells, the model (Figure 4, see steps 1–4) suggests that once the complex is formed co-translationally, PDCD2 escorts uS5 until it is incorporated into pre-40S particles in the nucleolus. In support of this model, the PDCD2–uS5 complex can be found in the cytoplasm, the nucleus, and the nucleolus in human cells [38]. Furthermore, upon depletion of PDCD2, reduced levels of uS5 are detected in the nucleolus [38]. Interestingly, the chaperoning function of Tsr4 appears to be restricted to the cytoplasm in *S. cerevisiae* [56], suggesting some differences in the mechanism of action of yeast Tsr4 and human PDCD2. Despite the essential role of PDCD2/Tsr4 in chaperoning nascent uS5, the mechanism of uS5 nuclear import and whether PDCD2/Tsr4 is required for this process remain unclear. Yet, recent work from the Pertschy lab (Brigitte Pertschy, personal communication) suggests the presence of two independent nuclear localization signals (NLS) in *S. cerevisiae* uS5, one located in the arginine-rich region between residues 10 through 28, which overlap with the Tsr4 binding site, and a second NLS between residues 76 through 145 that interacts with the importin Pse1 (human IPO5). Although we now have a reasonable understanding of how PDCD2/Tsr4 promotes uS5 incorporation into pre-40S particles by forming a stable complex with uS5, the mechanism by which this complex is disassembled remains to be determined. As structures of the eukaryotic ribosome show a connection between the N- and C-terminal region of uS5 [14,61], Black et al. [56] proposed the possibility that an intramolecular interaction between the N- and C-terminal regions of uS5 could destabilize the interaction between the N-terminal region of uS5 and PDCD2/Tsr4.

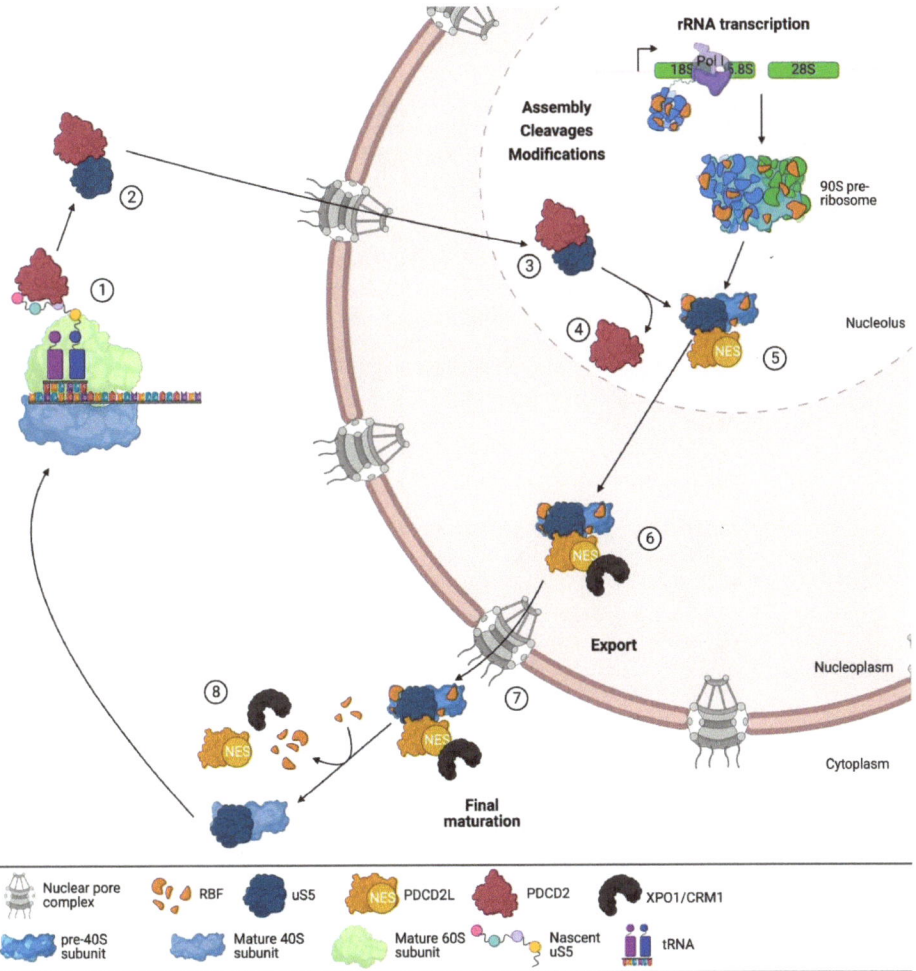

Figure 4. Model of how human PDCD2 and PDCD2L contribute to 40S ribosomal subunit biogenesis via interaction with uS5. (1) PDCD2 binds nascent uS5 co-translationally. (2) Interaction between uS5 and PDCD2 takes place in the cytoplasm and the (3) nucleolus. The role of PDCD2 in the nuclear import of uS5 remains to be determined. (4) In the nucleolus, PDCD2 would promote the incorporation of uS5 into the pre-40S ribosomal subunit. (5) PDCD2L binds to pre-40S subunits in the nucleolus via interaction with uS5. (6) The leucine-rich NES of PDCD2L promotes the recruitment of XPO1/CMR1 to the pre-40S particles. (7) Once pre-40S particles are exported to the cytoplasm, (8) PDCD2L and XPO1/CRM1 would dissociate from 40S precursors.

6. Conserved Role of PDCD2 in Stem Cell Biology and Embryonic Development

As mentioned previously, several studies using different multicellular model organisms report important roles for PDCD2 and homologs in stem cells and embryonic development. Zfrp8, the ortholog of PDCD2 in fruit flies, is important for the maintenance of two types of stem cells found in the *Drosophila* ovary: germ stem cells (GSC) and follicle stem cells (FSC). When Zfrp8 function is lost, GSCs and FSCs stop dividing, while differentiated cells show no growth phenotype [62]. Hematopoietic stem cells (HSC) are also greatly affected by the loss of Zfrp8 in *Drosophila*, as a Zfrp8 deficiency impedes self-renewal of HSCs but has no effect on pluripotent precursors [63]. In humans, PDCD2 was shown to be important for hematopoietic stem/progenitor cells viability and essential for erythroid differentiation and development [64]. Hematopoiesis is also impaired by the knockdown of PDCD2 in zebrafish embryos. Specifically, loss of *pdcd2* expression prevents HSC emergence/initiation and maintenance, in addition to causing erythroid differentiation arrest [65]. In human cells and zebrafish, ineffective hematopoiesis resulting from PDCD2 deficiency is associated with cell cycle defects [64,65]. As mentioned earlier, embryonic stem cells' viability and proliferation are dependent on PDCD2 in mice. A *PDCD2* deletion leads to embryonic development defects, with fertilized eggs attaining morula or blastocyst stages but not developing further [55]. *zfrp8*-null *Drosophila* also show abnormal development phenotypes consisting of developmental delay and lethality at larval stages [66]. In the silkworm (*Bombyx mori*), it was recently shown that BmZfrp8, the PDCD2 homolog, is expressed at different days, with a peak of expression in the middle of embryonic development [67], and this led the authors to suggest that BmZfrp8 is essential for the regulation of growth and development. Collectively, the aforementioned studies underline the importance of PDCD2 and its orthologs in stem cells' survival and embryonic development. More studies are therefore needed to elucidate the molecular mechanisms underlying the critical role of PDCD2 in stem cell biology and development and whether this role depends on the function of PDCD2 as a dedicated RP chaperone.

7. PDCD2L: A Paralog of Human PDCD2 That Associates with uS5

As will be discussed below, the ancestral duplication of the *PDCD2L* gene appears to be beneficial to organisms, as PDCD2L and PDCD2 paralogs participate in complementary functions involved in ribosome biogenesis in human cells [39]. As for PDCD2, PDCD2L physically associates with uS5. Specifically, affinity purification of GFP-PDCD2L from human cells coupled with mass spectrometry (AP-MS) revealed uS5 to be a strong interacting protein [39], and, reciprocally, PDCD2L copurifies with uS5 [68]. Interestingly, PDCD2 and PDCD2L show a mutually exclusive interaction with uS5, as the AP-MS assays of PDCD2L do not identify PDCD as a binding partner and vice versa [39]. However, and in contrast to PDCD2, the analysis of PDCD2L interactions revealed its association with several late 40S maturation factors [39]. Furthermore, one of the last precursors of the mature 18S rRNA, namely the 18S-E pre-rRNA, also specifically copurifies with PDCD2L, strongly suggesting that PDCD2L associates with late pre-40S particles [39]. Interestingly, analysis of the PDCD2L amino acid sequence revealed the presence of a leucine-rich nuclear export signal (NES), which was confirmed to act as a functional NES since (i) PDCD2L associates with XPO1/CRM1 and (ii) mutations in the PDCD2L NES prevents its association with XPO1/CRM1 and cause the accumulation of PDCD2L in the nucleus [39]. Notably, human cells deficient in PDCD2L show a marked accumulation of free 60S ribosomal subunits, a hallmark of 40S subunit deficiency. The absence of PDCD2L alone does not affect the maturation of 18S rRNA precursors, however, but clearly makes human cells more sensitive to the depletion of PDCD2, as the absence of both PDCD2 and PDCD2L exacerbates the ribosome biogenesis defects associated with the single PDCD2 depletion [39]. Together, the current data suggest that PDCD2L could act as a protein adapter for XPO1/CRM1 in the nuclear export of the pre-40S subunit. Yet, because PDCD2L is not essential for the export of 40S subunit precursors [39], it likely shares that function with one or several other ribosome maturation factors.

In contrast to the *PDCD2* and *TSR4* genes, *PDCD2L* is not essential for cell viability. As shown by the cancer dependency map, most cancer cell lines survive a deletion of *PDCD2L* [35]. However, PDCD2L seems to be required for embryonic development in mice, as *Pdcd2l*-null embryos were resorbed at mid and late gestation and no homozygous offsprings were born from heterozygous breeding [69]. In agreement with these results, knockdown of *trus*, the ortholog of *PDCD2L* in *Drosophila melanogaster*, causes a high rate of lethality at the third instar larval stage [69]. These results therefore suggest an important role for PDCD2L during embryogenesis. Further studies will be necessary in order to determine whether the role of PDCD2L during development depends on its association with uS5 and its function in ribosome biogenesis.

In summation, the available data indicate that the sequential interaction of PDCD2 and PDCD2L with uS5 contributes to different steps in 40S ribosomal subunit production, which is consistent with the additive ribosome biogenesis defects caused by the double depletion of PDCD2 and PDCD2L compared to single depletions [39]. Accordingly, as shown in Figure 4, we propose a working model of how PDCD2 and PDCD2L function in ribosome biogenesis via their mutually exclusive association with uS5. Specifically, published work supports a model (Figure 4) wherein human PDCD2 functions as a dedicated chaperone by recruiting uS5 co-translationally and by facilitating its incorporation into nucleolar pre-40S particles. Subsequently, PDCD2L would associate with nucleolar pre-40S particles via binding to uS5 and contribute to the efficient nuclear export of 40S precursors.

8. uS5 Arginine Methylation and uS5–PRMT3 Complex

An outstanding question in the field of ribosome function and regulation is the biological role of RP post-translational modifications (PTMs). Indeed, whereas RPs are subject to a variety of PTMs [70–73], few RP-modifying enzymes have been identified and studied to date. Arginine is the predominant methylated amino acid in both the eukaryotic 40S and 60S ribosomal subunits [74]. Methylation of RPs at arginines is evolutionarily conserved [75–77] and fluctuates during the cell cycle [78]. Although the methylation of arginine residues is not expected to alter the net charge of RPs, it can, however, change protein hydrophobicity and influence interactions with proteins and nucleic acids, thereby affecting functional properties such as stability, subcellular localization, complex assembly/disassembly, etc. [79]. Yet, the functional roles of RP methylation remain poorly understood. Arginine methylation is catalyzed by protein arginine methyltransferases (PRMTs), an evolutionarily conserved family of enzymes divided into two major classes depending on the type of dimethylarginine they generate: type I PRMTs modify proteins through the catalysis of asymmetric dimethylarginine, whereas type II PRMTs catalyze the formation of symmetric dimethylarginine [79,80]. Studies in the fission yeast *S. pombe* have led to the identification of PRMT3 as the first eukaryotic RP methyltransferase via arginine methylation of uS5 [17]. uS5 methylation by PRMT3 was subsequently demonstrated in human cells and in mice [19,81]. Although PRMT3 has no homolog in *S. cerevisiae* [80], budding yeast uS5 is arginine methylated by Hmt1 (homolog of human PRMT1) [18], and uS5 arginine methylation levels were found to increase during the stationary phase of *S. cerevisiae* [82]. In all of these species, uS5 was shown to be methylated on arginine residues located in its N-terminal RG-rich region [18,19,83].

PRMT3 is a primarily cytosolic type I arginine methyltransferase which possesses, in addition to its methyltransferase domain, a C2H2 zinc finger domain [84]. Although the zinc finger domain of PRMT3 is not required for methylation of an artificial substrate in vitro, it is necessary for the recognition of substrates in cell extracts [85]. Accordingly, the zinc finger domain of PRMT3 is necessary for binding uS5 in yeast and human cells [19,83]. As the structure of the PRMT3–uS5 complex has not yet been determined, we used AlphaFold-Multimer to visualize a predicted model of the human complex (Figure 5A). Here again, the best confident relaxed structure with the highest pLDDT score is presented, and the alternative predicted models showed highly similar pLDDT values. Surprisingly, direct physical contacts between the zinc finger domain of PRMT3 and residues of uS5 are not predicted in the model of the PRMT3–uS5 complex (Figure 5B), suggesting that the single zinc finger domain of PRMT3 is critical to fold its N-terminus into a structure that stably recognizes uS5. With regard to uS5 methylation by PRMT3, the model nicely predicts the arrangement of the uS5 RG-rich region proximal to the catalytic center of PRMT3 (Figure 5C) with a conserved glutamic acid residue involved in catalysis [83] located near arginine residues known to be methylated in uS5 (Figure 5D). Although arginine-methylated versions of uS5 appear to be part of actively translating ribosomes [81], it remains unclear when and where uS5 gets methylated by PRMT3: before or after incorporation into pre-40S ribosomal subunits. Whereas *S. pombe* and human uS5 form a complex with PRMT3 that is sufficiently stable to be easily isolated by affinity purification [17,19,39,83], only a small proportion of PRMT3 appears to co-sediment with the free 40S subunit [17,19].

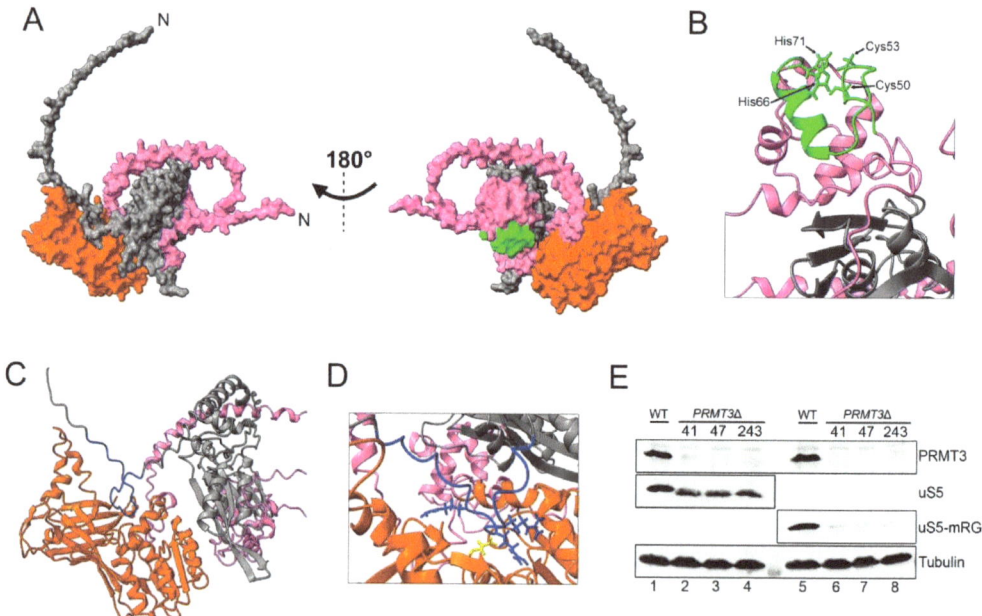

Figure 5. Predicted model of the human uS5–PRMT3 complex. (**A**) AlphaFold-Multimer [60] prediction of the human uS5 (grey)–PRMT3 (magenta) complex, left, and rotated 180 degrees, right. The C2H2 zinc finger (aa 48–71) and methyltransferase (aa 217–531) domains are shown in green and red, respectively. (**B**) The zinc finger domain of PRMT3 does not contact uS5. Shown is the zinc finger (green) of PRMT3 (magenta) with critical cysteine and histidine residues. Human uS5 is shown in grey. (**C**) The RG-rich region of human uS5 (aa 34–52, shown in dark blue) is located proximally to the catalytic center of the methyltransferase domain (red) of PRMT3. (**D**) Internal view of the PRMT3 methyltransferase domain with the Glu-338 critical for catalysis shown in yellow and arginine residues 42, 44, and 46 of uS5 shown in dark blue. (**E**) Western blot analysis using total extracts prepared from three independent clonal lines of HeLa cells deleted for *PRMT3* (lanes 2–4 and 6–8) and wild-type (lanes 1 and 5) HeLa cells. Lanes 1–4 were analyzed for total uS5, while lanes 5–8 were analyzed for arginine-methylated uS5 (uS5-mRG).

The biological significance of uS5 arginine methylation by PRMT3 remains poorly understood. In the past decade, there has been increasing interest in the idea that decoration of RPs with various modifications could customize ribosomes to translate a subset of functionally related mRNAs [86,87]. Our data in yeast and human cells show that the absence of uS5 methylation in *PRMT3* knockout cells results in a global shift in uS5 migration as analyzed by SDS-PAGE (Figure 5E, compare lanes 2–4 to lane 1) [83]. Although these results do not exclude the idea that uS5 methylation could contribute to ribosome heterogeneity, they suggest that the large majority of uS5 gets arginine methylated by PRMT3. Several studies have investigated the functional role of uS5 methylation via genetic alteration of *PRMT3* in various model organisms. In *S. pombe*, the deletion of *rmt3* results in increased levels of free 60S subunits, although pre-rRNAs processing is not disturbed [17]. Interestingly, the imbalance in free ribosomal subunits in *rmt3*-null *S. pombe* is not the consequence of deficient uS5 methylation but is rather due to the absence of uS5–Rmt3 interaction, as a methyltransferase-dead version of Rmt3 rescues the increased levels of free 60S subunits observed in *rmt3*-null cells [83]. Consistent with the view that the PRMT3–uS5 interaction is functionally important, studies in human cells indicate that PRMT3 stabilizes uS5 by inhibiting its ubiquitination and degradation by the proteasome [88]. Conversely, siRNA-mediated depletion of uS5 in humans cells considerably reduces the total cellular abundance of PRMT3 [68], suggesting that the uS5–PRMT3 interaction reciprocally stabilizes PRMT3. In mice, a targeted insertion in intron 14 of *PRMT3* that is predicted to remove the last 34 amino acids results in embryos that are smaller in size, but intriguingly, this size difference is lost by weaning age [81]. In contrast to the imbalance in free ribosomal subunits observed in *rmt3*-deleted fission yeast cells, mouse embryonic fibroblasts cultured from *PRMT3* mutant embryos showed a normal level of free ribosomal subunits, 80S monosomes, and polysomes [81]. The impact of the deletion of the *PRMT3* homolog was also studied in the plant *Arabidopsis thaliana*. In this species, the absence of AtPRMT3 alters the polysome profile and affects ribosome biogenesis [89]. More recently, the same group reported that the function of AtPRMT3 in ribosome biogenesis is primarily mediated by the physical interaction with RPS2B (one of four proteins encoded by genes orthologs of uS5 in *Arabidopsis*), but independent of its methyltransferase activity [90], consistent with previous findings using fission yeast [83]. Although the underlying mechanism by which the AtPRMT3–RPS2B interaction contributes to ribosome biogenesis remains to be defined, the authors propose that AtPRMT3 acts as a chaperone for RPS2B, preventing non-specific interactions of RPS2B and promoting its incorporation into pre-ribosomes [90].

9. ZNF277: The Newest Member among Conserved uS5-Associated Proteins

The discovery of PDCD2 and PDCD2L as novel uS5-associated proteins [38,39] beyond PRMT3 stimulated a comprehensive analysis of the human uS5 interactome. In addition to PRMT3, PDCD2, PDCD2L, RPs, and 40S maturation factors (MFs), we have recently identified a poorly characterized zinc finger protein, ZNF277, among the top 10% of uS5-associated proteins [68]. Importantly, the *uS5* mRNA is specifically enriched in ZNF277 precipitates [68], suggesting that ZNF277 is recruited co-translationally by nascent uS5, a frequent feature of dedicated RP chaperones [91,92]. A complex between ZNF277 and uS5 is also supported by independent studies that used high-throughput affinity purifications coupled with mass spectrometry in human cells [93] as well as analysis in *Drosophila* [94] and *C. elegans* [95]. These findings therefore support the existence of an evolutionarily conserved physical connection between ZNF277 and uS5.

Human ZNF277 contains five C2H2 zinc finger motifs, two featuring the typical amino acid consensus sequence (C-x(2,4)-C-x(3)-[LIVMFYWC]-x(8)-H-x(3,5)-H) and three containing atypical consensus motifs. Similar to PRMT3, the interaction between uS5 and ZNF277 depends on the integrity of its zinc finger domains, especially the two most C-terminal zinc fingers of ZNF277 [68]. Furthermore, current data support the view that ZNF277 and PRMT3 compete for uS5 binding: overexpression of wild-type PRMT3 in human cells inhibited the formation of the ZNF277–uS5 complex, whereas knockdown of ZNF277 resulted in increased levels of uS5 in PRMT3 precipitates [68]. These results therefore suggest that PRMT3 and ZNF277 have a common binding site on uS5. Although current proteomics data indicate that the PRMT3–uS5 complex is more abundant compared to the ZNF277–uS5 complex in human embryonic kidney cells [38,68], this stoichiometry may be different in other cell types.

To date, the molecular and cellular function of ZNF277 remains unclear. The homolog of human ZNF277 in mice, Zfp277, was shown to function as a transcriptional regulator [96] and to impact cellular proliferation and senescence [97]. In human cells, the depletion of ZNF277 does not appear to affect ribosome profiles as determined by polysome assays, despite direct physical interaction with uS5 and the localization of a fraction of ZNF277 to nucleoli in human cells [68]. Recent work in the nematode C. elegans shows that the homolog of human ZNF277, ZTF-7, as well as 40S RPs are required for the nucleolar depletion of the RNA exosome after a cold shock [95]. As the RNA exosome is a key complex required for pre-rRNA processing [7], these results suggest that ZNF277 may be involved in the regulation of ribosome biogenesis. It is also interesting to note that, similarly to PRMT3 conservation, S. cerevisiae does not appear to code for a protein with homology to ZNF277, whereas a ZNF277 homolog is found in the S. pombe genome.

10. Conclusions and Outlook

Since the initial discovery of the first non-ribosomal uS5-associated protein in fission yeast almost twenty years ago [17], studies have now identified a set of four evolutionarily conserved uS5-interacting partners: PDCD2, PDCD2L, PRMT3, and ZNF277. In this review, we highlighted the complex and conserved regulatory network responsible for monitoring the availability and the folding of uS5 for the formation of 40S ribosomal subunits. To our knowledge, uS5 is the RP with the greatest extent of conserved associated proteins outside of the ribosome. While PDCD2 and PDCD2L (and their homologs) contribute to the function of uS5 inside the ribosome, the functional significance of the PRMT3–uS5 (as well as uS5 arginine methylation by PRMT3) and ZNF277–uS5 complexes remains to be established. As *ZNF277* overexpression is associated with improved prognosis of human cancers according to the Human Protein Atlas Project [98], whereas *PRMT3* overexpression appears to be associated with poor prognosis, uncovering the process by which ZNF277 and PRMT3, two C2H2-type zinc finger proteins, compete for uS5 binding is likely to have relevance to cancer biology. As the PRMT3–uS5 complex is exclusively cytosolic [68], why would cells benefit from retaining a fraction of uS5 in the cytoplasm? Could this allow for the repair of damaged ribosomes, exchange between methylated and unmethylated uS5 in 40S subunits, or produce uS5-deficient ribosomes, as was recently shown for eS26 in yeast [99]?

Further studies will also be required to clarify the role of uS5 in the nuclear export of pre-40S particles [9,11,32] and whether PDCD2L functions as an adaptor protein for the CRM1-mediated export of pre-40S subunits [39] in specific cell types, thereby explaining its critical role during embryonic development [69]. Establishing the functional contribution of PDCD2 in stem cell biology and embryonic development and whether the critical role of PDCD2 in these processes is linked to its role in ribosome biogenesis via uS5 will also be very interesting. Ultimately, the ecosystem of uS5-associated proteins is complicated by the fact that these five proteins likely also form different trimers with uS5 acting as a bridging protein: PRMT3–uS5–PDCD2, PRMT3–uS5–PDCD2L, ZNF277–uS5–PDCD2, and ZNF277–uS5–PDCD2L [68]. Further research will therefore be essential to understand how these complexes coexist, cooperate, or antagonize each other.

Author Contributions: Writing of original draft preparation, review and editing by A.-M.L.-V., Z.M.H. and F.B.; analysis of protein complex prediction using AlphaFold Multimer by M.A. and F.B.; figures and visualization by A.-M.L.-V., Z.M.H. and F.B. All authors have read and agreed to the published version of the manuscript.

Funding: Studies on ribosome biogenesis and ribosomal proteins in the Bachand lab are supported by a grant from the Canadian Institutes for health research (CIHR) to F.B. Z.M.H. is supported by a studentship from the Université de Sherbrooke.

Institutional Review Board Statement: Not applicable for studies not involving humans or animals.

Informed Consent Statement: Not applicable for studies not involving humans.

Data Availability Statement: Not applicable.

Acknowledgments: We thank Brigitte Pertshy for sharing unpublished work. Studies on ribosome biogenesis and ribosomal proteins in the Bachand lab are supported by a grant from the Canadian Institutes for Health Research (CIHR) to F.B.; Z.M.H. is supported by a studentship from the Université de Sherbrooke.

Conflicts of Interest: The authors declare no conflict of interest.

References

1. Anger, A.M.; Armache, J.P.; Berninghausen, O.; Habeck, M.; Subklewe, M.; Wilson, D.N.; Beckmann, R. Structures of the human and Drosophila 80S ribosome. *Nature* **2013**, *497*, 80–85. [CrossRef]
2. Petrov, A.S.; Bernier, C.R.; Hsiao, C.; Norris, A.M.; Kovacs, N.A.; Waterbury, C.C.; Stepanov, V.G.; Harvey, S.C.; Fox, G.E.; Wartell, R.M.; et al. Evolution of the ribosome at atomic resolution. *Proc. Natl. Acad. Sci. USA* **2014**, *111*, 10251–10256. [CrossRef] [PubMed]
3. Khatter, H.; Myasnikov, A.G.; Natchiar, S.K.; Klaholz, B.P. Structure of the human 80S ribosome. *Nature* **2015**, *520*, 640–645. [CrossRef] [PubMed]
4. Melnikov, S.; Manakongtreecheep, K.; Soll, D. Revising the Structural Diversity of Ribosomal Proteins Across the Three Domains of Life. *Mol. Biol. Evol.* **2018**, *35*, 1588–1598. [CrossRef] [PubMed]
5. Pelletier, J.; Thomas, G.; Volarevic, S. Ribosome biogenesis in cancer: New players and therapeutic avenues. *Nat. Rev. Cancer* **2018**, *18*, 51–63. [CrossRef]
6. Kos, M.; Tollervey, D. Yeast pre-rRNA processing and modification occur cotranscriptionally. *Mol. Cell* **2010**, *37*, 809–820. [CrossRef]
7. Dorner, K.; Ruggeri, C.; Zemp, I.; Kutay, U. Ribosome biogenesis factors-from names to functions. *EMBO J.* **2023**, *42*, e112699. [CrossRef]
8. Henras, A.K.; Plisson-Chastang, C.; O'Donohue, M.F.; Chakraborty, A.; Gleizes, P.E. An overview of pre-ribosomal RNA processing in eukaryotes. *Wiley Interdiscip. Rev. RNA* **2015**, *6*, 225–242. [CrossRef]
9. Ferreira-Cerca, S.; Poll, G.; Gleizes, P.E.; Tschochner, H.; Milkereit, P. Roles of eukaryotic ribosomal proteins in maturation and transport of pre-18S rRNA and ribosome function. *Mol. Cell* **2005**, *20*, 263–275. [CrossRef]
10. Ferreira-Cerca, S.; Poll, G.; Kuhn, H.; Neueder, A.; Jakob, S.; Tschochner, H.; Milkereit, P. Analysis of the in vivo assembly pathway of eukaryotic 40S ribosomal proteins. *Mol. Cell* **2007**, *28*, 446–457. [CrossRef]
11. O'Donohue, M.F.; Choesmel, V.; Faubladier, M.; Fichant, G.; Gleizes, P.E. Functional dichotomy of ribosomal proteins during the synthesis of mammalian 40S ribosomal subunits. *J. Cell Biol.* **2010**, *190*, 853–866. [CrossRef] [PubMed]
12. Poll, G.; Braun, T.; Jakovljevic, J.; Neueder, A.; Jakob, S.; Woolford, J.L., Jr.; Tschochner, H.; Milkereit, P. rRNA maturation in yeast cells depleted of large ribosomal subunit proteins. *PLoS ONE* **2009**, *4*, e8249. [CrossRef] [PubMed]
13. Uhlen, M.; Fagerberg, L.; Hallstrom, B.M.; Lindskog, C.; Oksvold, P.; Mardinoglu, A.; Sivertsson, A.; Kampf, C.; Sjostedt, E.; Asplund, A.; et al. Proteomics. Tissue-based map of the human proteome. *Science* **2015**, *347*, 1260419. [CrossRef] [PubMed]
14. Behrmann, E.; Loerke, J.; Budkevich, T.V.; Yamamoto, K.; Schmidt, A.; Penczek, P.A.; Vos, M.R.; Burger, J.; Mielke, T.; Scheerer, P.; et al. Structural snapshots of actively translating human ribosomes. *Cell* **2015**, *161*, 845–857. [CrossRef] [PubMed]
15. Jumper, J.; Evans, R.; Pritzel, A.; Green, T.; Figurnov, M.; Ronneberger, O.; Tunyasuvunakool, K.; Bates, R.; Zidek, A.; Potapenko, A.; et al. Highly accurate protein structure prediction with AlphaFold. *Nature* **2021**, *596*, 583–589. [CrossRef]
16. Chowdhury, M.N.; Jin, H. The RGG motif proteins: Interactions, functions, and regulations. *Wiley Interdiscip. Rev. RNA* **2023**, *14*, e1748. [CrossRef]
17. Bachand, F.; Silver, P.A. PRMT3 is a ribosomal protein methyltransferase that affects the cellular levels of ribosomal subunits. *EMBO J.* **2004**, *23*, 2641–2650. [CrossRef]
18. Lipson, R.S.; Webb, K.J.; Clarke, S.G. Rmt1 catalyzes zinc-finger independent arginine methylation of ribosomal protein Rps2 in Saccharomyces cerevisiae. *Biochem. Biophys. Res. Commun.* **2010**, *391*, 1658–1662. [CrossRef]
19. Swiercz, R.; Person, M.D.; Bedford, M.T. Ribosomal protein S2 is a substrate for mammalian PRMT3 (protein arginine methyltransferase 3). *Biochem. J.* **2005**, *386*, 85–91. [CrossRef]
20. Antoine, M.; Reimers, K.; Wirz, W.; Gressner, A.M.; Muller, R.; Kiefer, P. Identification of an unconventional nuclear localization signal in human ribosomal protein S2. *Biochem. Biophys. Res. Commun.* **2005**, *335*, 146–153. [CrossRef]
21. Agarwal, D.; Kamath, D.; Gregory, S.T.; O'Connor, M. Modulation of decoding fidelity by ribosomal proteins S4 and S5. *J. Bacteriol.* **2015**, *197*, 1017–1025. [CrossRef] [PubMed]
22. Piepersberg, W.; Bock, A.; Wittmann, H.G. Effect of different mutations in ribosomal protein S5 of *Escherichia coli* on translational fidelity. *Mol. Gen. Genet.* **1975**, *140*, 91–100. [CrossRef] [PubMed]
23. Piepersberg, W.; Bock, A.; Yaguchi, M.; Wittmann, H.G. Genetic position and amino acid replacements of several mutations in ribosomal protein S5 from Escherichia coli. *Mol. Gen. Genet.* **1975**, *143*, 43–52. [CrossRef] [PubMed]
24. Rosset, R.; Gorini, L. A ribosomal ambiguity mutation. *J. Mol. Biol.* **1969**, *39*, 95–112. [CrossRef] [PubMed]
25. Culver, G.M. Meanderings of the mRNA through the ribosome. *Structure* **2001**, *9*, 751–758. [CrossRef] [PubMed]

26. Kurkcuoglu, O.; Doruker, P.; Sen, T.Z.; Kloczkowski, A.; Jernigan, R.L. The ribosome structure controls and directs mRNA entry, translocation and exit dynamics. *Phys. Biol.* **2008**, *5*, 046005. [CrossRef]
27. Llacer, J.L.; Hussain, T.; Saini, A.K.; Nanda, J.S.; Kaur, S.; Gordiyenko, Y.; Kumar, R.; Hinnebusch, A.G.; Lorsch, J.R.; Ramakrishnan, V. Translational initiation factor eIF5 replaces eIF1 on the 40S ribosomal subunit to promote start-codon recognition. *eLife* **2018**, *7*, e39273. [CrossRef]
28. Dong, J.; Hinnebusch, A.G. uS5/Rps2 residues at the 40S ribosome entry channel enhance initiation at suboptimal start codons in vivo. *Genetics* **2022**, *220*, iyab176. [CrossRef]
29. Shcherbakov, D.; Teo, Y.; Boukari, H.; Cortes-Sanchon, A.; Mantovani, M.; Osinnii, I.; Moore, J.; Juskeviciene, R.; Brilkova, M.; Duscha, S.; et al. Ribosomal mistranslation leads to silencing of the unfolded protein response and increased mitochondrial biogenesis. *Commun. Biol.* **2019**, *2*, 381. [CrossRef]
30. Moore, J.; Akbergenov, R.; Nigri, M.; Isnard-Petit, P.; Grimm, A.; Seebeck, P.; Restelli, L.; Frank, S.; Eckert, A.; Thiam, K.; et al. Random errors in protein synthesis activate an age-dependent program of muscle atrophy in mice. *Commun. Biol.* **2021**, *4*, 703. [CrossRef]
31. Steffen, K.K.; McCormick, M.A.; Pham, K.M.; MacKay, V.L.; Delaney, J.R.; Murakami, C.J.; Kaeberlein, M.; Kennedy, B.K. Ribosome deficiency protects against ER stress in Saccharomyces cerevisiae. *Genetics* **2012**, *191*, 107–118. [CrossRef] [PubMed]
32. Perreault, A.; Bellemer, C.; Bachand, F. Nuclear export competence of pre-40S subunits in fission yeast requires the ribosomal protein Rps2. *Nucleic Acids Res.* **2008**, *36*, 6132–6142. [CrossRef] [PubMed]
33. Cramton, S.E.; Laski, F.A. String of pearls encodes Drosophila ribosomal protein S2, has Minute-like characteristics, and is required during oogenesis. *Genetics* **1994**, *137*, 1039–1048. [CrossRef] [PubMed]
34. Marygold, S.J.; Roote, J.; Reuter, G.; Lambertsson, A.; Ashburner, M.; Millburn, G.H.; Harrison, P.M.; Yu, Z.; Kenmochi, N.; Kaufman, T.C.; et al. The ribosomal protein genes and Minute loci of Drosophila melanogaster. *Genome Biol.* **2007**, *8*, R216. [CrossRef]
35. Tsherniak, A.; Vazquez, F.; Montgomery, P.G.; Weir, B.A.; Kryukov, G.; Cowley, G.S.; Gill, S.; Harrington, W.F.; Pantel, S.; Krill-Burger, J.M.; et al. Defining a Cancer Dependency Map. *Cell* **2017**, *170*, 564–576.e516. [CrossRef]
36. Wang, M.; Hu, Y.; Stearns, M.E. RPS2: A novel therapeutic target in prostate cancer. *J. Exp. Clin. Cancer Res. CR* **2009**, *28*, 6. [CrossRef]
37. Cheng, J.; Lau, B.; Thoms, M.; Ameismeier, M.; Berninghausen, O.; Hurt, E.; Beckmann, R. The nucleoplasmic phase of pre-40S formation prior to nuclear export. *Nucleic Acids Res.* **2022**, *50*, 11924–11937. [CrossRef]
38. Landry-Voyer, A.M.; Bergeron, D.; Yague-Sanz, C.; Baker, B.; Bachand, F. PDCD2 functions as an evolutionarily conserved chaperone dedicated for the 40S ribosomal protein uS5 (RPS2). *Nucleic Acids Res.* **2020**, *48*, 12900–12916. [CrossRef]
39. Landry-Voyer, A.M.; Bilodeau, S.; Bergeron, D.; Dionne, K.L.; Port, S.A.; Rouleau, C.; Boisvert, F.M.; Kehlenbach, R.H.; Bachand, F. Human PDCD2L Is an Export Substrate of CRM1 That Associates with 40S Ribosomal Subunit Precursors. *Mol. Cell. Biol.* **2016**, *36*, 3019–3032. [CrossRef]
40. Burroughs, A.M.; Aravind, L. Analysis of two domains with novel RNA-processing activities throws light on the complex evolution of ribosomal RNA biogenesis. *Front. Genet.* **2014**, *5*, 424. [CrossRef]
41. Lutterbach, B.; Sun, D.; Schuetz, J.; Hiebert, S.W. The MYND motif is required for repression of basal transcription from the multidrug resistance 1 promoter by the t(8;21) fusion protein. *Mol. Cell. Biol.* **1998**, *18*, 3604–3611. [CrossRef] [PubMed]
42. Melnick, A.M.; Westendorf, J.J.; Polinger, A.; Carlile, G.W.; Arai, S.; Ball, H.J.; Lutterbach, B.; Hiebert, S.W.; Licht, J.D. The ETO protein disrupted in t(8;21)-associated acute myeloid leukemia is a corepressor for the promyelocytic leukemia zinc finger protein. *Mol. Cell. Biol.* **2000**, *20*, 2075–2086. [CrossRef] [PubMed]
43. Owens, G.P.; Hahn, W.E.; Cohen, J.J. Identification of mRNAs associated with programmed cell death in immature thymocytes. *Mol. Cell. Biol.* **1991**, *11*, 4177–4188. [CrossRef] [PubMed]
44. Fan, C.W.; Chan, C.C.; Chao, C.C.; Fan, H.A.; Sheu, D.L.; Chan, E.C. Expression patterns of cell cycle and apoptosis-related genes in a multidrug-resistant human colon carcinoma cell line. *Scand. J. Gastroenterol.* **2004**, *39*, 464–469. [CrossRef]
45. Kawakami, T.; Furukawa, Y.; Sudo, K.; Saito, H.; Takami, S.; Takahashi, E.; Nakamura, Y. Isolation and mapping of a human gene (PDCD2) that is highly homologous to Rp8, a rat gene associated with programmed cell death. *Cytogenet. Cell Genet.* **1995**, *71*, 41–43. [CrossRef]
46. Baron, B.W.; Zeleznik-Le, N.; Baron, M.J.; Theisler, C.; Huo, D.; Krasowski, M.D.; Thirman, M.J.; Baron, R.M.; Baron, J.M. Repression of the PDCD2 gene by BCL6 and the implications for the pathogenesis of human B and T cell lymphomas. *Proc. Natl. Acad. Sci. USA* **2007**, *104*, 7449–7454. [CrossRef]
47. Kusam, S.; Munugalavadla, V.; Sawant, D.; Dent, A. BCL6 cooperates with CD40 stimulation and loss of p53 function to rapidly transform primary B cells. *Int. J. Cancer J. Int. Cancer* **2009**, *125*, 977–981. [CrossRef]
48. Liu, H.; Wang, M.; Liang, N.; Guan, L. PDCD2 sensitizes HepG2 cells to sorafenib by suppressing epithelial-mesenchymal transition. *Mol. Med. Rep.* **2019**, *19*, 2173–2179. [CrossRef]
49. Steinemann, D.; Gesk, S.; Zhang, Y.; Harder, L.; Pilarsky, C.; Hinzmann, B.; Martin-Subero, J.I.; Calasanz, M.J.; Mungall, A.; Rosenthal, A.; et al. Identification of candidate tumor-suppressor genes in 6q27 by combined deletion mapping and electronic expression profiling in lymphoid neoplasms. *Genes Chromosomes Cancer* **2003**, *37*, 421–426. [CrossRef]
50. Wang, W.; Song, X.W.; Bu, X.M.; Zhang, N.; Zhao, C.H. PDCD2 and NCoR1 as putative tumor suppressors in gastric gastrointestinal stromal tumors. *Cell. Oncol.* **2016**, *39*, 129–137. [CrossRef]

51. Yang, X.; Lee, Y.; Fan, H.; Sun, X.; Lussier, Y.A. Identification of common microRNA-mRNA regulatory biomodules in human epithelial cancers. *Chin. Sci. Bull.* **2010**, *55*, 3576–3589. [CrossRef] [PubMed]
52. Yang, Y.; Jin, Y.; Du, W. Programmed cell death 2 functions as a tumor suppressor in osteosarcoma. *Int. J. Clin. Exp. Pathol.* **2015**, *8*, 10894–10900. [PubMed]
53. Fukae, J.; Sato, S.; Shiba, K.; Sato, K.; Mori, H.; Sharp, P.A.; Mizuno, Y.; Hattori, N. Programmed cell death-2 isoform1 is ubiquitinated by parkin and increased in the substantia nigra of patients with autosomal recessive Parkinson's disease. *FEBS Lett.* **2009**, *583*, 521–525. [CrossRef] [PubMed]
54. Tan, W.; Schauder, C.; Naryshkina, T.; Minakhina, S.; Steward, R. Zfrp8 forms a complex with fragile-X mental retardation protein and regulates its localization and function. *Dev. Biol.* **2016**, *410*, 202–212. [CrossRef]
55. Mu, W.; Munroe, R.J.; Barker, A.K.; Schimenti, J.C. PDCD2 is essential for inner cell mass development and embryonic stem cell maintenance. *Dev. Biol.* **2010**, *347*, 279–288. [CrossRef]
56. Black, J.J.; Musalgaonkar, S.; Johnson, A.W. Tsr4 Is a Cytoplasmic Chaperone for the Ribosomal Protein Rps2 in Saccharomyces cerevisiae. *Mol. Cell. Biol.* **2019**, *39*, e00019–e00094. [CrossRef]
57. Rossler, I.; Embacher, J.; Pillet, B.; Murat, G.; Liesinger, L.; Hafner, J.; Unterluggauer, J.J.; Birner-Gruenberger, R.; Kressler, D.; Pertschy, B. Tsr4 and Nap1, two novel members of the ribosomal protein chaperOME. *Nucleic Acids Res.* **2019**, *47*, 6984–7002. [CrossRef]
58. Li, Z.; Lee, I.; Moradi, E.; Hung, N.J.; Johnson, A.W.; Marcotte, E.M. Rational extension of the ribosome biogenesis pathway using network-guided genetics. *PLoS Biol.* **2009**, *7*, e1000213. [CrossRef]
59. Minakhina, S.; Naryshkina, T.; Changela, N.; Tan, W.; Steward, R. Zfrp8/PDCD2 Interacts with RpS2 Connecting Ribosome Maturation and Gene-Specific Translation. *PLoS ONE* **2016**, *11*, e0147631. [CrossRef]
60. Evans, R.; O'Neill, M.; Pritzel, A.; Antropova, N.; Senior, A.; Green, T.; Žídek, A.; Bates, R.; Blackwell, S.; Yim, J.; et al. Protein complex prediction with AlphaFold-Multimer. *BioRxiv* **2022**. [CrossRef]
61. Ben-Shem, A.; Garreau de Loubresse, N.; Melnikov, S.; Jenner, L.; Yusupova, G.; Yusupov, M. The structure of the eukaryotic ribosome at 3.0 A resolution. *Science* **2011**, *334*, 1524–1529. [CrossRef]
62. Minakhina, S.; Changela, N.; Steward, R. Zfrp8/PDCD2 is required in ovarian stem cells and interacts with the piRNA pathway machinery. *Development* **2014**, *141*, 259–268. [CrossRef] [PubMed]
63. Minakhina, S.; Steward, R. Hematopoietic stem cells in Drosophila. *Development* **2010**, *137*, 27–31. [CrossRef] [PubMed]
64. Kokorina, N.A.; Granier, C.J.; Zakharkin, S.O.; Davis, S.; Rabson, A.B.; Sabaawy, H.E. PDCD2 knockdown inhibits erythroid but not megakaryocytic lineage differentiation of human hematopoietic stem/progenitor cells. *Exp. Hematol.* **2012**, *40*, 1028–1042.e1023. [CrossRef] [PubMed]
65. Kramer, J.; Granier, C.J.; Davis, S.; Piso, K.; Hand, J.; Rabson, A.B.; Sabaawy, H.E. PDCD2 controls hematopoietic stem cell differentiation during development. *Stem Cells Dev.* **2013**, *22*, 58–72. [CrossRef] [PubMed]
66. Minakhina, S.; Druzhinina, M.; Steward, R. Zfrp8, the Drosophila ortholog of PDCD2, functions in lymph gland development and controls cell proliferation. *Development* **2007**, *134*, 2387–2396. [CrossRef]
67. Abbas, M.N.; Liang, H.; Kausar, S.; Dong, Z.; Cui, H. Zinc finger protein RP-8, the Bombyx mori ortholog of programmed cell death 2, regulates cell proliferation. *Dev. Comp. Immunol.* **2020**, *104*, 103542. [CrossRef]
68. Dionne, K.L.; Bergeron, D.; Landry-Voyer, A.M.; Bachand, F. The 40S ribosomal protein uS5 (RPS2) assembles into an extra-ribosomal complex with human ZNF277 that competes with the PRMT3-uS5 interaction. *J. Biol. Chem.* **2019**, *294*, 1944–1955. [CrossRef]
69. Houston, B.J.; Oud, M.S.; Aguirre, D.M.; Merriner, D.J.; O'Connor, A.E.; Okutman, O.; Viville, S.; Burke, R.; Veltman, J.A.; O'Bryan, M.K. Programmed Cell Death 2-Like (Pdcd2l) Is Required for Mouse Embryonic Development. *G3* **2020**, *10*, 4449–4457. [CrossRef]
70. Lee, S.W.; Berger, S.J.; Martinovic, S.; Pasa-Tolic, L.; Anderson, G.A.; Shen, Y.; Zhao, R.; Smith, R.D. Direct mass spectrometric analysis of intact proteins of the yeast large ribosomal subunit using capillary LC/FTICR. *Proc. Natl. Acad. Sci. USA* **2002**, *99*, 5942–5947. [CrossRef]
71. Louie, D.F.; Resing, K.A.; Lewis, T.S.; Ahn, N.G. Mass spectrometric analysis of 40 S ribosomal proteins from Rat-1 fibroblasts. *J. Biol. Chem.* **1996**, *271*, 28189–28198. [CrossRef] [PubMed]
72. Odintsova, T.I.; Muller, E.C.; Ivanov, A.V.; Egorov, T.A.; Bienert, R.; Vladimirov, S.N.; Kostka, S.; Otto, A.; Wittmann-Liebold, B.; Karpova, G.G. Characterization and analysis of posttranslational modifications of the human large cytoplasmic ribosomal subunit proteins by mass spectrometry and Edman sequencing. *J. Protein Chem.* **2003**, *22*, 249–258. [CrossRef] [PubMed]
73. Xue, S.; Barna, M. Specialized ribosomes: A new frontier in gene regulation and organismal biology. *Nat. Rev. Mol. Cell Biol.* **2012**, *13*, 355–369. [CrossRef] [PubMed]
74. Chang, F.N.; Navickas, I.J.; Chang, C.N.; Dancis, B.M. Methylation of ribosomal proteins in HeLa cells. *Arch. Biochem. Biophys.* **1976**, *172*, 627–633. [CrossRef] [PubMed]
75. Kruiswijk, T.; Kunst, A.; Planta, R.J.; Mager, W.H. Modification of yeast ribosomal proteins. Methylation. *Biochem. J.* **1978**, *175*, 221–225. [CrossRef]
76. Lhoest, J.; Lobet, Y.; Costers, E.; Colson, C. Methylated proteins and amino acids in the ribosomes of Saccharomyces cerevisiae. *Eur. J. Biochem.* **1984**, *141*, 585–590. [CrossRef]
77. Ramagopal, S. Covalent modifications of ribosomal proteins in growing and aggregation-competent dictyostelium discoideum: Phosphorylation and methylation. *Biochem. Cell Biol.* **1991**, *69*, 263–268. [CrossRef]

78. Chang, F.N.; Navickas, I.J.; Au, C.; Budzilowicz, C. Identification of the methylated ribosomal proteins in HeLa cells and the fluctuation of methylation during the cell cycle. *Biochim. Biophys. Acta* **1978**, *518*, 89–94. [CrossRef]
79. Xu, J.; Richard, S. Cellular pathways influenced by protein arginine methylation: Implications for cancer. *Mol. Cell* **2021**, *81*, 4357–4368. [CrossRef]
80. Bachand, F. Protein Arginine Methyltransferases: From Unicellular Eukaryotes to Humans. *Eukaryot. Cell* **2007**, *6*, 889–898. [CrossRef]
81. Swiercz, R.; Cheng, D.; Kim, D.; Bedford, M.T. Ribosomal protein rpS2 is hypomethylated in PRMT3-deficient mice. *J. Biol. Chem.* **2007**, *282*, 16917–16923. [CrossRef]
82. Ladror, D.T.; Frey, B.L.; Scalf, M.; Levenstein, M.E.; Artymiuk, J.M.; Smith, L.M. Methylation of yeast ribosomal protein S2 is elevated during stationary phase growth conditions. *Biochem. Biophys. Res. Commun.* **2014**, *445*, 535–541. [CrossRef] [PubMed]
83. Perreault, A.; Gascon, S.; D'Amours, A.; Aletta, J.M.; Bachand, F. A methyltransferase-independent function for Rmt3 in ribosomal subunit homeostasis. *J. Biol. Chem.* **2009**, *284*, 15026–15037. [CrossRef] [PubMed]
84. Tang, J.; Gary, J.D.; Clarke, S.; Herschman, H.R. PRMT 3, a type I protein arginine N-methyltransferase that differs from PRMT1 in its oligomerization, subcellular localization, substrate specificity, and regulation. *J. Biol. Chem.* **1998**, *273*, 16935–16945. [CrossRef] [PubMed]
85. Frankel, A.; Clarke, S. PRMT3 is a distinct member of the protein arginine N-methyltransferase family. Conferral of substrate specificity by a zinc-finger domain. *J. Biol. Chem.* **2000**, *275*, 32974–32982. [CrossRef] [PubMed]
86. Ferretti, M.B.; Karbstein, K. Does functional specialization of ribosomes really exist? *RNA* **2019**, *25*, 521–538. [CrossRef] [PubMed]
87. Genuth, N.R.; Barna, M. The Discovery of Ribosome Heterogeneity and Its Implications for Gene Regulation and Organismal Life. *Mol. Cell* **2018**, *71*, 364–374. [CrossRef]
88. Choi, S.; Jung, C.R.; Kim, J.Y.; Im, D.S. PRMT3 inhibits ubiquitination of ribosomal protein S2 and together forms an active enzyme complex. *Biochim. Biophys. Acta* **2008**, *1780*, 1062–1069. [CrossRef]
89. Hang, R.; Liu, C.; Ahmad, A.; Zhang, Y.; Lu, F.; Cao, X. Arabidopsis protein arginine methyltransferase 3 is required for ribosome biogenesis by affecting precursor ribosomal RNA processing. *Proc. Natl. Acad. Sci. USA* **2014**, *111*, 16190–16195. [CrossRef]
90. Hang, R.; Wang, Z.; Yang, C.; Luo, L.; Mo, B.; Chen, X.; Sun, J.; Liu, C.; Cao, X. Protein arginine methyltransferase 3 fine-tunes the assembly/disassembly of pre-ribosomes to repress nucleolar stress by interacting with RPS2B in arabidopsis. *Mol. Plant* **2021**, *14*, 223–236. [CrossRef]
91. Pausch, P.; Singh, U.; Ahmed, Y.L.; Pillet, B.; Murat, G.; Altegoer, F.; Stier, G.; Thoms, M.; Hurt, E.; Sinning, I.; et al. Co-translational capturing of nascent ribosomal proteins by their dedicated chaperones. *Nat. Commun.* **2015**, *6*, 7494. [CrossRef] [PubMed]
92. Pillet, B.; Mitterer, V.; Kressler, D.; Pertschy, B. Hold on to your friends: Dedicated chaperones of ribosomal proteins: Dedicated chaperones mediate the safe transfer of ribosomal proteins to their site of pre-ribosome incorporation. *Bioessays* **2017**, *39*, 1–12. [CrossRef] [PubMed]
93. Huttlin, E.L.; Ting, L.; Bruckner, R.J.; Gebreab, F.; Gygi, M.P.; Szpyt, J.; Tam, S.; Zarraga, G.; Colby, G.; Baltier, K.; et al. The BioPlex Network: A Systematic Exploration of the Human Interactome. *Cell* **2015**, *162*, 425–440. [CrossRef] [PubMed]
94. Giot, L.; Bader, J.S.; Brouwer, C.; Chaudhuri, A.; Kuang, B.; Li, Y.; Hao, Y.L.; Ooi, C.E.; Godwin, B.; Vitols, E.; et al. A Protein Interaction Map of Drosophila melanogaster. *Science* **2003**, *302*, 1727–1736. [CrossRef]
95. Xu, T.; Liao, X.; Huang, M.; Zhu, C.; Huang, X.; Jin, Q.; Xu, D.; Fu, C.; Chen, X.; Feng, X.; et al. A ZTF-7/RPS-2 complex mediates the cold-warm response in C. elegans. *PLoS Genet.* **2023**, *19*, e1010628. [CrossRef]
96. Negishi, M.; Saraya, A.; Mochizuki, S.; Helin, K.; Koseki, H.; Iwama, A. A novel zinc finger protein Zfp277 mediates transcriptional repression of the Ink4a/arf locus through polycomb repressive complex 1. *PLoS ONE* **2010**, *5*, e12373. [CrossRef]
97. Xie, G.; Peng, Z.; Liang, J.; Larabee, S.M.; Drachenberg, C.B.; Yfantis, H.; Raufman, J.P. Zinc finger protein 277 is an intestinal transit-amplifying cell marker and colon cancer oncogene. *JCI Insight* **2022**, *7*, e150894. [CrossRef]
98. Uhlen, M.; Zhang, C.; Lee, S.; Sjostedt, E.; Fagerberg, L.; Bidkhori, G.; Benfeitas, R.; Arif, M.; Liu, Z.; Edfors, F.; et al. A pathology atlas of the human cancer transcriptome. *Science* **2017**, *357*, eaan2507. [CrossRef]
99. Yang, Y.M.; Karbstein, K. The chaperone Tsr2 regulates Rps26 release and reincorporation from mature ribosomes to enable a reversible, ribosome-mediated response to stress. *Sci. Adv.* **2022**, *8*, eabl4386. [CrossRef]

Disclaimer/Publisher's Note: The statements, opinions and data contained in all publications are solely those of the individual author(s) and contributor(s) and not of MDPI and/or the editor(s). MDPI and/or the editor(s) disclaim responsibility for any injury to people or property resulting from any ideas, methods, instructions or products referred to in the content.

Article

RPS27a and RPL40, Which Are Produced as Ubiquitin Fusion Proteins, Are Not Essential for p53 Signalling

Matthew John Eastham †, Andria Pelava †, Graeme Raymond Wells, Nicholas James Watkins * and Claudia Schneider *

Biosciences Institute, The Medical School, Newcastle University, Newcastle upon Tyne NE2 4HH, UK
* Correspondence: nick.watkins@ncl.ac.uk (N.J.W.); claudia.schneider@ncl.ac.uk (C.S.);
 Tel.: +44-191-208-7708 (C.S.)
† These authors contributed equally to this work.

Abstract: Two of the four human ubiquitin-encoding genes express ubiquitin as an N-terminal fusion precursor polypeptide, with either ribosomal protein (RP) RPS27a or RPL40 at the C-terminus. RPS27a and RPL40 have been proposed to be important for the induction of the tumour suppressor p53 in response to defects in ribosome biogenesis, suggesting that they may play a role in the coordination of ribosome production, ubiquitin levels and p53 signalling. Here, we report that RPS27a is cleaved from the ubiquitin-RP precursor in a process that appears independent of ribosome biogenesis. In contrast to other RPs, the knockdown of either RPS27a or RPL40 did not stabilise the tumour suppressor p53 in U2OS cells. Knockdown of neither protein blocked p53 stabilisation following inhibition of ribosome biogenesis by actinomycin D, indicating that they are not needed for p53 signalling in these cells. However, the knockdown of both RPS27a and RPL40 in MCF7 and LNCaP cells robustly induced p53, consistent with observations made with the majority of other RPs. Importantly, RPS27a and RPL40 are needed for rRNA production in all cell lines tested. Our data suggest that the role of RPS27a and RPL40 in p53 signalling, but not their importance in ribosome biogenesis, differs between cell types.

Citation: Eastham, M.J.; Pelava, A.; Wells, G.R.; Watkins, N.J.; Schneider, C. RPS27a and RPL40, Which Are Produced as Ubiquitin Fusion Proteins, Are Not Essential for p53 Signalling. *Biomolecules* **2023**, *13*, 898. https://doi.org/10.3390/biom13060898

Academic Editor: Leonard B. Maggi, Jr.

Received: 13 April 2023
Revised: 24 May 2023
Accepted: 26 May 2023
Published: 28 May 2023

Copyright: © 2023 by the authors. Licensee MDPI, Basel, Switzerland. This article is an open access article distributed under the terms and conditions of the Creative Commons Attribution (CC BY) license (https:// creativecommons.org/licenses/by/ 4.0/).

Keywords: ribosomal proteins; ribosome; ubiquitin; p53; ribosome biogenesis

1. Introduction

The production of ribosomes, the most energetically consuming process in the cell, is up-regulated in cancer, down-regulated in differentiated cells and blocked by many forms of cellular stress [1–3]. As expected for such an important pathway, ribosome production is controlled by and also controls the major signalling pathways in the cell, such as mTOR, c-Myc and p53 [4]. Even though ribosome production appears to be upregulated in cancer, ribosome biogenesis defects are also linked to multiple types of cancer [5–7]. Defects in ribosome biogenesis are further found in more than 20 genetic diseases, termed ribosomopathies, which include Diamond Blackfan anaemia and Treacher– Collins syndrome [8,9]. Interestingly, in several cases, some or all of the ribosomopathy symptoms have been shown to be p53-dependent [10,11]. Therefore, understanding how ribosome production is coupled to cellular signalling is key to understanding how defects in ribosome production cause disease.

Ribosomes consist of two ribosomal subunits (small (SSU, 40S) and large (LSU, 60S)) that contain four ribosomal (r)RNAs and about 80 ribosomal proteins. The 18S, 5.8S and 28S rRNAs are transcribed as part of a long precursor rRNA (pre-rRNA) by RNA polymerase I in the nucleolus. The 47S pre-rRNA undergoes extensive endo- and exonucleolytic processing to generate the mature rRNAs [12,13]. The fourth rRNA, the 5S rRNA, is transcribed by RNA polymerase III and joins the LSU as the 5S RNP, together with the ribosomal proteins RPL5 (also known as uL18) and RPL11 (uL5) [14]. The pre-rRNA is bound by a myriad of trans-acting factors and assembles with the ribosomal proteins to generate the functional ribosomal subunits [15,16]. Surprisingly, the majority of the ribosomal proteins are over-produced [17]. Ribosomal proteins are unstable outside the ribosome, and the excess proteins are turned over by the proteasome.

Two eukaryotic ribosomal proteins, RPS27a (eS31; UBA80 gene) and RPL40 (eL40; UBA52 gene), are both encoded as C-terminal fusion proteins with ubiquitin. There are four ubiquitin-encoding genes in humans. Two are multi-copy ubiquitin repeats (UBB and UBC) and two are ubiquitin–ribosomal protein fusions (Figure 1A) [18]. The ubiquitin moiety is needed for Rps31 (yeast RPS27a) and Rpl40 production in yeast [19,20]. Ubiquitin, which makes up 0.5–1% of the total protein in the cell, is an 8.6 kDa polypeptide and a common post-translational modification, which attaches covalently to one or more lysine residues in the target protein [18]. There are multiple lysine residues in ubiquitin that can be used for linkage to proteins. Different ubiquitin linkage types have distinct functions, ranging from the modification of protein function to the degradation of the target protein [21]. The multicopy ubiquitin repeats and the ubiquitin–ribosomal protein fusions all need to be processed by yet-to-be-defined deubiquitinases to release the individual polypeptides. Interestingly, the deubiquitinase USP16 has recently been demonstrated to remove a regulatory mono-ubiquitin from an internal lysine within RPS27a. This likely represents a novel quality-control step in pre-40S maturation, but USP16 does not appear to process the RPS27a precursor fusion protein [22]. Importantly, the ubiquitin moiety would be expected to interfere with ribosome biogenesis and translation if it is not cleaved from the ribosomal protein [23]. Indeed, ubiquitin release has been shown to be required for the maturation and function of both the 40S and 60S subunits in yeast [19,24]. Interestingly, both genes encoding the ubiquitin–ribosomal protein fusions are preferentially over-expressed during hepatoma cell apoptosis [25].

Ribosome biogenesis directly controls p53 homeostasis [10,26–29]. The 5S RNP binds to and inhibits the E3 ubiquitin ligase MDM2. MDM2 suppresses p53 transcriptional activity and targets p53, through ubiquitination, for proteasomal degradation [30]. Defects in ribosome biogenesis result in 5S RNP accumulation and binding/inhibition of MDM2, leading to both p53 stabilisation and activation [26–29]. Multiple ribosomal proteins have been shown to bind MDM2, including RPL5, RPL11 and RPS27a, and to regulate its activity [10,31–36]. However, much of this work is based on protein over-expression, and only in a few cases has the endogenous protein been shown to bind to MDM2 [31–33,36]. RPS27a is one example where the endogenous protein has been shown to bind directly to MDM2 and suppress its activity [32]. Indeed, as seen with RPL5 and RPL11, knockdown of RPS27a inhibited actinomycin D (ActD; inhibits RNA pol I) and 5 fluorouracil (5FU; inhibits pre-rRNA processing) from activating p53 in U2OS cells [32]. However, the knockdown of most other ribosomal proteins causes a ribosome biogenesis defect that leads to p53 activation/stabilisation via the 5S RNP [28,29,37]. Tagged RPL40, expressed from a transfected plasmid, has also been shown to bind MDM2 and the data indicate that RPL40 regulates p53 together with the long non-coding (lnc)RNA LUCAT1 [38]. However, in several conflicting reports, the knockdown of either RPS27 or RPL40 was shown to induce p53 in HCT116 cells [29] and A549 cells [28,39]. This data suggests that RPS27a and RPL40 are not essential for p53 activation/stabilisation in response to defects in ribosome biogenesis in all cell types.

As ubiquitin and ribosome biogenesis are central to the regulation of the major cellular signalling pathways, especially p53 regulation, it is tempting to speculate that these fusion proteins couple the ubiquitin pathway to ribosome biogenesis and p53 signalling. It is further possible that the cleavage event separating ubiquitin from the ribosomal protein, as previously shown in yeast, may also be a key event in human ribosomal maturation.

Here, we have set up an in vivo system using stably transfected U2OS cells to investigate whether processing of the ubiquitin-ribosomal protein precursor protein is coupled to ribosome biogenesis. Comparing three different cell lines commonly used to study p53 signalling, we have also examined the roles of RPS27a and RPL40 in human ribosome biogenesis and cellular signalling to shed light on previous conflicting reports.

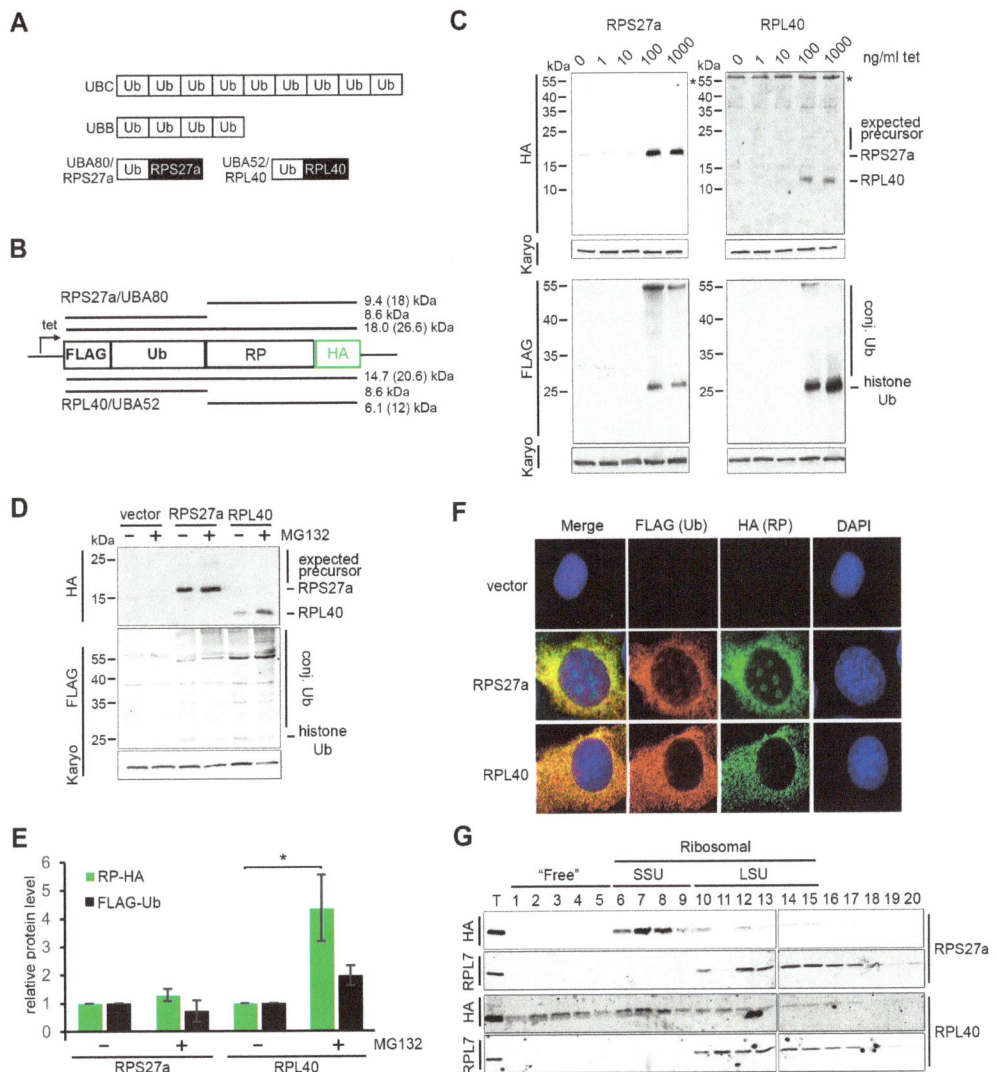

Figure 1. The stability and function of RPL40, but not RPS27a, are affected by the affinity tag. (**A**) Schematic representation of the organisation of the poly-ubiquitin (Ub) and ubiquitin–ribosomal protein fusion precursors. (**B**) Schematic representation of the ubiquitin (Ub)-ribosomal protein (RP) expression construct with N-terminal FLAG tag and C-terminal HA tag under the control of a tetracycline (tet)-regulated promoter. The bars and molecular masses above and below represent the expected fusion protein, and cleaved products, from the RPS27a and RPL40 cDNAs. Note that the HA-tagged ribosomal proteins appear slightly larger than expected (**C**), likely due to the basic nature of the amino acid sequence in each case and the gel-based mass for the HA-tagged ribosomal protein and expected mass for the fusion protein are shown in brackets. (**C**) U2OS cells stably expressing

either RPS27a or RPL40 ubiquitin fusion precursor proteins under the control of a tetracycline-regulated promoter were incubated with 0–1000 ng/mL tetracycline (tet), as indicated above each lane. Proteins from these cells were harvested after 18 h and analysed by western blotting using antibodies that recognise the HA-tag, the FLAG-tag or karyopherin (Karyo; loading control), as indicated on the left of each panel. The positions of the various proteins and expected ubiquitin–ribosomal protein fusions are indicated on the right of the panels. * indicates a non-specific protein band detected by the anti-HA antibody. (**D**) U2OS cells stably expressing either the RPS27a or the RPL40 ubiquitin fusion precursor protein or the empty pcDNA5 vector were incubated for 18 h with 1000 ng/mL tetracycline in the absence (−) or presence (+) of 25 µM MG132. Proteins from these cells were analysed by Western blotting as described in panel (**C**). (**E**) The ribosomal protein-HA (RP-HA, green bars) and FLAG-Ub (black bars) Western blot signals from panel (**D**) were quantitated and normalised to karyopherin and plotted. Quantification is based on 3 independent repeats. Error bars indicate SEM. * < 0.05. (**F**) U2OS cells stably expressing either the RPS27a or the RPL40 ubiquitin fusion precursor protein or the empty pcDNA5 vector were analysed by immunofluorescence using antibodies that recognise the FLAG-tag (FLAG (Ub); red) or the HA-tag (HA (RP), green) or DAPI to visualise DNA (blue). (**G**) Whole-cell extracts from U2OS cells stably expressing either the RPS27a or the RPL40 ubiquitin fusion precursor protein were separated on 10–40% glycerol gradients. Fractions were analysed by western blotting, using an anti-HA antibody to detect the HA-tagged ribosomal protein and an anti-RPL7 antibody to visualise LSU complexes. Positions of free, non-RNP associated proteins ("Free"), 40S (SSU) and 60S (LSU) (pre-)ribosomal complexes are indicated.

2. Materials and Methods

2.1. Cell Culture, RNAi and the Generation of Stable Cell Lines

U2OS cells were grown according to standard protocols at 37 °C with 5% CO_2 in DMEM supplemented with 10% foetal calf serum and 1% penicillin/streptomycin. MCF7 cells were cultured using RPMI1640 media, 10% foetal calf serum and 1% penicillin/streptomycin. LNCaP cells were grown in RPMI1640 media with L-Glutamine, 25 mM HEPES, 10% foetal calf serum and 1% penicillin/streptomycin. U2OS Flp-In T-REx cell lines were a generous gift from Prof. Laurence Pelletier [40]. For the Flp-In system, the cDNAs for RPS27a and RPL40 were cloned into a pcDNA5 vector to enable expression of the proteins with an N-terminal 2xFLAG-PreScission protease site-His6 (FLAG) tag and a 2xHA tag at the C-terminus. These plasmids, or the empty pcDNA5 vector, were transfected into Flp-In T-REx U2OS cells and cells that had stably integrated the plasmid into their genome were selected using Hygromycin B, according to the manufacturer's instructions (Invitrogen/Thermo Fisher). Expression of tagged proteins was induced by the addition of 0–1000 ng/mL tetracycline for 18–48 h prior to harvesting.

For RNAi-mediated knockdowns, the cells were transfected with siRNA duplexes (50 nM) using Lipofectamine RNAiMAX reagent according to the manufact- urer's instructions (Invitrogen/Thermo Fisher) as described previously [41]. For RPL40 and RPL7 knockdowns, Dharmacon smartpools were used. Individual siRNAs (Eurofins MWG) were used for RPL5 (5′-GGUUGGCCUGACAAAUUAUdTdT-3′ [42]), RPS19 (5′-GAUGGCGGCCGCAAACUGUCAdTdT-3′ [43]) and RPS27a (5′-CAGACAUUAUUGUGG-CAAAdTdT-3′ or 5′-UUAGUCGCCUUCGUCGAGAdTdT-3′ [32]). Knockdown cells and cells treated with the GL2 control siRNA targeting firefly luciferase (5′-CGUACGCGGAAU-ACUUCGAdTdT-3′ [44]) were harvested 48 h after transfection. Cells were incubated for 18 h with low levels (5 ng/mL) of actinomycin D (ActD) to block ribosome biogenesis or 25 µM of MG132 (in ethanol) to inhibit the proteasome.

2.2. Western Blotting

Total cellular protein was separated by SDS polyacrylamide gel electrophoresis (SDS-PAGE) and transferred to a nitrocellulose membrane. Proteins were detected using ECL (Figure 1C, "FLAG") or fluorescently labelled secondary antibodies and the LI-COR Odyssey system (all other Figures), and levels were determined using Image Quant software (G.E. Healthcare). Antibodies to detect the affinity tags on the fusion proteins were purchased from Sigma (anti-FLAG, F7425) or Berkeley Ab Company (anti-HA, MMS-101P), respectively. Other antibodies were: Anti-RPL7 (Abcam; ab72550), anti-p53 (Santa Cruz Biotechnology; sc-126), anti-karyopherin (loading control, Santa Cruz Biotechnology; sc11367), anti-RPL5 (Cell Signalling Technology; #14568). Anti-RPS19 antibodies were a kind gift from Phil Mason (Washington University School of Medicine, St. Louis, MO, USA).

2.3. Immunofluorescence

For immunofluorescence, U2OS T-REx Flp-In cells expressing the tagged protein of interest were plated on coverslips in a 24-well plate and treated with 1000 ng/mL tetracycline for 18 h. Cells were fixed in PBS containing 4% paraformaldehyde and then permeabilised in PBS/0.1% Triton. After blocking for 1–2 h using PBS/0.1% Triton/10% FCS solution, cells were incubated in the same solution containing the primary antibody for 1–2 h, followed by washes with PBS, then incubation with the secondary antibody for 1–2 h. After washing with PBS, and one wash with 0.01 ng/mL DAPI (4′,6-diamidino-2-phenylindole) diluted in PBS, the coverslips were mounted on a glass slide using Moviol. The cells were visualised using a Zeiss Axiovert 200M inverted microscope and analysed using the Axiovert software. For primary antibodies, a rabbit anti-FLAG antibody (Sigma Aldrich; F7425) and a mouse anti-HA antibody (Berkeley Ab Company; MMS-101P) were used. The secondary antibodies anti-Rabbit Alexa Fluor 555 (A-31572) and anti-Mouse Alexa Fluor 647 (A-31570) were both purchased from Invitrogen.

2.4. Gradient Analysis

For glycerol gradient analysis, U2OS T-REx Flp-In cells expressing the tagged protein of interest were treated with 1000 ng/mL tetracycline for 18 h. Whole-cell extracts prepared by sonication (approximately 8×10^6 cells) were loaded on 10–40% glycerol gradients and separated by centrifugation (1.5 h at 52,000 rpm at 4 °C) using a swTi60 rotor (Beckman L7-80). The resultant fractions were analysed by SDS-PAGE and western blotting as described above.

2.5. RNA Analysis

RNA was extracted from cells using TRI reagent (Sigma-Aldrich) and separated on a 1.2% agarose-glyoxal gel and transferred to a nylon membrane by capillary blotting. DNA oligonucleotide probes specific to ITS1 (hybridising between site 2a and site 2, described previously [41]), RNase P (5′-CCTTCCCAAGGGACATGGGAGTGGAGTG-3′), the mature 18S rRNA (5′-GGGCGGTGGCTCGCCTCGCG-3′) and the mature 28S rRNA (5′-TGGTCCGTGTTTCAAGACGGGT-3′) were 5′-labelled using T4 polynucleotide kinase and ^{32}P γATP. The probe used for ITS2 was generated by random-primed labelling from a PCR product (primers: forward 5′-GTGCGCGGCTGGGGGTTCCCTCGCAGG-3′ and reverse 5′-CCGGCACCCTTCCCCTTCCGGACC-3′) using ^{32}P dATP as described previously [45]. All bands were detected using a PhosphorImager.

For RT-PCR, the RNA samples were treated with DNase Turbo before cDNAs were generated using oligo-dT primer and Superscript III (Invitrogen) reverse transcriptase. The specific cDNA fragments were then amplified using GoTaq polymerase (Promega) and the following primer pairs: GAPDH, 5′-GGTCGGAGTCAACGGATTTGGTCG-3′ and 5′-CGTTGTCATACCAGGAAATG-AGCTTGAC-3′; RPL40, 5′-AGGAGGGTATCCCA-CCTGACCAGC-3′ and 5′-CGAGCATAGCACTTGCGGCAGATC-3′; RPS27a, 5′-CCCTCGA-GGTTGAACCCTCGG-3′ and 5′-GCCATTCTCATCCACCTTATAATATTTCAGG-3′. PCR products were separated by agarose gel electrophoresis using SYBR Safe stain (Invitrogen) and visualised using a PhosphorImager.

3. Results

3.1. Both the RPS27a and RPL40 Ubiquitin Fusion Precursors Are Efficiently and Rapidly Cleaved In Vivo

RPS27a and RPL40 are co-expressed as C-terminal fusions with ubiquitin and, as such, represent two of the four ubiquitin genes in mammals (Figure 1A; [18]). It is not yet completely clear when and how the fusion proteins are processed. It is possible that the separation of the ubiquitin and ribosomal protein moieties could take place during ribosome biogenesis. Since ubiquitin is important in ribosome biogenesis/function and excess ribosomal protein turnover, it may provide a means to co-regulate ribosomal protein production with ubiquitin levels in the cell [18]. To investigate this, we generated U2OS cells stably expressing either the RPS27a or RPL40 ubiquitin fusion proteins under the control of a tetracycline-regulated promoter. In each case, a FLAG-tag was added to the N-terminus of ubiquitin, and an HA-tag was added to the C-terminus of the ribosomal protein (Figure 1B).

These cells were first treated with a range of tetracycline concentrations (0–1000 ng/mL) to induce protein expression, and 18 h later, the cells were harvested, and expression of the ubiquitin–ribosomal protein fusions analysed by western blotting (Figure 1C). While commercially available antibodies were reported to detect RPS27a and RPL40 in prostate cancer cell lines [46], they did not function when tested in our hands. Therefore, levels of the endogenous ribosomal proteins could not be assessed. Cells expressing the RPS27a fusion protein produced a single, HA-tagged protein with an apparent molecular weight of ~18 kDa. We also observed a range of proteins conjugated to FLAG-tagged ubiquitin, with the most prominent band at ~25 kDa likely representing histone-ubiquitin. Importantly, the ~18 kDa band detected by the anti-HA antibody was not detected by the anti-FLAG antibody, indicating that this is the already cleaved RPS27a protein lacking the ubiquitin moiety. The HA-tagged RPS27a, which represents the cleaved ribosomal protein, appeared larger than predicted (9.4 kDa from the amino acid sequence). We assume that the highly basic nature of the sequence affects the migration of the protein through the polyacrylamide gel. Only background signals for both the anti-HA and anti-FLAG antibodies were detected in cells treated with tetracycline concentrations lower than 100 ng/mL. Upon treatment with 100 or 1000 ng/mL tetracycline, cells expressing RPL40 also produced a single, HA-tagged protein, albeit at lower levels compared to RPS27a, with an apparent molecular weight of about 12 kDa, and a similar distribution of FLAG-tagged, ubiquitinated proteins. Again, the HA-tagged RPL40, which is also a very basic protein, appeared larger than predicted from its amino acid sequence (6.2 kDa). No higher-molecular weight precursor proteins (~26.6 kDa for RPS27a or ~20.6 kDa for RPL40–considering the apparent MW of the two ribosomal proteins from the gel), detectable by both anti-HA and anti-FLAG antibodies were seen suggesting that processing of both ribosomal protein-fusion proteins is efficient.

Many ribosomal proteins are produced in excess, and the free proteins are quickly degraded by the proteasome [17]. The apparent lack of ubiquitin-fusion precursors upon tetracycline treatment could indicate rapid turnover of the excess non-cleaved precursor proteins (Figure 1C). However, it is also possible that the ubiquitin component might be cleaved before the proteasomal degradation of the excess ribosomal protein so that the ubiquitin can still be integrated into the ubiquitin pool. To test this, U2OS cells expressing tagged ubiquitin-RPS27a or RPL40, or the pcDNA5 empty vector, were treated for 18 h with 25 µM MG132, a proteasomal inhibitor [47]. A total of 1000 ng/mL tetracycline was added to the cells at the same time so that MG132 affected the expressed fusion proteins (Figure 1D).

Interestingly, the levels of HA-tagged RPS27a did not significantly change when the proteasome was blocked by treatment with MG132 (Figure 1E). In contrast, the levels of HA-tagged RPL40 were significantly increased after MG132 treatment. These results suggest that HA-RPL40, but not HA-RPS27a, is turned over by the proteasome. It should be noted that the level of overexpression may also affect the stability of the proteins. Due to the lack of functional antibodies, it was not possible to assess the ratios between HA-tagged and endogenous proteins in either case. However, previous studies using the tetracycline induction system for other ribosomal proteins, e.g., RPL5 and RPL11, found that proteins were expressed at a maximum of 20% of the endogenous protein levels [26]. We would therefore assume similar relative levels of expression for RPS27a and RPL40. Ubiquitin levels were not significantly affected by treatment with MG132 in either case or upon expression of the empty vector, indicating that ubiquitin is produced as normal. No obvious accumulation of either expected ubiquitin-fusion precursor was detected after inhibition of the proteasome by MG132 (Figure 1D), apart from a very faint ~25 kDa band seen for RPS27a, which may represent low levels of the fusion protein. These data suggest that the processing of the ubiquitin-fusion proteins occurs efficiently and, in the case of RPL40, before degradation of the excess cleaved protein.

3.2. The Function of RPL40, but Not RPS27a, Is Compromised by the Affinity Tag

The observed proteasomal degradation of HA-RPL40 suggests that the function of HA-RPL40 could be affected by the affinity tag used in this study, which might, in turn, affect its stability (Figure 1D,E).

To test if the affinity tag impacts on the cellular localisation of RPS27a or RPL40, immunofluorescence was performed (Figure 1F). U2OS cells containing either the pcDNA5 empty vector or constructs expressing the RPS27a or RPL40 fusion proteins were treated with 1000 ng/mL tetracycline for 18 h before staining. An anti-FLAG antibody was used to stain FLAG-tagged ubiquitin, and an anti-HA antibody was used to detect cleaved HA-tagged ribosomal proteins. U2OS cells containing the pcDNA5 empty vector showed a clear DAPI-staining marking on the nucleus but only a background signal when the anti-FLAG or the anti-HA antibodies were used (Figure 1F). In cells expressing the fusion proteins, FLAG-tagged ubiquitin was found mainly in the cytoplasm, with some traces in the nucleus but not in the nucleolus. This agrees with previous data showing that FLAG-tagged ubiquitin was found conjugated to both cytoplasmic and nuclear proteins (Figure 1C,D). HA-tagged RPS27a localised in both the cytoplasm and the nucleolus, while HA-tagged RPL40 was found mainly in the cytoplasm, together with a weak nucleoplasmic signal seen (Figure 1F). Overall, the localisation of the tagged proteins is consistent with what was previously seen for the endogenous RPS27a (https://www.proteinatlas.org/ENSG00000143947-RPS27A/subcellular; accessed on 11 April 2023) and RPL40 https://www.proteinatlas.org/ENSG00000221983-UBA52/subcellular, accessed on 11 April 2023) proteins in U2OS cells in the Human Protein Atlas resource project (https://www.proteinatlas.org/; accessed on 11 April 2023) [48].

Glycerol gradient analyses to separate free proteins from those associated with RNP complexes were performed to assess the integration of HA-tagged RPS27a and RPL40 into (pre-)ribosomal particles. For this, whole cell extracts from U2OS cells expressing HA-tagged RPS27a or RPL40 after the addition of tetracycline for 18 h were separated using a 10%–40% glycerol gradient and analysed by Western blotting (Figure 1G). HA-tagged ribosomal proteins were detected in the individual fractions using an anti-HA antibody, while an antibody against the endogenous ribosomal protein RPL7 was used as a marker for (pre-)60S complexes. HA-tagged RPS27a was mainly found in small ribosomal subunit (SSU) complexes (fractions 6–9), as expected, and no HA-tagged RPS27a was sedimented in the free protein fractions (fractions 1–5). While a small amount of HA-tagged RPL40 was detected in the expected large ribosomal subunit (LSU) complexes (fractions 10–15), the majority of the protein sedimented together with the SSU complexes (fractions 6–9) or in the free, non-ribosomal complex fractions (fractions 1–5).

The glycerol gradient, therefore, suggests that the affinity tag does not appear to affect the function of RPS27a. In contrast, the integration of HA-tagged RPL40 into (pre-)60S particles is impaired. Due to a lack of a functional antibody, we cannot determine whether this is due to the affinity tag on RPL40 or due to the protein being overexpressed, or both. Based on these observations, the cell line expressing HA-RPL40 was excluded from further analyses. However, the gradient data may provide an explanation for the observed instability of the HA-tagged RPL40 protein (Figure 1D,E). After cleavage from the fusion protein and without appropriate integration into (pre-)ribosomal complexes, HA-RPL40 is likely prone to degradation by the proteasome.

3.3. RPS27a Is Separated from Its Precursor in a Process Independent from Ribosome Biogenesis

To test whether ribosome biogenesis is important for the cleavage of the ubiquitin-RPS27a fusion protein, we treated cells expressing affinity-tagged RPS27a, alongside cells containing the empty pcDNA5 vector, with 1000 ng/mL tetracycline for 18 h to induce protein expression. Low levels (5 ng/mL) of Actinomycin D (ActD) were added to block rRNA transcription by RNA polymerase I and thus abolish ribosome production. ActD and tetracycline were added at the same time so that all the tagged proteins were produced after ribosome production had been blocked. In this situation, newly synthesised ribosomal proteins for both subunits are unstable [17]. Treatment of cells with ActD resulted in a five-fold decrease in HA-tagged RPS27a but no significant change in FLAG-ubiquitin levels (Figure 2A,B). Again, no ubiquitin–ribosomal protein fusion precursor band was observed.

To further analyse the role of ribosome biogenesis in the processing of the ubiquitin-RPS27a fusion protein, we tested the effect of depleting either a small (RPS19) or large (RPL7) ribosomal subunit protein on the accumulation of HA-RPS27a and FLAG-ubiquitin. Cells were simultaneously transfected with the respective siRNA to mediate protein knockdown and treated with 1000 ng/mL tetracycline to induce expression of the fusion protein for 48 h (Figure 2C,E). The knockdown efficiency of RPS19 and RPL7 was monitored by Western blotting. Northern blotting was performed to demonstrate that the knockdown of each ribosomal protein causes a strong decrease in the levels of the mature 18S or 28S rRNAs, respectively (Figure 2D).

Knockdown of RPS19 resulted in an about three-fold decrease in HA-tagged RPS27a levels, while knockdown of RPL7 had no impact (Figure 2C,E). Neither knockdown resulted in the accumulation of the ubiquitin-RPS27a precursor or impacted FLAG-ubiquitin production. Our data, therefore, show that HA-RPS27a is dependent on SSU, but not LSU production, while the production of ubiquitin appears independent of ribosome biogenesis. These observations suggest that cleavage of the ubiquitin-RPS27a precursor does not require ribosome biogenesis and that ubiquitin and HA-RPS27a accumulate independently of one another.

3.4. Cells Expressing the Affinity-Tagged Ubiquitin-RPS27a Precursor Exhibit Higher p53 Levels, Which Do Not Change in Response to ActD Treatment

Overexpression of RPS27a and its ubiquitin-fusion precursor has been shown to inhibit MDM2-mediated p53 degradation in U2OS cells [32]. To test whether this is also the case in our in vivo system, we treated cells expressing HA-tagged RPS27a, alongside those containing the empty pcDNA5 vector, with 1000 ng/mL tetracycline for 18 h to induce protein expression. Low levels (5 ng/mL) of Actinomycin D (ActD) were added at the same time, which was previously demonstrated to cause p53 stabilisation through blocking ribosome biogenesis [26].

Over-expression of RPS27a resulted in a significant five-fold p53 increase compared to the cells containing the empty pcDNA5 vector (Figure 2F). Treatment with ActD in U2OS cells containing the empty vector resulted in a slightly lower but also significant two-fold p53 increase compared to the non-treated cells. Interestingly, ActD treatment in U2OS cells expressing HA-tagged RPS27a did not result in a significant change in p53 levels as compared to the non-ActD-treated U2OS cells expressing the HA-tagged ribosomal protein. Likewise, p53 levels did not significantly differ between ActD-treated cells expressing HA-tagged RPS27a or the empty vector. These results confirm that the expression of HA-tagged RPS27a results in p53 stabilisation, which is consistent with the previous study [32]. However, the data also indicate that over-expression of RPS27a does not further enhance p53 stabilisation seen after ActD-induced ribosome biogenesis defects.

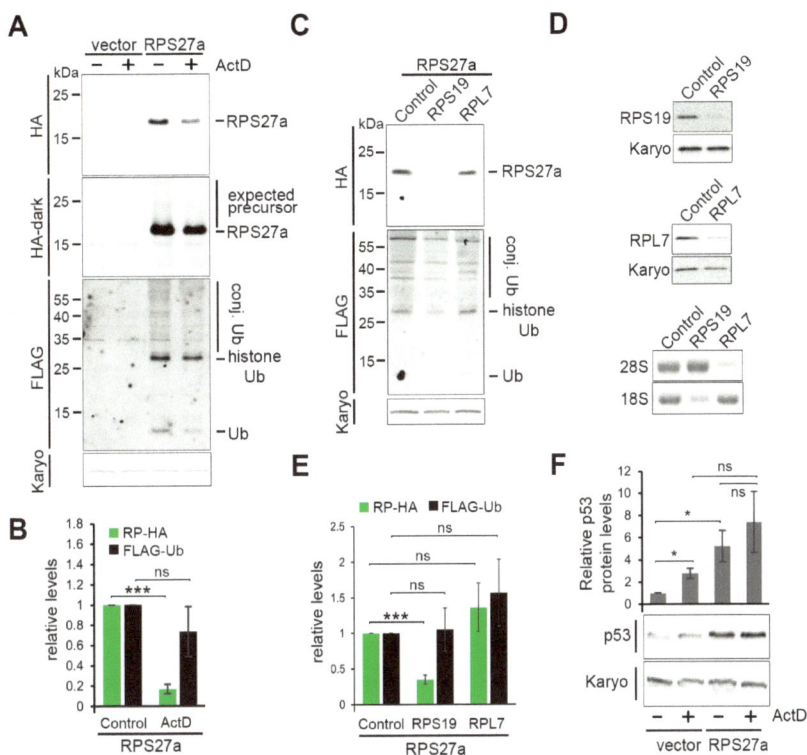

Figure 2. Processing of ubiquitin from the Ub-RPS27a precursor occurs independently of ribosome biogenesis. (**A**) U2OS cells stably expressing the RPS27a ubiquitin fusion precursor protein or the empty pcDNA5 vector, were incubated for 18 h with 1000 ng/mL tetracycline in the absence (-) or presence (+) of 5 ng/mL ActD. Proteins from these cells were analysed by western blotting using antibodies that recognise the HA-tag, the FLAG-tag or karyopherin (Karyo; loading control), as indicated on the left of each panel. HA-dark is a stronger exposure of the HA-signal to facilitate visualisation of the expected precursor. The positions of the various proteins, and expected ubiquitin-ribosomal protein fusions, are indicated on the right of the panels. (**B**) The ribosomal protein-HA (RP, green bars) and FLAG-Ub (black bars) western blot signals from (**A**) were quantitated and normalised to karyopherin and plotted. (**C**) The RPS27a ubiquitin fusion precursor protein was expressed with 1000 ng/mL tetracycline in U2OS cells transfected with the control siRNA or siRNAs targeting either RPS19 or RPL7 (indicated above each lane). Proteins from these cells were harvested after 48 h and analysed by western blotting as described in (**A**). (**D**) U2OS cells stably expressing the RPS27a ubiquitin fusion precursor protein, as described in (**C**), were transfected with siRNAs to deplete either RPS19 or RPL7. Total protein was analysed by western blotting using antibodies specific to the ribosomal proteins or karyopherin (Karyo; loading control), as indicated on the left of each panel. The impact of the siRNAs on mature rRNA levels was assessed by glyoxal/agarose gel electrophoresis and northern blotting using probes specific to the mature 18S and 28S rRNAs (as indicated on the left). (**E**) The ribosomal protein-HA (RP, green bars) and FLAG-Ub (black bars) western blot signals from (**C**) were quantitated and normalised to karyopherin and plotted. (**F**) Proteins from (**A**) were analysed by western blotting using antibodies that recognise p53 or karyopherin (Karyo; loading control), as indicated on the left. Western blot signals were quantitated, normalised to karyopherin and plotted. For all graphs, quantification is based on 3 independent repeats. Error bars indicate SEM. * < 0.05, *** < 0.001, ns > 0.05 (not statistically significant).

3.5. Knockdown of RPS27a or RPL40 Leads to p53 Stabilisation in MCF7 and LNCaP Cells, but Not in U2OS Cells

Having established the effect of overexpressing RPS27a on p53 levels, we next investigated the impact of reducing the levels of RPS27a or RPL40 on p53 signalling in U2OS cells. Knockdown of many ribosomal proteins has been shown to induce p53 through defects in ribosome biogenesis and the 5S RNP [28,29,37]. Indeed, knockdown of either RPS27a or RPL40 caused significant p53 stabilisation in HCT116 cells [29] and A549 cells [28,39]. However, in the aforementioned study describing the effect of RPS27a overexpression in U2OS cells, RPS27a knockdown did not induce p53 but instead blocked p53 stabilisation in response to inhibiting ribosome biogenesis using either ActD or 5FU [32].

To shed light on these conflicting reports and to clarify whether knockdown of RPS27a and/or RPL40 inhibits or causes p53 accumulation, we transfected U2OS cells with siRNAs targeting RPS27a or RPL40 or with control siRNAs targeting firefly luciferase and analysed p53 levels by Western blotting (Figure 3). Since no functional commercial antibodies were available for RPS27a and RPL40, the efficiency of the knockdown was confirmed by RT-PCR (Figure 3A). In each case, a reduction in RT-PCR signal, and therefore ribosomal protein mRNA levels, relative to the levels of the GAPDH mRNA, was observed for the cells transfected with the siRNAs targeting the ribosomal protein mRNA relative to cells transfected with the control siRNA.

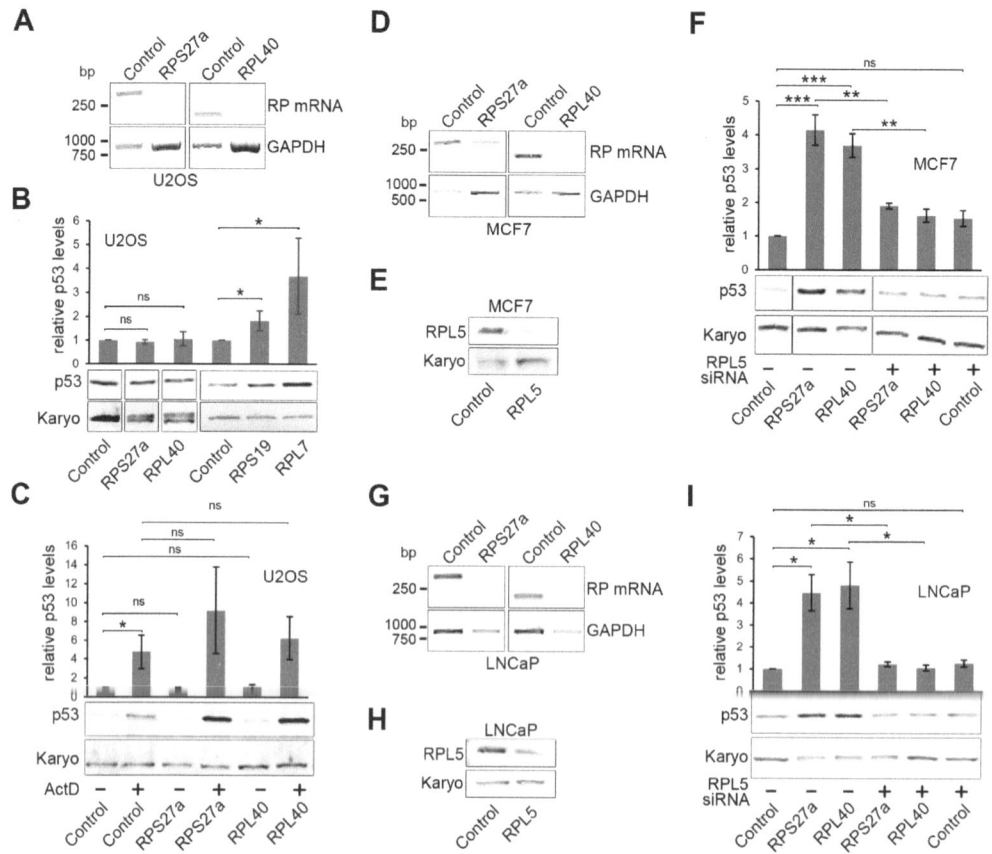

Figure 3. Knockdown of RPS27a and RPL40 results in p53 stabilisation in MCF7 and LNCaP cells, but not in U2OS cells. (**A,D,G**) RNA was extracted from U2OS (**A**), MCF7 (**D**) or LNCaP (**G**) cells depleted

of RPS27a or RPL40 by RNAi or control cells (indicated at the top of each lane), harvested 48 h post-transfection, and analysed by RT-PCR using primers specific for the RPS27a, RPL40 or GAPDH (loading control) mRNAs (indicated to the right of each panel). PCR products were separated by agarose gel electrophoresis and SYBR Safe-stained bands detected using a PhosphorImager. The positions of the DNA ladder markers are shown on the left of the panels. (**B**) RPS27a, RPL40, RPS19 or RPL7 were knocked down in U2OS cells and harvested after 48 h. Proteins isolated from these cells and control cells (indicated below each lane) were analysed by western blotting using antibodies that recognise p53 and karyopherin (Karyo; loading control). The p53 levels, relative to the control, were calculated for each lane and plotted. (**C**) RPS27a and RPL40 were knocked down in U2OS cells for 48 h and either untreated (−) or treated (+) with 5 ng/mL ActD for 18 h (added 30 h post-transfection). Proteins isolated from these cells and control cells (indicated below each lane) were analysed by western blotting using antibodies that recognise p53 and karyopherin (Karyo; loading control). The p53 levels, relative to the control -ActD lane, were calculated for each lane and plotted. (**E,H**) MCF7 (**E**) or LNCaP (**H**) cells were transfected with the control siRNA or a siRNA targeting RPL5, and proteins were isolated after 48 h and analysed by western blotting using antibodies that recognise RPL5 or karyopherin (Karyo; loading control). (**F,I**) RPS27a or RPL40 were knocked down in MCF7 (**F**) or LNCaP (**I**) cells for 48 h, either alone or together with RPL5, and the levels of p53 determined by western blotting as described in (**B**). The p53 levels, relative to those seen with control cells, were calculated for each lane and plotted. The siRNAs used are indicated at the bottom of the lanes and the protein detected by the antibody on the left of each panel. For all graphs, quantification is based on 3 independent repeats. Error bars indicate SEM. * < 0.05, ** < 0.01, *** < 0.001, ns > 0.05 (not statistically significant).

Confirming the earlier observation [32], the knockdown of RPS27a did not result in p53 accumulation in U2OS cells (Figure 3B). The same result was observed for RPL40. Notably, the knockdown of RPS19 or RPL7 caused a significant ~two-fold or ~four-fold increase in p53 levels, respectively, which is consistent with previous studies in HCT116 cells [29] and A549 cells [28].

The fact that the knockdown of RPS27a and RPL40 does not induce p53 in U2OS cells suggests that, as previously reported for RPS27a [32], they may be involved in p53 signalling. If this is the case, their knockdown should block p53 stabilisation upon ActD treatment. To test whether RPS27a or RPL40 is important for p53 stabilisation, U2OS cells were transfected with either the RP-targeting siRNAs or the control siRNA for 48 h. Cells were also treated with low levels (5 ng/mL) of ActD (for 18 h, added 30 h post-transfection) to induce p53, and p53 levels were determined by western blotting (Figure 3C). ActD treatment resulted in a significant, four-fold to five-fold increase in p53 levels. Surprisingly, the knockdown of neither RPS27a nor RPL40 had any impact on p53 stabilisation by ActD. Therefore, our data show that RPS27a and RPL40 are not needed for p53 stabilisation in U2OS cells upon ActD-induced defects in ribosome biogenesis, which is different from the previous report [32].

Next, we wanted to further investigate whether the impact of RPS27a or RPL40 knockdown on p53 signalling could indeed be cell-type dependent. For this, we chose MCF7 (breast cancer) and LNCaP (prostate cancer) cancer cell lines, both of which have wild-type p53 and are routinely used to study p53-dependent cellular signalling. Knockdown efficiency for RPS27a and RPL40 in MCF7 and LNCaP cells was determined by RT-PCR (Figure 3D,G), while the impact on p53 signalling was analysed by western blotting (Figure 3F,I).

In both MCF7 and LNCaP cells, knockdown of either ribosomal protein induced a robust, three-fold to five-fold p53 stabilisation (Figure 3F,I). This demonstrates that as in HCT116 [29] and A549 [28,39] cells, the knockdown of RPS27a or RPL40 stabilises p53 in both MCF7 and LNCaP cells. Knockdown of ribosomal proteins generally induces p53 through the 5S RNP and p53 stabilisation is blocked by the co-depletion of either of the 5S RNP proteins, RPL5 or RPL11 [26,27]. To determine whether this is also the case with RPS27a and RPL40, we co-depleted the 5S RNP protein RPL5 (Figure 3F,I) with either

RPS27a or RPL40 in both MCF7 and LNCaP cells. Treatment of the cells with siRNAs depleting RPL5 resulted in a reduction of RPL5 levels to 40–50% of those treated with control siRNAs (Figure 3E,H). Importantly, co-depletion of RPL5 significantly decreased p53 stabilisation caused by the knockdown of either RPS27a or RPL40 (Figure 3F,I). Note that in both MCF7 and LNCaP cells, the knockdown of RPL5 alone did not significantly increase p53 levels as expected since RPL5 is essential for p53 signalling. We also did not observe a decrease in p53 levels with knockdown of just RPL5, as we had earlier published in U2OS cells [26], which may reflect that RPL5 is less important for maintaining p53 levels in MCF7 and LNCaP cells under non-stress conditions.

Taken together, our data demonstrate that the knockdown of RPS27a or RPL40 stabilises p53 via the 5S RNP-MDM2 pathway in both MCF7 and LNCaP cells but not in U2OS cells. We further conclude that neither RPS27a nor RPL40 are needed for p53 signalling in response to ribosome biogenesis defects in either cell line.

3.6. RPS27a and RPL40 Are Important for Ribosome Production in U2OS, MCF7 and LNCaP Cells

We next asked if the knockdown of RPS27a and RPL40 may affect different stages of pre-rRNA processing in U2OS, MCF7 and LNCaP cells (Figure 4) and whether this may influence whether p53 is stabilised upon protein knockdown or not. This analysis is important since the knockdown of RPS27a in HeLa and HCT116 cells has already shown somewhat different effects on pre-rRNA processing. RPS27a was shown to be important for intermediate stages of 18S rRNA maturation in HeLa cells (mainly 26S and 21S accumulation upon knockdown) [49] but needed for intermediate and earlier stages (21S/21SC and 41S pre-rRNA accumulation upon knockdown) in HCT116 cells [29].

U2OS, MCF7 and LNCaP cells were transfected with siRNAs designed to knockdown RPS27a and RPL40, RNA was extracted and the effect on mature 18S and 28S rRNA levels and pre-rRNA processing was analysed by glyoxal agarose gel electrophoresis followed by northern blotting.

Northern blot analysis revealed that knockdown of RPS27a resulted in a significant drop in mature 18S rRNA levels, relative to RNase P RNA, compared to control knockdown cells for each cell line, with no significant change in 28S rRNA levels (Figure 4B–D). Interestingly, the reduction in 18S levels also coincided with an increase in the detection of a shorter product, presumably an 18S degradation intermediate, in U2OS cells (Figure 4B, mature 18S probe). Knockdown of RPL40 resulted in a major decrease in mature 28S rRNA levels (Figure 4B–D), with mature 28S rRNA levels dropping to about 50% of those seen in control cells in U2OS and MCF7 cells (Figure 4B,C). In LNCaP cells, the impact on 28S rRNA levels upon RPL40 knockdown was less severe, although statistically significant (Figure 4D).

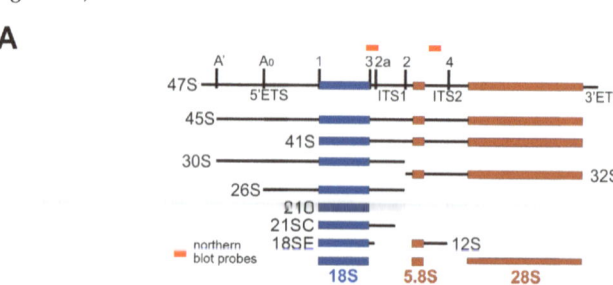

Figure 4. *Cont.*

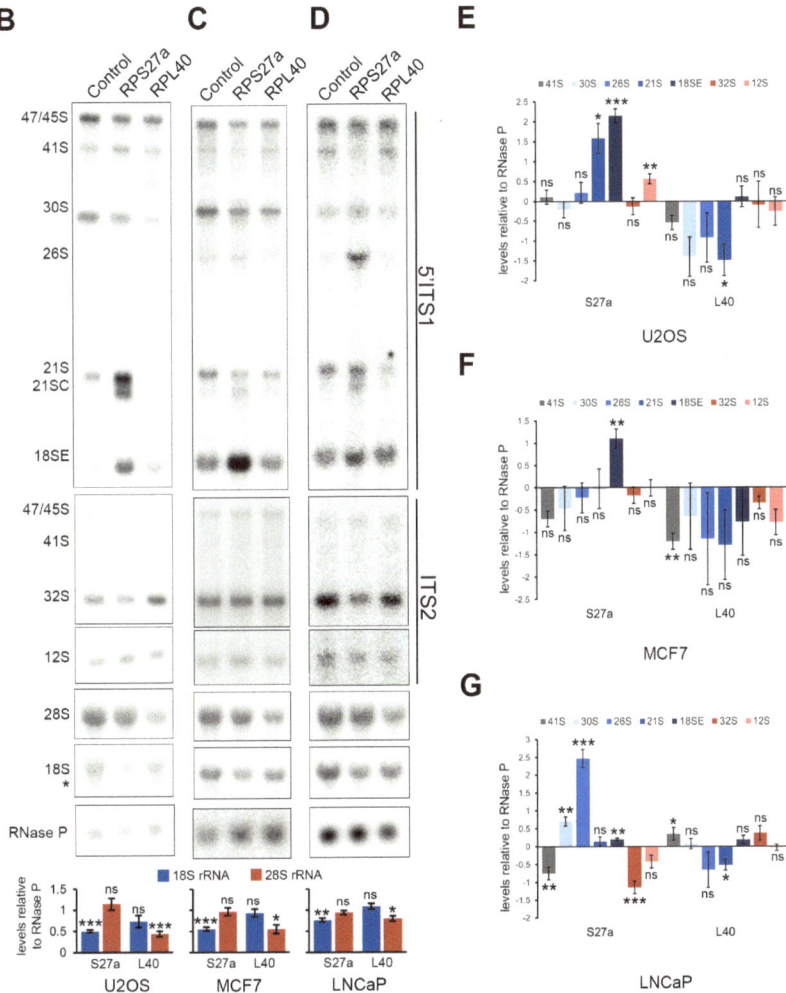

Figure 4. RPS27a and RPL40 are needed for pre-rRNA processing in U2OS, MCF7 and LNCaP cells. (**A**) Schematic representation of the pre-rRNA processing intermediates with the external transcribed spacers (ETS) and internal transcribed spacers (ITS) and processing/cleavage sites indicated. The relative positions of the northern blot probes are indicated above the 47S pre-rRNA. (**B–D**) RNA was extracted from control U2OS (**B**), MCF7 (**C**) and LNCaP (**D**) cells or cells depleted by RNAi of RPS27a or RPL40 for 48 h, separated by glyoxal/agarose gel electrophoresis and then analysed by northern blotting using probes specific to the 5′ end of ITS1 (5′ITS1), ITS2, the mature 18S and 28S rRNAs, or RNase P (loading control). The identities of the RNAs and pre-rRNAs are indicated on the left. * indicates a putative 18S degradation intermediate in U2OS cells (**B**). The levels of the mature 18S and 28S rRNAs after knockdown of RPS27a and RPL40, respectively, were determined and plotted relative to RNase P levels in the control cells. (**E–G**) The levels of SSU and LSU precursors after knockdown of RPS27a and RPL40 in U2OS (**B**), MCF7 (**C**) and LNCaP (**D**) cells, respectively, were determined. Bar charts show log2 ratios of precursors relative to RNase P levels in the control cells. For all graphs, quantification is based on 3 independent repeats. Error bars indicate SEM. * < 0.05, ** < 0.01, *** < 0.001, ns > 0.05 (not statistically significant).

Notably, while depletion of RPS27a always resulted in a defect in the mid to late stages of 18S rRNA processing, the change in pre-rRNA intermediate levels were different in each cell line tested (Figure 4B–D). In U2OS cells, there were no changes in 41S, 30S or 26S pre-rRNA levels, but significant increases in 21S and 18SE pre-rRNA levels, accompanied by the appearance of the aberrant 21SC pre-rRNA (Figure 4B, 5′ITS1 probe and Figure 4E). In MCF7 cells, on the other hand, depletion of RPS27a resulted in an apparent but not significant reduction in 41S levels and no significant changes in the 30S, 26S or 21S pre-rRNAs. However, the appearance of 21SC and, most strikingly, a significant increase in the levels of the 18SE pre-rRNA were observed (Figure 4C, 5′ITS1 probe and Figure 4F). In LNCaP cells, knockdown of RPS27a resulted in a significant reduction in 41S pre-rRNA levels, significant accumulation of the 30S, 26S and 18SE precursors and the appearance of 21SC, while 21S levels were unchanged (Figure 4D, 5′ITS1 probe and Figure 4G). Surprisingly, the knockdown of RPS27a in U2OS cells also resulted in a significant accumulation of the 12S precursor, while RPS27a knockdown in LNCaP cells caused a significant reduction in 32S levels and an apparent reduction in the levels of the 12S pre-rRNA that are both linked to LSU production. However, these defects did not appear to be strong enough to result in significant changes in the levels of the mature 28S rRNA, and they were not seen in MCF7 cells (Figure 4B–D, ITS2 probe and Figure 4E–G). Contrary to the strong impact of the RPL40 knockdown on mature 28S rRNA levels, Northern blotting revealed no significant change in the levels of the LSU pre-rRNAs in either cell line (Figure 4B–D, ITS2 probe and Figure 4E–G). However, RPL40 knockdown surprisingly led to a significant reduction of the 21S SSU precursor in U2OS and LNCaP cells. In contrast, levels of the 41S pre-rRNA were significantly decreased in MCF7 cells and increased in LNCaP cells (Figure 4B–D, 5′ITS1 probe and Figure 4E–G).

Taken together, our data indicate that RPS27a and RPL40 are essential for the production of their respective (18S or 28S) mature rRNA in all cell lines tested. However, the accumulation of specific SSU and LSU precursors seen upon knockdown varies between different cell types.

4. Discussion

The ribosomal proteins RPS27a and RPL40 are produced as fusion proteins with an N-terminal ubiquitin in all eukaryotes, and their genes represent two of the four ubiquitin genes in the mammalian genome [18]. To investigate their production and function in human cells, we generated U2OS cells stably expressing either the RPS27a or RPL40 ubiquitin fusion proteins. In each case, a FLAG-tag was added to the N-terminus of ubiquitin, and an HA-tag added to the C-terminus of the ribosomal protein. While the affinity-tag on either protein did not appear to interfere with their cellular localisation, HA-tagged RPL40 was unstable and showed impaired integration into (pre-)ribosomal complexes. The cell line expressing the RPL40 fusion protein was therefore excluded from further analysis. A previous study on RPL40 in hepatoma cells utilised a C-terminal RFP tag on RPL40 in their fusion construct, which also did not affect the cellular localisation of the cleaved RPL40 protein [25]. However, the integration of RPL40-RFP into (pre-)ribosomal complexes was not assessed. Notably, adding an HA-tag to the N-terminus of yeast Rpl40 did not interfere with its assembly into ribosomes [50], suggesting that an N-terminal HA tag on RPL40 may be more suitable for future analysis in human cells.

To our surprise, we never saw the accumulation of the ubiquitin–ribosomal fusion proteins, not even in the case of RPS27a, when ribosome biogenesis was blocked. Blocking ribosome biogenesis also only affected the production of the ribosomal protein but not ubiquitin. This strongly suggests that the processing of the RPS27a fusion protein, and presumably also the RPL40 fusion protein, is independent of ribosome production, as has been proposed in yeast [19,51,52]. Indeed, the apparently rapid nature of this processing event suggests that cleavage takes place co-translationally on the ribosome. The identification of the enzyme/deubiquitinase responsible is needed before this processing event can be studied in more detail. While we have shed new light on the production and function of the two ribosomal proteins in ribosome production and cellular signalling, it remains unclear why these proteins are produced, throughout eukaryotes, as ubiquitin fusion proteins that require post-translational processing.

We also demonstrate that RPS27 and RPL40 are essential for the production of their respective subunits in all cell lines tested. Surprisingly, and consistent with a previous study on RPS27a in U2OS cells [32], the knockdown of either protein did not cause p53 stabilisation in U2OS cells. In addition, our experiments revealed that neither RPS27a nor RPL40 is needed for p53 stabilisation in U2OS cells upon ActD-induced defects in ribosome biogenesis, which is different from the previous report on RPS27a [32]. However, and again consistent with the previous study [32], p53 stabilisation was observed upon expression of the RPS27a fusion protein. Taken together, the combined data suggest that RPS27a may play a role in p53 signalling, at least in U2OS cells, beyond its being an essential component of the ribosome. Further experiments involving the use of other cell lines (see below) are essential to clarify its role in cell signalling and how this is linked to ribosome biogenesis.

Contrary to what was seen in U2OS cells, knockdown of RPS27a and RPL40 led to p53 stabilisation in MCF7 and LNCaP cells, consistent with previous studies in HCT116 cells and A549 cells and what was seen for the majority of other ribosomal proteins. [28,29,39] We cannot explain why the knockdown of either RPS27a or RPL40 does not induce p53 in U2OS cells. This could be due to cell-line-specific differences in the expression levels of RPS27a and RPL40, but a comparison of the available expression data from cell lines collated in the Human Protein Atlas resource project (https://www.proteinatlas.org/; accessed on 11 April 2023) [48] does not support this hypothesis.

Interestingly, our data on pre-rRNA processing defects seen upon RP27a and RPL40 knockdown is somewhat similar to what was observed in yeast. Depletion of Rps31 (yeast RPS27a) showed some defects in the mid to early steps of yeast 18S maturation but blocked the final processing of the cytoplasmic 20S precursor [51,53] (equivalent of human 18SE). Likewise, yeast Rpl40 depletion resulted in only a minor depletion of the mature LSU rRNAs [54]. However, in yeast, an increase in the initial transcript, 35S (47S in humans), and a slight delay in ITS2 processing [50] was seen upon Rpl40 depletion.

It is quite striking that while both RPS27a and RPL40 knockdowns had an impact on the levels of their respective mature rRNA in all cell lines tested, only RPS27a knockdown caused a strong accumulation of mostly mid to late SSU pre-rRNAs. In contrast, RPL40 depletion had no significant impact on mid to late LSU precursor RNA levels. Again, our data is somewhat in agreement with the data in yeast which showed minimal impact on pre-rRNA levels upon Rpl40 depletion [50,54]. It has also been shown that Rpl40 assembles with pre-60S complexes late in the cytoplasm, which is likely similar in human cells after most pre-rRNA cleavages have already occurred [50]. Moreover, earlier work in HCT116 cells had revealed that the knockdown of a few other LSU ribosomal proteins (e.g., RPL27a (uL15) and RPL26 (uL24)) also significantly impacted mature 28S rRNA levels with minimal change in pre-rRNA levels [29]. In these instances, and likely similar to what we observed upon RPL40 knockdown in three different cell lines, we believe that the ribosomal proteins are more important for the stability of the mature LSU than for any of the individual pre-rRNA processing steps.

Finally, we observed differences in the pre-rRNA processing defects seen upon knockdown of RPS27a and RPL40 between U2OS, MCF7 and LNCaP cells and also compared to those previously published in HeLa and HCT116 cells [29,49]. Firstly, the significant SSU or LSU processing defects seen with respect to their own subunit varied between different cell lines. RPS27a knockdown surprisingly also caused significant defects in LSU maturation and changes in the 41S pre-rRNA levels, while both 41S and 21S SSU precursor levels were significantly altered upon RPL40 depletion, and this was again different in different cells. These could reflect cell-line-specific differences in the functions of these proteins in ribosome biogenesis. This is likely due to the numerous and distinct mutations in genes encoding both ribosomal proteins and ribosome biogenesis factors, which may directly impact pre-rRNA processing [55,56]. However, analysis of the mutations in the available exome sequencing databases revealed no obvious explanation for the cell-line specific data we observed, both with respect to p53 signalling and ribosome biogenesis. We and others have previously reported that mammalian pre-rRNA processing pathways vary slightly between different cell lines or tissues, for example, with respect to 5′ETS and ITS1 processing [57,58]. However, we have not, with the exception of the RPS27a and RPL40 knockdowns described here, observed significant cell-line specific differences in the pre-rRNA processing defects seen with other ribosomal protein knockdowns (our unpublished data). Further characterisation of the genetic background of each cell line will help to understand the impact of individual ribosomal protein knockdowns on both ribosome biogenesis and p53-dependent cellular signalling.

Author Contributions: Conceptualisation, N.J.W. and C.S.; methodology, M.J.E., A.P. and G.R.W.; investigation, M.J.E., A.P. and G.R.W.; writing—original draft preparation, N.J.W. and C.S.; writing—review and editing, M.J.E., G.R.W., N.J.W. and C.S.; visualisation, M.J.E., A.P., N.J.W. and C.S.; supervision, N.J.W. and C.S.; project administration, N.J.W. and C.S.; funding acquisition, N.J.W. and C.S. All authors have read and agreed to the published version of the manuscript.

Funding: This research was funded by the BBSRC/MRC (grant number BB/R00143X/1), MRC (grant number MR/N013840/1), Royal Society (grant number UF150691) and DBA UK/DBAF.

Institutional Review Board Statement: Not applicable.

Informed Consent Statement: Not applicable.

Data Availability Statement: Not applicable.

Acknowledgments: We would like to thank Phil Mason for the gift of anti-RPS19 antibodies and Laurence Pelletier for the gift of the U2OS Flp-In TRex cells.

Conflicts of Interest: The authors declare no conflict of interest. The funders had no role in the design of the study; in the collection, analyses, or interpretation of data; in the writing of the manuscript; or in the decision to publish the results.

References

1. Warner, J.R. The economics of ribosome biosynthesis in yeast. *Trends Biochem. Sci.* **1999**, *24*, 437–440. [CrossRef]
2. Teng, T.; Thomas, G.; Mercer, C.A. Growth control and ribosomopathies. *Curr. Opin. Genet. Dev.* **2013**, *23*, 63–71. [CrossRef]
3. Lempiainen, H.; Shore, D. Growth control and ribosome biogenesis. *Curr. Opin. Cell Biol.* **2009**, *21*, 855–863. [CrossRef]
4. Gentilella, A.; Kozma, S.C.; Thomas, G. A liaison between mTOR signaling, ribosome biogenesis and cancer. *Biochim. Biophys. Acta* **2015**, *1849*, 812–820. [CrossRef]
5. Penzo, M.; Montanaro, L.; Trere, D.; Derenzini, M. The Ribosome Biogenesis-Cancer Connection. *Cells* **2019**, *8*, 55. [CrossRef]
6. Pelletier, J.; Thomas, G.; Volarevic, S. Ribosome biogenesis in cancer: New players and therapeutic avenues. *Nat. Rev. Cancer* **2018**, *18*, 51–63. [CrossRef] [PubMed]
7. Bustelo, X.R.; Dosil, M. Ribosome biogenesis and cancer: Basic and translational challenges. *Curr. Opin. Genet. Dev.* **2018**, *48*, 22–29. [CrossRef] [PubMed]
8. Freed, E.F.; Bleichert, F.; Dutca, L.M.; Baserga, S.J. When ribosomes go bad: Diseases of ribosome biogenesis. *Mol. Biosyst.* **2010**, *6*, 481–493. [CrossRef] [PubMed]
9. Aspesi, A.; Ellis, S.R. Rare ribosomopathies: Insights into mechanisms of cancer. *Nat. Rev. Cancer* **2019**, *19*, 228–238. [CrossRef]
10. Pelava, A.; Schneider, C.; Watkins, N.J. The importance of ribosome production, and the 5S RNP-MDM2 pathway, in health and disease. *Biochem. Soc. Trans.* **2016**, *44*, 1086–1090. [CrossRef]
11. Fumagalli, S.; Thomas, G. The role of p53 in ribosomopathies. *Semin. Hematol.* **2011**, *48*, 97–105. [CrossRef] [PubMed]
12. Mullineux, S.T.; Lafontaine, D.L. Mapping the cleavage sites on mammalian pre-rRNAs: Where do we stand? *Biochimie* **2012**, *94*, 1521–1532. [CrossRef]
13. Bohnsack, K.E.; Bohnsack, M.T. Uncovering the assembly pathway of human ribosomes and its emerging links to disease. *EMBO J.* **2019**, *38*, e100278. [CrossRef]
14. Ciganda, M.; Williams, N. Eukaryotic 5S rRNA biogenesis. *Wiley Interdiscip. Rev. RNA* **2011**, *2*, 523–533. [CrossRef] [PubMed]
15. Cerezo, E.; Plisson-Chastang, C.; Henras, A.K.; Lebaron, S.; Gleizes, P.E.; O'Donohue, M.F.; Romeo, Y.; Henry, Y. Maturation of pre-40S particles in yeast and humans. *Wiley Interdiscip. Rev. RNA* **2019**, *10*, e1516. [CrossRef]
16. Woolford, J.L., Jr.; Baserga, S.J. Ribosome biogenesis in the yeast Saccharomyces cerevisiae. *Genetics* **2013**, *195*, 643–681. [CrossRef]
17. Lam, Y.W.; Lamond, A.I.; Mann, M.; Andersen, J.S. Analysis of nucleolar protein dynamics reveals the nuclear degradation of ribosomal proteins. *Curr. Biol.* **2007**, *17*, 749–760. [CrossRef]
18. Park, C.W.; Ryu, K.Y. Cellular ubiquitin pool dynamics and homeostasis. *BMB Rep.* **2014**, *47*, 475–482. [CrossRef]

19. Lacombe, T.; Garcia-Gomez, J.J.; de la Cruz, J.; Roser, D.; Hurt, E.; Linder, P.; Kressler, D. Linear ubiquitin fusion to Rps31 and its subsequent cleavage are required for the efficient production and functional integrity of 40S ribosomal subunits. *Mol. Microbiol.* **2009**, *72*, 69–84. [CrossRef]
20. Martin-Villanueva, S.; Fernandez-Pevida, A.; Kressler, D.; de la Cruz, J. The Ubiquitin Moiety of Ubi1 Is Required for Productive Expression of Ribosomal Protein eL40 in Saccharomyces cerevisiae. *Cells* **2019**, *8*, 850. [CrossRef] [PubMed]
21. Song, L.; Luo, Z.Q. Post-translational regulation of ubiquitin signaling. *J. Cell Biol.* **2019**, *218*, 1776–1786. [CrossRef]
22. Montellese, C.; van den Heuvel, J.; Ashiono, C.; Dorner, K.; Melnik, A.; Jonas, S.; Zemp, I.; Picotti, P.; Gillet, L.C.; Kutay, U. USP16 counteracts mono-ubiquitination of RPS27a and promotes maturation of the 40S ribosomal subunit. *Elife* **2020**, *9*, 1776–1786. [CrossRef]
23. Kobayashi, M.; Oshima, S.; Maeyashiki, C.; Nibe, Y.; Otsubo, K.; Matsuzawa, Y.; Nemoto, Y.; Nagaishi, T.; Okamoto, R.; Tsuchiya, K.; et al. The ubiquitin hybrid gene UBA52 regulates ubiquitination of ribosome and sustains embryonic development. *Sci. Rep.* **2016**, *6*, 36780. [CrossRef] [PubMed]
24. Martin-Villanueva, S.; Fernandez-Pevida, A.; Fernandez-Fernandez, J.; Kressler, D.; de la Cruz, J. Ubiquitin release from eL40 is required for cytoplasmic maturation and function of 60S ribosomal subunits in Saccharomyces cerevisiae. *FEBS J.* **2020**, *287*, 345–360. [CrossRef] [PubMed]
25. Han, X.J.; Lee, M.J.; Yu, G.R.; Lee, Z.W.; Bae, J.Y.; Bae, Y.C.; Kang, S.H.; Kim, D.G. Altered dynamics of ubiquitin hybrid proteins during tumor cell apoptosis. *Cell Death Dis.* **2012**, *3*, e255. [CrossRef] [PubMed]
26. Sloan, K.E.; Bohnsack, M.T.; Watkins, N.J. The 5S RNP couples p53 homeostasis to ribosome biogenesis and nucleolar stress. *Cell Rep.* **2013**, *5*, 237–247. [CrossRef]
27. Donati, G.; Peddigari, S.; Mercer, C.A.; Thomas, G. 5S ribosomal RNA is an essential component of a nascent ribosomal precursor complex that regulates the Hdm2-p53 checkpoint. *Cell Rep.* **2013**, *4*, 87–98. [CrossRef]
28. Hannan, K.M.; Soo, P.; Wong, M.S.; Lee, J.K.; Hein, N.; Poh, P.; Wysoke, K.D.; Williams, T.D.; Montellese, C.; Smith, L.K.; et al. Nuclear stabilization of p53 requires a functional nucleolar surveillance pathway. *Cell Rep.* **2022**, *41*, 111571. [CrossRef]
29. Nicolas, E.; Parisot, P.; Pinto-Monteiro, C.; de Walque, R.; De Vleeschouwer, C.; Lafontaine, D.L. Involvement of human ribosomal proteins in nucleolar structure and p53-dependent nucleolar stress. *Nat. Commun.* **2016**, *7*, 11390. [CrossRef]
30. Wade, M.; Wang, Y.V.; Wahl, G.M. The p53 orchestra: Mdm2 and Mdmx set the tone. *Trends Cell Biol.* **2010**, *20*, 299–309. [CrossRef]
31. Dai, M.S.; Zeng, S.X.; Jin, Y.; Sun, X.X.; David, L.; Lu, H. Ribosomal protein L23 activates p53 by inhibiting MDM2 function in response to ribosomal perturbation but not to translation inhibition. *Mol. Cell. Biol.* **2004**, *24*, 7654–7668. [CrossRef]
32. Sun, X.X.; DeVine, T.; Challagundla, K.B.; Dai, M.S. Interplay between ribosomal protein S27a and MDM2 protein in p53 activation in response to ribosomal stress. *J. Biol. Chem.* **2011**, *286*, 22730–22741. [CrossRef]
33. Yadavilli, S.; Mayo, L.D.; Higgins, M.; Lain, S.; Hegde, V.; Deutsch, W.A. Ribosomal protein S3: A multi-functional protein that interacts with both p53 and MDM2 through its KH domain. *DNA Repair* **2009**, *8*, 1215–1224. [CrossRef] [PubMed]
34. Kim, T.H.; Leslie, P.; Zhang, Y. Ribosomal proteins as unreveled caretakers for cellular stress and genomic instability. *Oncotarget* **2014**, *5*, 860–871. [CrossRef]
35. He, X.; Li, Y.; Dai, M.S.; Sun, X.X. Ribosomal protein L4 is a novel regulator of the MDM2-p53 loop. *Oncotarget* **2016**, *7*, 16217–16226. [CrossRef]
36. Bursac, S.; Brdovcak, M.C.; Pfannkuchen, M.; Orsolic, I.; Golomb, L.; Zhu, Y.; Katz, C.; Daftuar, L.; Grabusic, K.; Vukelic, I.; et al. Mutual protection of ribosomal proteins L5 and L11 from degradation is essential for p53 activation upon ribosomal biogenesis stress. *Proc. Natl. Acad. Sci. USA* **2012**, *109*, 20467–20472. [CrossRef]
37. Fumagalli, S.; Ivanenkov, V.V.; Teng, T.; Thomas, G. Suprainduction of p53 by disruption of 40S and 60S ribosome biogenesis leads to the activation of a novel G2/M checkpoint. *Genes Dev.* **2012**, *26*, 1028–1040. [CrossRef] [PubMed]
38. Zhou, Q.; Hou, Z.; Zuo, S.; Zhou, X.; Feng, Y.; Sun, Y.; Yuan, X. LUCAT1 promotes colorectal cancer tumorigenesis by targeting the ribosomal protein L40-MDM2-p53 pathway through binding with UBA52. *Cancer Sci.* **2019**, *110*, 1194–1207. [CrossRef] [PubMed]
39. Li, H.; Zhang, H.; Huang, G.; Bing, Z.; Xu, D.; Liu, J.; Luo, H.; An, X. Loss of RPS27a expression regulates the cell cycle, apoptosis, and proliferation via the RPL11-MDM2-p53 pathway in lung adenocarcinoma cells. *J. Exp. Clin. Cancer Res.* **2022**, *41*, 33. [CrossRef]
40. Al-Hakim, A.K.; Bashkurov, M.; Gingras, A.C.; Durocher, D.; Pelletier, L. Interaction proteomics identify NEURL4 and the HECT E3 ligase HERC2 as novel modulators of centrosome architecture. *Mol. Cell. Proteom.* **2012**, *11*, M111.014233. [CrossRef]
41. Sloan, K.E.; Mattijssen, S.; Lebaron, S.; Tollervey, D.; Pruijn, G.J.; Watkins, N.J. Both endonucleolytic and exonucleolytic cleavage mediate ITS1 removal during human ribosomal RNA processing. *J. Cell Biol.* **2013**, *200*, 577–588. [CrossRef]
42. Müller, J.S.; Burns, D.T.; Griffin, H.; Wells, G.R.; Zendah, R.A.; Munro, B.; Schneider, C.; Horvath, R. RNA exosome mutations in pontocerebellar hypoplasia alter ribosome biogenesis and p53 levels. *Life Sci. Alliance* **2020**, *3*, e202000678. [CrossRef]
43. Idol, R.A.; Robledo, S.; Du, H.Y.; Crimmins, D.L.; Wilson, D.B.; Ladenson, J.H.; Bessler, M.; Mason, P.J. Cells depleted for RPS19, a protein associated with Diamond Blackfan Anemia, show defects in 18S ribosomal RNA synthesis and small ribosomal subunit production. *Blood Cells Mol. Dis.* **2007**, *39*, 35–43. [CrossRef]
44. Elbashir, S.M.; Harborth, J.; Weber, K.; Tuschl, T. Analysis of gene function in somatic mammalian cells using small interfering RNAs. *Methods* **2002**, *26*, 199–213. [CrossRef]
45. Turner, A.J.; Knox, A.A.; Prieto, J.L.; McStay, B.; Watkins, N.J. A novel small-subunit processome assembly intermediate that contains the U3 snoRNP, nucleolin, RRP5, and DBP4. *Mol. Cell. Biol.* **2009**, *29*, 3007–3017. [CrossRef] [PubMed]

46. Zhang, R.; Yang, Y.; Huang, H.; Li, T.; Ye, L.; Lin, L.; Wei, Y. UBC Mediated by SEPT6 Inhibited the Progression of Prostate Cancer. *Mediat. Inflamm.* **2021**, *2021*, 7393029. [CrossRef] [PubMed]
47. Oh, C.; Park, S.; Lee, E.K.; Yoo, Y.J. Downregulation of ubiquitin level via knockdown of polyubiquitin gene Ubb as potential cancer therapeutic intervention. *Sci. Rep.* **2013**, *3*, 2623. [CrossRef] [PubMed]
48. Uhlen, M.; Fagerberg, L.; Hallstrom, B.M.; Lindskog, C.; Oksvold, P.; Mardinoglu, A.; Sivertsson, A.; Kampf, C.; Sjostedt, E.; Asplund, A.; et al. Proteomics. Tissue-based map of the human proteome. *Science* **2015**, *347*, 1260419. [CrossRef]
49. O'Donohue, M.F.; Choesmel, V.; Faubladier, M.; Fichant, G.; Gleizes, P.E. Functional dichotomy of ribosomal proteins during the synthesis of mammalian 40S ribosomal subunits. *J. Cell Biol.* **2010**, *190*, 853–866. [CrossRef]
50. Fernandez-Pevida, A.; Rodriguez-Galan, O.; Diaz-Quintana, A.; Kressler, D.; de la Cruz, J. Yeast ribosomal protein L40 assembles late into precursor 60 S ribosomes and is required for their cytoplasmic maturation. *J. Biol. Chem.* **2012**, *287*, 38390–38407. [CrossRef]
51. Fernandez-Pevida, A.; Martin-Villanueva, S.; Murat, G.; Lacombe, T.; Kressler, D.; de la Cruz, J. The eukaryote-specific N-terminal extension of ribosomal protein S31 contributes to the assembly and function of 40S ribosomal subunits. *Nucleic Acids Res.* **2016**, *44*, 7777–7791. [CrossRef]
52. Finley, D.; Bartel, B.; Varshavsky, A. The tails of ubiquitin precursors are ribosomal proteins whose fusion to ubiquitin facilitates ribosome biogenesis. *Nature* **1989**, *338*, 394–401. [CrossRef]
53. Ferreira-Cerca, S.; Poll, G.; Gleizes, P.E.; Tschochner, H.; Milkereit, P. Roles of eukaryotic ribosomal proteins in maturation and transport of pre-18S rRNA and ribosome function. *Mol. Cell* **2005**, *20*, 263–275. [CrossRef]
54. Poll, G.; Braun, T.; Jakovljevic, J.; Neueder, A.; Jakob, S.; Woolford, J.L., Jr.; Tschochner, H.; Milkereit, P. rRNA maturation in yeast cells depleted of large ribosomal subunit proteins. *PLoS ONE* **2009**, *4*, e8249. [CrossRef] [PubMed]
55. Iorio, F.; Knijnenburg, T.A.; Vis, D.J.; Bignell, G.R.; Menden, M.P.; Schubert, M.; Aben, N.; Goncalves, E.; Barthorpe, S.; Lightfoot, H.; et al. A Landscape of Pharmacogenomic Interactions in Cancer. *Cell* **2016**, *166*, 740–754. [CrossRef] [PubMed]
56. Tate, J.G.; Bamford, S.; Jubb, H.C.; Sondka, Z.; Beare, D.M.; Bindal, N.; Boutselakis, H.; Cole, C.G.; Creatore, C.; Dawson, E.; et al. COSMIC: The Catalogue Of Somatic Mutations In Cancer. *Nucleic Acids Res.* **2019**, *47*, D941–D947. [CrossRef] [PubMed]
57. Sloan, K.E.; Bohnsack, M.T.; Schneider, C.; Watkins, N.J. The roles of SSU processome components and surveillance factors in the initial processing of human ribosomal RNA. *RNA* **2014**, *20*, 540–550. [CrossRef]
58. Wang, M.; Anikin, L.; Pestov, D.G. Two orthogonal cleavages separate subunit RNAs in mouse ribosome biogenesis. *Nucleic Acids Res.* **2014**, *42*, 11180–11191. [CrossRef]

Disclaimer/Publisher's Note: The statements, opinions and data contained in all publications are solely those of the individual author(s) and contributor(s) and not of MDPI and/or the editor(s). MDPI and/or the editor(s) disclaim responsibility for any injury to people or property resulting from any ideas, methods, instructions or products referred to in the content.

Article

Ribosomal Protein S12 Hastens Nucleation of Co-Transcriptional Ribosome Assembly

Margaret L. Rodgers [1,2,*], Yunsheng Sun [1] and Sarah A. Woodson [1,*]

1. Thomas C. Jenkins Department of Biophysics, Johns Hopkins University, Baltimore, MD 21218, USA
2. The Laboratory of Biochemistry and Genetics, The National Institute of Diabetes and Digestive and Kidney Diseases, The National Institutes of Health, Bethesda, MD 20892, USA
* Correspondence: margaret.rodgers@nih.gov (M.L.R.); swoodson@jhu.edu (S.A.W.); Tel.: +1-301-496-6724 (M.L.R.); +1-410-516-2015 (S.A.W.)

Abstract: Ribosomal subunits begin assembly during transcription of the ribosomal RNA (rRNA), when the rRNA begins to fold and associate with ribosomal proteins (RPs). In bacteria, the first steps of ribosome assembly depend upon recognition of the properly folded rRNA by primary assembly proteins such as S4, which nucleates assembly of the 16S 5′ domain. Recent evidence, however, suggests that initial recognition by S4 is delayed due to variable folding of the rRNA during transcription. Here, using single-molecule colocalization co-transcriptional assembly (smCoCoA), we show that the late-binding RP S12 specifically promotes the association of S4 with the pre-16S rRNA during transcription, thereby accelerating nucleation of 30S ribosome assembly. Order of addition experiments suggest that S12 helps chaperone the rRNA during transcription, particularly near the S4 binding site. S12 interacts transiently with the rRNA during transcription and, consequently, a high concentration is required for its chaperone activity. These results support a model in which late-binding RPs moonlight as RNA chaperones during transcription in order to facilitate rapid assembly.

Keywords: ribosome assembly; RNA chaperones; single-molecule fluorescence; ribosomal protein S12; co-transcriptional RNA folding

Citation: Rodgers, M.L.; Sun, Y.; Woodson, S.A. Ribosomal Protein S12 Hastens Nucleation of Co-Transcriptional Ribosome Assembly. *Biomolecules* 2023, 13, 951. https://doi.org/10.3390/biom13060951

Academic Editors: Brigitte Pertschy and Ingrid Rössler

Received: 27 April 2023
Revised: 1 June 2023
Accepted: 4 June 2023
Published: 6 June 2023

Copyright: © 2023 by the authors. Licensee MDPI, Basel, Switzerland. This article is an open access article distributed under the terms and conditions of the Creative Commons Attribution (CC BY) license (https://creativecommons.org/licenses/by/4.0/).

1. Introduction

In all kingdoms of life, the structures of ribosomal subunits derive from the three-dimensional organization of the rRNA in complex with more than 20 unique ribosomal proteins (RPs) [1]. The RPs not only stabilize the rRNA in its native conformation, but also induce conformational changes in the rRNA that favor the next steps of assembly [2–5]. This linkage between rRNA folding and RP binding produces a hierarchy of protein addition that ensures the cooperativity of assembly (Figure 1a) [6,7]. In this hierarchy, primary assembly proteins associate with the naked rRNA, whereas secondary and tertiary assembly proteins only join the complex after a primary assembly protein has bound. This hierarchy is not strict, however, as 30S and 50S assembly can proceed via alternative paths [3,8,9].

In cells, RPs begin to associate with the pre-rRNA as it is being transcribed [10]. Footprinting experiments on refolded rRNA showed that the path of RP assembly at 30 °C in vitro is aligned with the 5′ to 3′ direction of transcription [11], although assembly can begin in any domain [4]. During transcription, assembly is likely nucleated by primary assembly RPs uS17, uS20, and uS4 that bind the 16S 5′ domain, since this domain is transcribed first. Protein uS4 (S4 hereafter) binds a five-way helix junction (5WJ) in the 16S rRNA, stabilizes RNA tertiary interactions throughout the 16S 5′ domain [12], and nucleates further assembly of the 5′ and central domains [7,13–15]. After S4 binds the 5WJ, it induces a conformational change in 16S h18 [16,17], an element of the 30S decoding site that is also stabilized by RP uS12 (S12) (Figure S1).

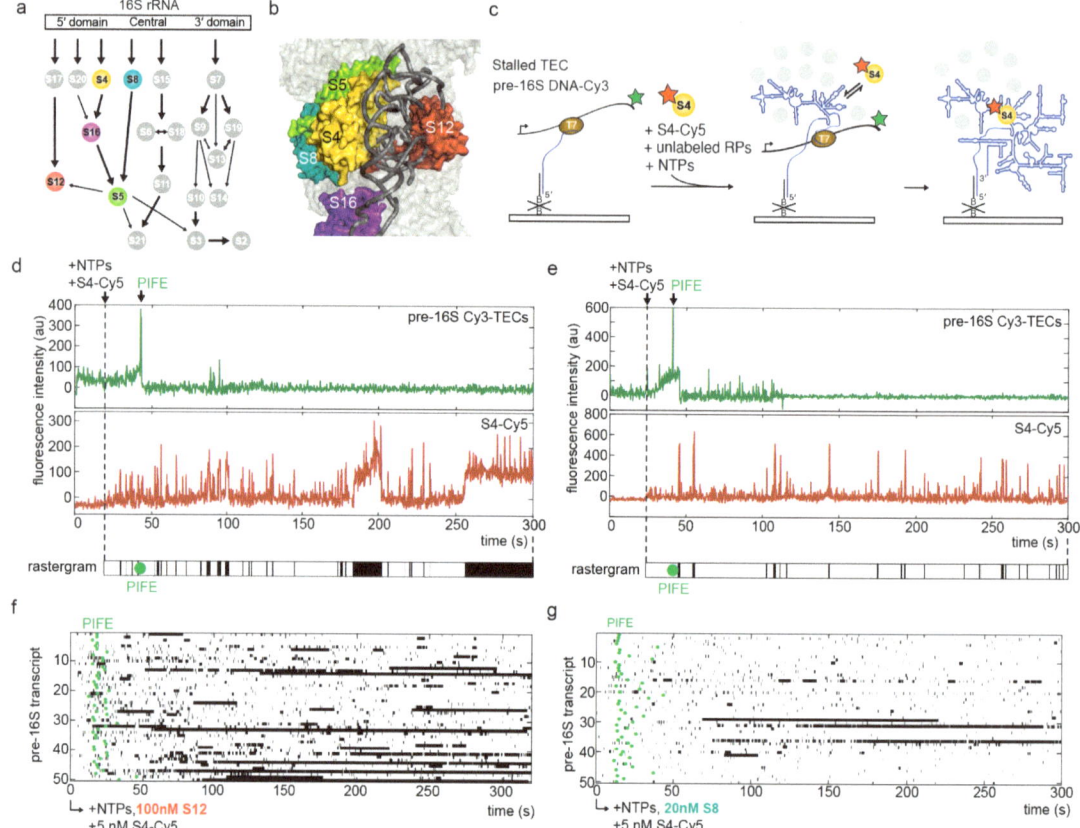

Figure 1. Single-molecule system to study the association of S4 in the presence of other RPs. (**a**) Nomura 30S ribosome assembly map highlighting RPs used in this study. (**b**) RPs surrounding the S4 binding site in the mature 30S ribosome. The 5WJ recognized by S4 is shown as a dark grey ribbon. PDB: 4V4A. (**c**) Schematic of the single-molecule system used to study the binding dynamics of S4 during and after transcription of the pre-16S rRNA. Green star, Cy3; red star, Cy5. (**d**,**e**) Sample raw single-molecule traces illustrating PIFE at the end of transcription (green; top) and colocalization of S4-Cy5 with the transcript (red; bottom). A rastergram for each transcript is shown below each plot of Cy5 intensity. This simplified annotation is used to visualize the timing of S4-Cy5 binding (black bars). PIFE (green circle) indicates the end of transcription. See Methods for details. (**f**,**g**) Rastergrams for 50 randomly selected pre-16S transcripts during and after transcription in the presence of (**f**) 100 nM unlabeled S12 and (**g**) 20 nM unlabeled S8. See Figure S2 for data with other RPs.

We previously used single-molecule fluorescence co-localization co-transcriptional assembly (smCoCoA) to visualize association of RP S4 with the rRNA during transcription [18]. Although native S4-rRNA complexes are stable at 20 mM $MgCl_2$ ($t^{1/2} \gg 10$ min) [5,13,19], most S4 binding events during transcription are unstable (<2 s) due to variable folding of the rRNA during transcription [18,20]. Interestingly, the likelihood of stable S4 binding during transcription increased when other RPs, including uS5 (S5), uS8 (S8), S12, bS16 (S16), uS17 (S17), and bS20 (S20), were also present.

These observations suggested that ribosomal proteins can accelerate rRNA folding and assembly in cells. Yet, the mechanism of this chaperone effect remains unknown. S17 and S20 bind to different helix junctions in the 16S 5′ domain, whereas S5, S8, S12, and S16 bind near S4 in the mature 30S subunit (Figure 1a,b). Therefore, these RPs may facilitate

S4 binding by altering the conformational landscape of the 5′ domain. Additionally, RPs and other RNA chaperones could restructure the S4 binding site by unfolding an incorrect secondary structure, or by transiently stabilizing the tertiary structure of helix junctions [21].

Here, we use our co-transcriptional assembly assay to understand how additional RPs accelerate proper recognition of the S4 binding site, which is a key first step in 30S ribosome assembly. The results show that the conserved late-binding protein S12 promotes S4 association during transcription by acting on the nascent rRNA. S12 was previously found to have RNA chaperone activity, and may help the rRNA refold [22]. Interestingly, S12 binds the same rRNA 5WJ as S4, but on the opposite side. Thus, certain RPs may perform the dual functions of stabilizing native rRNA structure and accelerating rRNA refolding.

2. Materials and Methods

2.1. Preparation of Fluorescent DNA Templates

Pre-16S DNA template for transcription was prepared by PCR using a reverse primer containing a fluorescently labeled nucleotide, as previously described [18]. Fluorescent labeling was carried out by reacting an internal C6-amino-modified T (IDT, Coralville, IA, USA) with Cy3-NHS mono reactive dye (Lumiprobe Corp., Hunt Valley, MD, USA), as follows: 1 mg Cy3-NHS mono reactive dye was dissolved in 33 µL DMSO and added to a reaction mixture containing 10 nmol oligonucleotide, adjusted to 100 µL final volume with 100 mM sodium bicarbonate pH 8.5. Labeling reactions were incubated at room temperature for 24 h. Reactions were purified using a Nucleospin column (Takara Bio, San Jose, CA, USA). Fluorescent DNA transcription templates were generated by PCR using Q5 high-fidelity polymerase (NEB, Ipswich, MA, USA). Following PCR, fluorescently labeled DNA templates were separated on 1% agarose, and isolated from the gel using a Nucleospin gel purification kit (Takara Bio).

2.2. Protein Purification and Fluorescent Labeling

Unlabeled ribosomal proteins uS8 (S8), uS5 (S5), S12, and bS16 (S16) were expressed and purified as previously described [23,24]. T7 RNA polymerase (RNAP) was recombinantly expressed in E. coli BL21 (DE3) cells as previously described [25], and natively purified on a P11 phosphocellulose column followed by a Blue Dextran-Sepharose column, using a protocol developed for SP6 polymerase [26].

Fluorescently labeled S4:C32S,S189C was purified and labeled as previously described [5]. Briefly, S4:C32S,S189C was incubated for 3 h with a six-fold molar excess of dye in 80 mM K-HEPES pH 7.6, 1 M KCl, 1 mM TCEP, and 3 M urea at 20 °C. Unreacted dye was removed by cation exchange followed by dialysis against 80 mM K-HEPES pH 7.6, 1 M KCl, and 6 mM 2-mercaptoethanol (BME).

The S12:A48C expression plasmid was a kind gift from the Noller lab. Fluorescent S12:A48C was prepared and fluorescently labeled as previously described, but with a few modifications [27]. Briefly, S12:A48C was expressed in BL21 (DE3) cells. Upon lysis, S12 was found primarily in inclusion bodies. The inclusion bodies were dissolved in 10 mL buffer A (6 M guanidinium-HCl, 1 M KCl, 6 mM BME), cleared by centrifugation, and the soluble protein was dialyzed against 1 L buffer B (20 mM Na-acetate pH = 5.6, 1 M KCl, 6 M urea, 6 mM BME), 1 L buffer C (20 mM Na-acetate pH = 5.6, 500 mM KCl, 6 M urea, 6 mM BME), and twice against 1 L buffer D (20 mM Na-acetate pH = 5.6, 100 mM KCl, 6 M urea, 6 mM BME), for 2 h at each buffer change. The dialyzed protein was cleared by centrifugation and filtered (0.45 µm), then applied to a UNO-S6 column (Bio-Rad, Hercules, CA, USA) and eluted with a 0–40% linear gradient of 1 M KCl in buffer D. Half of the protein was dialyzed against buffer 1 (80 mM HEPES, pH = 7.6, 20 mM $MgCl_2$, 1 M KCl, 6 mM BME) for 2 h, twice, flash frozen in aliquots and stored at −80 °C. The other half was dialyzed overnight against buffer 2 (80 mM HEPES pH = 7.5, 1 M KCl, 1 mM TCEP) for labeling.

For labeling with Cy5, S12 protein was warmed to 20 °C and diluted with buffer 1 to 1.7 mL total (40 µM S12 final). Cy5-maleimide mono-reactive dye (GE-Healthcare,

Chicago, IL, USA) was dissolved in DMSO to 20 mM and immediately added to the protein solution. The reaction was incubated at 20 °C in the dark for 2 h. BME was added to 0.5% v/v to quench the reaction. The labeled protein was purified on an UNO-S6 column with 0–60% linear gradient of 1 M KCl in 10 mM Tris-HCl pH = 6.3, 6 M urea, 0.01% Nikkol, and 6 mM BME, with elution at 460 mM KCl. The protein was dialyzed against buffer 1 and stored at −80 °C as above.

2.3. Single-Molecule Fluorescence Microscopy

Single-molecule smCoCoA experiments were carried out as previously described [18]. Briefly, single-molecule microscopy was performed on a custom-built prism-based total internal reflection fluorescence microscope. Cy3-labeled biomolecules were imaged using a green (532 nm) laser and Cy5-labeled biomolecules were imaged using a red (640 nm) laser.

Stalled transcription elongation complexes (TECs) were assembled at RT for 2 min in the following reaction: 50–100 nM Cy3-labeled DNA template, 40 mM Tris-HCl pH 7.5, 20 mM $MgCl_2$, 50 nM T7 RNAP, 200 µM GTP, 200 µM ATP, 50 µM UTP, 2 U RNasin Plus, and 100 nM biotinylated tether oligomer. Stalled TECs (20 µL) were diluted 1:10 in transcription buffer (40 mM Tris-HCl pH 7.5, 20 mM $MgCl_2$), immobilized on quartz slides passivated with DDS-Tween20 [28], and functionalized with streptavidin. Following immobilization, stalled TECs were washed with imaging buffer (40 mM Tris-HCl pH 7.5, 20 mM $MgCl_2$, 150 mM KCl, 1% w/v glucose, 165 U/mL glucose oxidase, 4 mM Trolox, 2 U RNasin Plus).

Before imaging, the restart imaging solution was assembled as follows: 40 mM Tris-HCl pH 7.5, 20 mM $MgCl_2$, 150 mM KCl, 5 nM Cy5-S4 or Cy5-S12, 1 mM ATP, 1 mM GTP, 1 mM CTP, 1 mM UTP, 1% w/v glucose, 165 U/mL glucose oxidase, 2170 U/mL catalase, 4 mM Trolox, and 2 U RNasin Plus, with additional RPs as stated. The restart imaging solution was injected to the slide chamber during imaging. Imaging was performed with alternating frames of green and red excitation every 100 ms for a total of ~3000 frames (5 min).

2.4. Analysis of Single-Molecule Data

Single-molecule movies were analyzed as previously described [18] using Imscroll software [29]. Briefly, Cy3-labeled TECs were selected as areas of interest (AOIs) at the beginning of the movie. Colocalized spots were determined by translating the Cy3-TEC AOI locations to the Cy5 channel using a mapping function, as previously described [29]. The intensities for AOIs in both channels were integrated over the duration of the movie to generate single-molecule time traces. PIFE and Cy5-S4 colocalization was not observed in control reactions without NTPs, confirming that these signals report on transcription elongation and Cy5-S4 binding to the transcript, respectively [18].

Single-molecule traces were examined for transcription of the pre-16S rRNA as indicated by protein-induced fluorescence enhancement (PIFE), as previously described [18]. Only traces exhibiting a single PIFE signal were included in the analysis of colocalized Cy5-labeled RP to ensure that the analysis was limited to single, full-length pre-16S transcripts. Binding intervals for Cy5-labeled RPs were generated as previously described [29]. Dwell times lasting for a single frame (0.2 s) were indistinguishable from nonspecific binding of S4 or S12 with the slide surface in the absence of the RNA and were not included in the analysis. Maximum likelihood estimation (MLE) was used to globally fit the unbinned kinetic data using single- and triple-exponential kinetic binding models, as in Equations (1) and (2), respectively, where x is the duration of the binding event; t_m is the minimum resolvable time interval in the experiment; t_x is the maximum time interval; τ, τ_1, τ_2, τ_3 represent characteristic lifetimes; and a_1 and a_2 are the amplitudes associated with the fitted lifetimes.

$$\frac{1}{\left(e^{-\frac{t_m}{\tau}} - e^{-\frac{t_x}{\tau}}\right)} \cdot \frac{1}{\tau} e^{-\frac{x}{\tau}} \qquad (1)$$

$$\frac{1}{a_1\left(e^{-\frac{tm}{\tau_1}} - e^{-\frac{tx}{\tau_1}}\right) + a_2\left(e^{-\frac{tm}{\tau_2}} - e^{-\frac{tx}{\tau_2}}\right) + (1 - a_1 - a_2)\left(e^{-\frac{tm}{\tau_3}} - e^{-\frac{tx}{\tau_3}}\right)} \cdot \left(\frac{a_1}{\tau_1}e^{-\frac{x}{\tau_1}} + \frac{a_2}{\tau_2}e^{-\frac{x}{\tau_2}} + (1 - a_1 - a_2)e^{-\frac{x}{\tau_3}}\right)$$ (2)

Errors in MLE parameters were estimated by bootstrapping the data to obtain a 95% confidence interval, as previously described [29]. Histograms were generated in MATLAB (the Mathworks) by unequal binning of the data to minimize empty bins and visualize the MLE fits. Error bars in the histogram represent the variance in a binomial distribution, described by Equation (3), where N is the number of observations and P is the event probability.

$$\sigma = \sqrt{NP(1-P)}$$ (3)

Cumulative density plots were generated in MATLAB and single-exponential fitting of association times was carried out using GraphPad Prism.

3. Results

3.1. A Single-Molecule System to Study the Chaperone Activity of Ribosomal Proteins

Previous single-molecule experiments showed that the presence of secondary and tertiary RPs increase the likelihood that protein S4 binds the 16S rRNA stably during transcription (Figure 1a,b; [18]). This effect was greatest for a combination of proteins that bind the 16S 5′ domain (S4, S12, S16, S17, S20) and the adjacent central domain (S5, S8). By contrast, protein uS9 (S9) that binds the 16S 3′ domain had no effect on the likelihood of stable S4 binding.

In order to understand how each RP acts on the rRNA, we used single-molecule colocalization assembly (smCoCoA) to simultaneously monitor transcription of the pre-rRNA and association of Cy5-labeled S4 with the nascent RNA (Figure 1c). We used the stability of S4-Cy5 binding events as a readout for proper folding of the pre-16S rRNA, and measured how S4 binding dynamics changed in the presence of individual RPs.

To observe synthesis of the pre-rRNA in real time, we first assembled stalled T7 RNA polymerase transcription elongation complexes (TECs) on a Cy3-labeled pre-16S DNA template. Stalled TECs were immobilized on a microscope slide using a biotinylated DNA oligonucleotide that was complementary to the 5′ end of the nascent RNA (Figure 1c). Transcription was restarted during imaging by adding NTPs together with S4-Cy5 and unlabeled RPs in order to monitor S4-Cy5 binding dynamics during elongation of the pre-rRNA (Figure 1c). The end of transcription was marked by protein-induced fluorescence enhancement (PIFE), which occurs as RNAP passes over the Cy3 fluorophore attached to the DNA template, as described previously [18].

We first examined whether RPs that bind near S4 in the mature ribosome contribute to S4-Cy5 recruitment (Figure 1a,b). S5, S8, S12, or S16 was added to the transcription reaction at a concentration that mimics 30S reconstitution conditions: Primary protein S8 was added at 20 nM, while tertiary proteins S5 and S12 were added at 100 nM. Most S4 binding events were brief (Figure 1d,e), consistent with previous findings that improper pre-rRNA folding during transcription results in unstable or non-specific S4 binding [18]. Interestingly, we observed that the presence of S12 increased the number of stable S4 binding events compared to S4 alone, or to S4 with other RPs, such as S8 (Figures 1f,g and S2).

3.2. S12 Increases Likelihood of Stable S4 Recruitment to Pre-16S rRNA during Transcription

In order to characterize the contribution of each RP on the binding dynamics of S4-Cy5, we measured the distribution of S4-Cy5 dwell times, which report on the stability of S4-rRNA complexes in the presence of another RP. Because the S4-Cy5 dwell time distribution is dominated by short events, the mean dwell time ~1 s was similar for S4 alone and S4 in the presence of S5, S8, or S16. However, the mean was slightly higher in the presence of 100 nM S12 (~2 s), and the longest dwell times were more represented with 100 nM S12 compared to S4 alone (Figure 2a).

Table 1. Lifetimes of S4 complexes. Lifetimes (τ) and amplitudes (a) of all binding events, from maximum likelihood triple-exponential fits of the unbinned dwell times for S4-Cy5 in the presence of RPs.

RP Added	N_{mol}	τ_1 (s)	τ_2 (s)	τ_3 (s) [2]	a_1 [1]	a_2	a_3 [2]
S4 alone	226	0.47 ± 0.01	3.83 ± 1.01	44 ± 11	0.957 ± 0.008	0.037 ± 0.007	0.003 ± 0.008
+100 nM S12	200	0.71 ± 0.02	4.78 ± 0.76	60 ± 13	0.893 ± 0.012	0.089 ± 0.011	0.018 ± 0.012
+20 nM S8	176	0.63 ± 0.02	5.79 ± 1.03	60 ± 21	0.941 ± 0.009	0.053 ± 0.009	0.006 ± 0.009
+100 nM S5	183	0.67 ± 0.01	4.43 ± 0.93	118 ± 29	0.978 ± 0.006	0.020 ± 0.006	0.002 ± 0.006
+50 nM S16	130	0.52 ± 0.02	2.35 ± 0.52	16 ± 3.8	0.923 ± 0.014	0.060 ± 0.015	0.016 ± 0.015

[1] All distributions are dominated by short events (<1 s). [2] The uncertainties in τ_3 and a_3 are large, owing to the few long-lived binding events relative to short events, and τ_3 is underestimated because some long-lived events extend beyond the end of the movie.

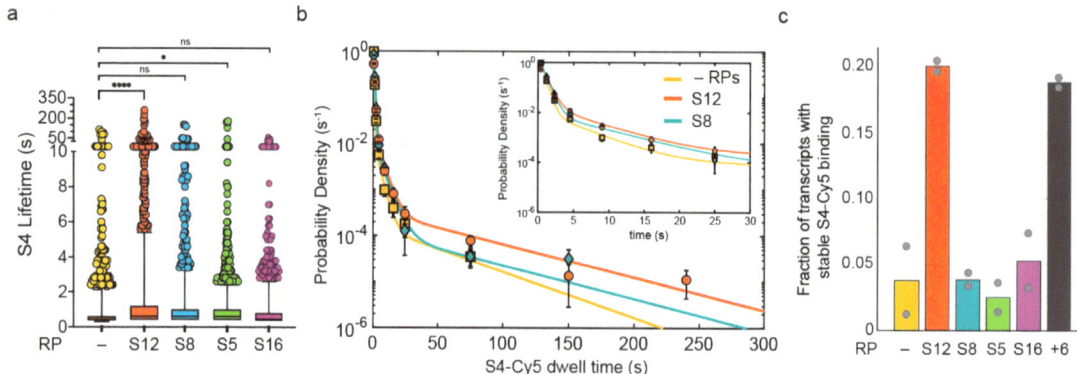

Figure 2. Stable binding of S4 is enhanced by S12. (**a**) Box plot of the distribution of S4-Cy5 dwell times in the presence of no other protein (−), 100 nM S12, 20 nM S8, 100 nM S5, and 50 nM S16. (****, $p \leq 0.0001$; *, $p \leq 0.05$; ns, $p > 0.05$; Student's t-test) (**b**) Maximum likelihood analysis of the unbinned dwell time distribution shown in (**a**). Triple-exponential fit is shown with a colored line, and fitting parameters are reported in Table 1. Centers of the binned dwell time data for S12, S8, and −RPs are shown as circles, diamonds, and squares, respectively. See Figure S3 for additional data. (**c**) Fraction of pre-16S TECs that experienced an S4 binding event lasting longer than 20 s, for experiments as in (**a**). For comparison, the grey bar (+6) indicates the effect of adding six RPs at the same time: 20 nM S20, 20 nM S17, 20 nM S8, 50 nM S16, 100 nM S5, 100 nM S12; data from Ref. [18]. Bars, average; grey circles, individual replicates. See Table 1 for the number of transcripts analyzed in each experiment.

Because S4 interacts much more tightly with the properly folded 5WJ than with unstructured RNA [15,18,19], an increase in stable S4-Cy5 binding indicates that the rRNA is becoming better folded, on average. Therefore, the proportion of stable S4-Cy5 complexes can be used as a readout for changes in the folding quality of the rRNA. We used maximum likelihood analysis of the dwell time distributions to determine the characteristic dwell times associated with S4-Cy5 binding in the presence of each RP, and their amplitudes (Figures 2b and S2). Consistent with previous work [18], we found that S4-Cy5 exhibits dwell times of $\tau_1 \sim 0.5$ s, $\tau_2 \sim 6$ s, and $\tau_3 > 20$ s (Table 1), which represent the inherent lifetimes of the different S4 binding modes. When only S4-Cy5 was added to the transcription reaction, the amplitude of the most stable binding mode, a_3, was very small (0.3%; Table 1).

When S12 was also present, however, a_3 was sixfold higher, indicating that S4-Cy5 binding was six times more likely to produce a stable complex.

We next examined the fraction of pre-16S transcripts that formed a long-lived S4-Cy5 complex (>20 s) at some point during each movie (Figure 2c). This fraction reports on the probability that the pre-16S rRNA is competent for native-like S4 binding, which can be reliably estimated from the single transcripts. The results showed that 100 nM S12 increased the fraction of competent pre-16S transcripts by fourfold compared to other RPs and S4 alone (red bar, Figure 2c), suggesting that S12 generates a more native-like binding site for S4.

3.3. S12 Only Interacts Transiently with Pre-16S in the Absence of Other Ribosomal Proteins

We considered that S12 must bind the rRNA to facilitate folding of the 5WJ and stable binding of S4. One explanation for this effect is that S12 binds the rRNA together with S4 to facilitate folding of the 5WJ (co-binding mechanism). Another explanation is that S12 transiently binds the rRNA and changes its folding path, leaving the rRNA in a more native-like structure that is competent to stably bind S4 (independent binding mechanism). To distinguish between these mechanisms, we sought to measure how S12 binds the rRNA on its own.

In order to characterize the binding dynamics of S12 directly, we labeled S12 with a Cy5 fluorophore and carried out smCoCoA experiments on pre-16S TECs in the same manner as for S4-Cy5 (Figure 3a). We expected that S12 alone should not interact stably with the rRNA, because addition of S12 to the 16S rRNA requires many other RPs, including S4, in the Nomura assembly map. Consistent with the Nomura map, the single-molecule results showed that S12 only binds pre-16S rRNAs transiently under the conditions tested (Figure 3b,c). We measured the dwell times of S12-Cy5 complexes and found that the characteristic dwell time was $\tau \sim 0.5$ s, similar to non-specific S4-Cy5 binding to pre-16S rRNA (Figure 3d).

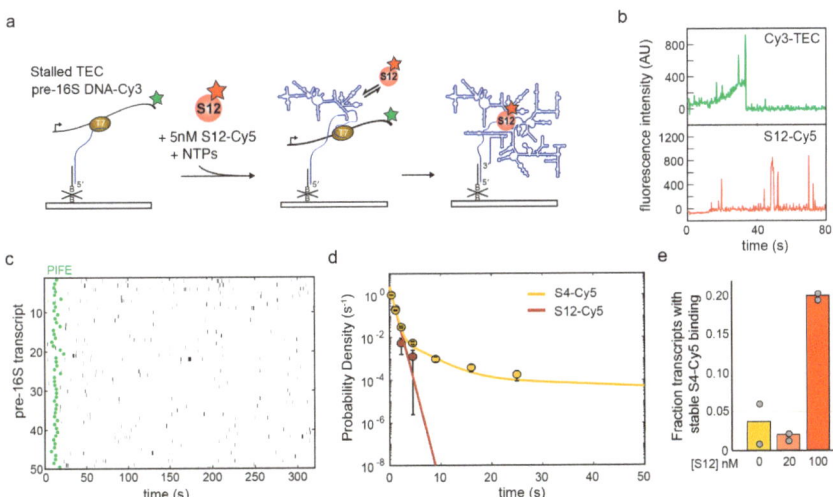

Figure 3. S12-Cy5 only interacts transiently with pre-16S transcripts in the absence of other RPs. (**a**) Single-molecule assay to measure S12 binding during transcription. (**b**) Example of a single-molecule trace illustrating transcription from a single Cy3-TEC (green) and colocalization of S12-Cy5 (red). (**c**) Rastergram of S12-Cy5 binding (black bars) to 50 randomly selected pre-16S TECs, as in Figure 1. (**d**) MLE analysis of S12-Cy5 binding to pre-16S transcripts (red) compared to S4-Cy5 (gold). The characteristic S12-Cy5 dwell time is $\tau = 0.46$ s; see Table 1 for S4-Cy5 parameters. (**e**) The fraction of pre-16S TECs with stable S4-Cy5 binding is dependent on S12 concentration, as 20 nM unlabeled S12 does not improve S4-Cy5 binding.

Since S12-Cy5 binds the rRNA only for a short time, we reasoned that the effect of S12 on S4-Cy5 binding may be concentration dependent for either co-binding or independent binding mechanisms. In order to examine the concentration dependence, we measured S4-Cy5 binding dynamics in the presence of 20 nM S12, and found that this amount of S12 was insufficient to improve S4 binding (Figure 3e). These results indicated that S12 must be present at a high concentration to facilitate the formation of stable S4-rRNA complexes through frequent, low-affinity interactions with the rRNA.

3.4. S12 Can Act on Pre-16S during and after Transcription to Increase Stable S4 Recruitment

In all aforementioned experiments, S12 was allowed to access the rRNA during and after transcription. Therefore, it was unclear if S12 helps the rRNA fold as it is synthesized or if S12 helps the rRNA refold after it has already misfolded.

In order to examine whether S12 acts during or after transcription, we varied the order in which S12 is added to the pre-rRNA in the single-molecule experiments. First, we immobilized unlabeled stalled pre-16S TECs and injected NTPs to restart transcription, as described above (Figure 4). For co-transcriptional S12 interactions (Figure 4a), we added 100 nM S12 together with the NTPs to allow S12 to help fold the rRNA during transcription.

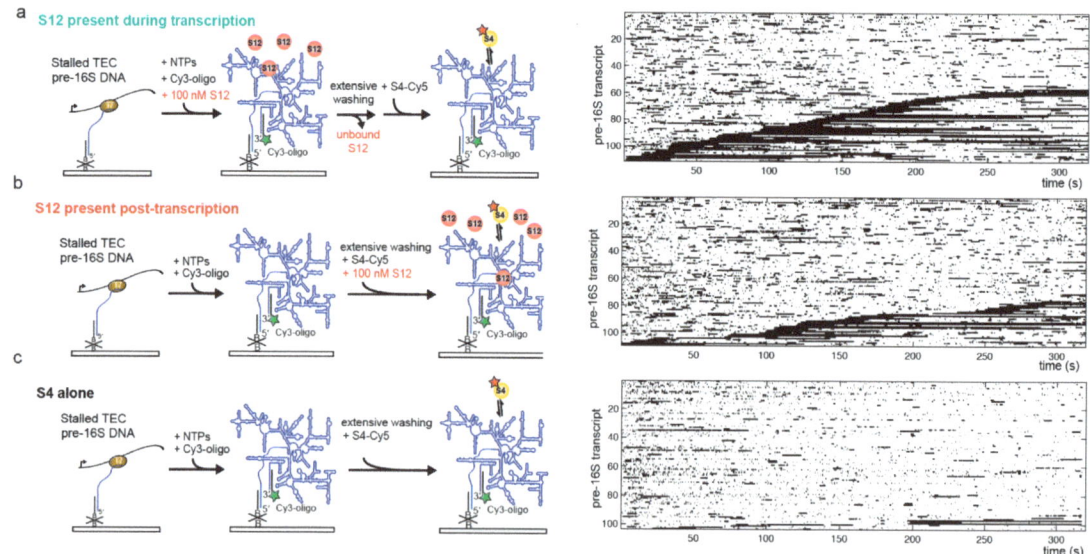

Figure 4. S12 is more effective when present during transcription. Strategy for visualizing S4-Cy5 binding to the pre-16S after transcription. Immobilized transcripts were detected by hybridization of a complementary Cy3-oligomer. (**a**) Binding to pre-16S transcribed in the presence of 100 nM S12. Rastergram of S4-Cy5 binding at right. Pre-16S transcripts were ordered by the start of the first S4-Cy5 binding event longer than 20 s. (**b**) Binding in the presence of 100 nM S12, as in (**a**). (**c**) Binding in the absence of S12, as in (**a**). See Figure S4 and Table S1 for binding lifetimes.

Following transcription, the slide was thoroughly washed to remove unbound S12 and NTPs from the reaction chamber. The co-transcriptionally folded pre-16S rRNAs were then labeled in situ with a Cy3-labeled oligomer that is complementary to the 3′ end of the pre-16S rRNA. Binding of S4-Cy5 to S12-treated transcripts was then monitored following injection of S4-Cy5 into the slide chamber during imaging (Figure 4a). To determine whether S12 can act after transcription, 100 nM S12 was added together with S4-Cy5 to previously transcribed pre-16S pre-rRNA (Figure 4b). Finally, we compared the results to a control in which S4-Cy5 was allowed to bind the pre-rRNA post-transcription in the absence of S12 (Figure 4c). We observed that adding S12 during transcription or after transcription increased the amount of stable S4-Cy5 binding compared to S4 alone (Figure 4). Moreover, S12 was able to enhance stable S4-Cy5 binding even after S12 was washed out (Figure 4a), supporting the idea that S12 acts on the pre-rRNA rather than on S4.

In order to determine whether S12 was more effective during or after transcription, we compared the fraction of pre-16S rRNAs competent to stably bind S4-Cy5 when S12 was added co-transcriptionally (co-txn) or post-transcriptionally (post-txn; Figure 5). The fraction of competent pre-16S rRNA increased nearly fivefold when S12 was added during transcription, compared with ~twofold when S12 was added post-transcription (Figure 5a). These data suggested that S12 more efficiently chaperones rRNA folding as it is synthesized compared to when the RNA has had time to misfold.

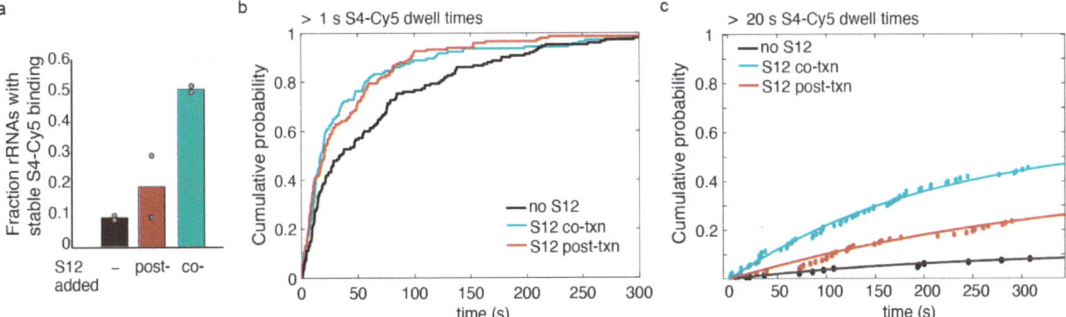

Figure 5. The presence of S12 influences S4 binding during transcription and after transcription. (a) Fraction of pre-16S rRNAs that bind S4 > 20 s in the absence of S12 (black bar; as in Figure 4c), in the presence of S12 after transcription (red bar; as in Figure 4b), and in the presence of S12 during transcription (teal bar; as in Figure 4a). Gray symbols indicate the values for independent replicates. (b) Cumulative probability plot of S4-Cy5 arrival times for specific events lasting >1 s. Association times were combined from two independent replicates. Association of S4-Cy5 is faster in the presence of S12 added during or after transcription. The cumulative density functions for S12 co-txn and S12 post-txn are statistically similar, and both are statistically different than no S12 (K–S test). (c) Cumulative probability plot of S4-Cy5 arrival times for stable events lasting >20 s. Apparent association times were fit with a single exponential function (lines). Stable association of S4-Cy5 is enhanced by the presence of S12 during and after transcription.

Next, we examined the kinetics of S4-Cy5 association with the rRNA when S12 was added, either during transcription (co-txn) or simultaneously with S4-Cy5 after transcription (post-txn). Previous experiments established that S4-Cy5 dwell times longer than 1 s mainly represent specific interactions with the 5WJ in the 16S rRNA [18].

Therefore, we first examined the first moment of S4-Cy5 binding >1 s for all pre-16S transcripts. The cumulative probability distributions of association times showed that S12 accelerated specific S4-Cy5 binding, whether S12 was present during or after transcription (Figure 5b). This suggests that S12 favors recognition of the 5WJ by S4, and that S12 need not be present with S4 to act.

Because the likelihood of stable S4-Cy5 binding increased dramatically when S12 was present during transcription, we also measured the apparent association time for S4-Cy5 binding >20 s, as above (Figure 5c). The cumulative probability distributions showed that, when S12 is present during transcription, S4-Cy5 binding reached a higher maximum probability compared to when S12 was added after transcription or not at all (Table 2). However, the apparent association rates of long-lived S4-Cy5 binding events, $k_{on,app}$, were similar within error between the different conditions, indicating that the effective on-rate had not changed in the presence of S12 (Figure 5c and Table 2). This is consistent with an RNA chaperone activity of S12, in which S12 accelerates proper folding of the rRNA, thereby generating a larger fraction of rRNAs that are competent to stably interact with S4.

Table 2. Exponential fit parameters for apparent association rates of long-lived S4-rRNA interactions. Cumulative density plots from Figure 5c were fit to a single-exponential rate equation, in which A is the amplitude of plateau and $k_{on,app}$ is the apparent association rate for S4-Cy5 binding events > 20 s.

Experiment	N_{mol}	$k_{on,app}$ (s^{-1})	A
S4 alone post-txn	153	0.0043 ± 0.0018	0.11 ± 0.02
S12 added co-txn	143	0.0047 ± 0.0004	0.59 ± 0.02
S12 added post-txn	151	0.0028 ± 0.0012	0.43 ± 0.09

4. Discussion

How the rRNA folds during and after transcription is critical for ribosome biogenesis, which is coupled with pre-rRNA synthesis in cells. Estimates of ribosome synthesis kinetics in E. coli [30] imply that the primary assembly RP S4 is recruited to the rRNA soon after its binding site is transcribed [31]. In vitro, however, S4 struggles to stably recognize its binding site due to improper folding of the newly made pre-16S rRNA [18]. This dichotomy implies that cells possess mechanisms for facilitating proper rRNA folding during transcription.

Here, we examined how RPs facilitate the binding of protein S4 to the 16S rRNA, which is one of the first complexes formed during 30S ribosome assembly. Surprisingly, we find that RP S12 alone can accelerate stable association of S4 (Figures 1 and 2), although S12 stably joins the complex much later during 30S assembly. Protein S12 only binds weakly with assembly intermediates [32] and pulse-chase mass spectrometry experiments revealed that S12 has a very slow on-rate during reconstitution relative to other RPs [3]. Consistent with these earlier results, we observe that S12 can only interact transiently with the pre-16S rRNA in single-molecule experiments (Figure 3).

There are at least two mechanisms by which S12 may facilitate S4-rRNA complex formation: S12 may co-bind the rRNA with S4 to stabilize the complex, or S12 may independently change the folding path of the S4 binding site, acting as a canonical RNA chaperone. In the mature subunit, S12 binds the 16S 5WJ opposite S4 (Figure S1). Therefore, the two proteins could cooperatively stabilize each other's interactions with the rRNA when they co-bind the rRNA. However, S12 is unlikely to improve the thermodynamics of S4 binding, as native S4-rRNA complexes are already very stable [13], and S4 binds before S12 in the assembly map. Moreover, it is unlikely that the two proteins occupy the same rRNA under the conditions of our experiments (≤ 1 per 200 s at 100 nM S12). Instead, our results show that S12 increases the effective rate of stable recruitment, rather than the lifetimes of the S4 complexes (Figures 2 and 5), by acting on the rRNA.

Protein S12 has been previously suggested to act as a general RNA chaperone that facilitates splicing of a group I intron and the hammerhead ribozyme [22]. Interestingly, S12 also binds other RNAs weakly and non-specifically, with only a slight binding preference for unstructured RNA [22]. Despite only interacting with the newly transcribed RNA for short periods of time (~0.5 s), S12 is able to elicit an effect on the rRNA folding path, leading to increased binding of S4. Because we observe that S12 must be present in a high concentration in order to facilitate S4 binding (Figure 3e), it is likely that there is a low probability that S12 successfully chaperones the RNA at each encounter. Similar to other RNA-binding proteins with chaperone activity [33], S12 possesses a broad RNA binding surface, with flexible loops and tails that could help refold the RNA during these short interactions [34]. It is unclear if S12 is acting as a chaperone by interacting with its normal binding site, or if S12 is helping to refold the entire rRNA through nonspecific interactions.

Our order-of-addition experiments suggest that S12 acts on the rRNA independently of S4, since the presence of S12 during transcription followed by its removal leads to even more stable S4 binding post-transcription than when S12 was present together with S4 (Figures 4 and 5). Similarly, S12 does not have to be present during group I intron splicing to improve the splicing efficiency [22]. Moreover, our data suggest that S12 may perform better on rRNAs during transcription than on refolded rRNAs (Figure 5). Altogether, the results support a model in which S12 binds to the rRNA while it is still relatively unstructured to help form native-like initial structures and prevent the rRNA from misfolding. Rapid and transient binding may prevent the accumulation of unproductive S12-rRNA complexes that could inhibit S4 binding.

During assembly in vivo, stable recruitment of S4 must occur soon after transcription of the 5WJ in order to nucleate assembly of the 5′ domain. In our minimal in vitro system, S4 is still unable to bind the rRNA stably during transcription, even in the presence of S12 (Figure 1). Furthermore, S12 only increases the fraction of rRNAs competent for S4 addition to ~20% (Figure 2c). While it is not known how many transcribed rRNAs are ultimately assembled into a mature 30S subunit, it is likely that cells contain redundant mechanisms to accelerate rRNA folding and facilitate binding of S4 and other RPs during transcription. These chaperone mechanisms could include other RPs not tested here, such as RP S1, which has been shown to have RNA chaperone activity [35–38]. There is evidence that general RNA chaperones function during ribosome assembly, including Hfq, CspA, CsdA, and other RNA binding proteins [33,39]. Future work examining other chaperones will be important in order to characterize the mechanism of the earliest steps in ribosome biogenesis during transcription.

Supplementary Materials: The following supporting information can be downloaded at: https://www.mdpi.com/article/10.3390/biom13060951/s1, Figure S1: Binding sites of S4 and S12 in E. coli 16S rRNA after [40].; Figure S2: Presence of RP S16 or S5 does not influence S4-Cy5 binding dynamics on pre-16S rRNAs.; Figure S3: Maximum likelihood estimation (MLE) analysis of S4-Cy5 dwell times in the presence of different RPs.; Figure S4: Probability density histograms and maximum likelihood estimation (MLE) fits of S4-Cy5 binding to pre-16S rRNAs post-transcription, as outlined in Figure 4; Table S1: MLE analysis of S4-Cy5 binding in the presence and absence of S12 during and after transcription.

Author Contributions: Conceptualization, M.L.R. and S.A.W.; methodology, M.L.R. and Y.S.; formal analysis, M.L.R. and Y.S.; writing—original draft preparation, M.L.R.; writing—review and editing, M.L.R., Y.S. and S.A.W.; funding acquisition, M.L.R. and S.A.W. All authors have read and agreed to the published version of the manuscript.

Funding: This research was funded by the National Institute of General Medical Sciences, K99/R00 GM140204 to M.L.R. and R35 GM136351 to S.A.W.

Institutional Review Board Statement: Not applicable.

Informed Consent Statement: Not applicable.

Data Availability Statement: The data presented in this study are available on request from the corresponding author.

Acknowledgments: We thank the Noller lab for the expression plasmid for single-cysteine S12 and all members of the Woodson lab for helpful comments on this manuscript. This research was supported in part by the Intramural Research Program of the NIH, The National Institute of Diabetes and Digestive and Kidney Diseases (NIDDK).

Conflicts of Interest: The authors declare no conflict of interest.

References

1. Noller, H.F. RNA Structure: Reading the Ribosome. *Science* **2005**, *309*, 1508–1514. [CrossRef] [PubMed]
2. Stern, S.; Powers, T.; Changchien, L.-M.; Noller, H.F. RNA-Protein Interactions in 30S Ribosomal Subunits: Folding and Function of 16S rRNA. *Science* **1989**, *244*, 783–790. [CrossRef] [PubMed]
3. Talkington, M.W.T.; Siuzdak, G.; Williamson, J.R. An assembly landscape for the 30S ribosomal subunit. *Nature* **2005**, *438*, 628–632. [CrossRef] [PubMed]
4. Adilakshmi, T.; Bellur, D.L.; Woodson, S.A. Concurrent nucleation of 16S folding and induced fit in 30S ribosome assembly. *Nature* **2008**, *455*, 1268–1272. [CrossRef] [PubMed]
5. Kim, H.; Abeysirigunawarden, S.C.; Chen, K.; Mayerle, M.; Ragunathan, K.; Luthey-Schulten, Z.; Ha, T.; Woodson, S.A. Protein-guided RNA dynamics during early ribosome assembly. *Nature* **2014**, *506*, 334–338. [CrossRef]

6. Held, W.A.; Mizushima, S.; Nomura, M. Reconstitution of *Escherichia coli* 30 S Ribosomal Subunits from Purified Molecular Components. *J. Biol. Chem.* **1973**, *248*, 5720–5730. [CrossRef]
7. Nowotny, V.; Nierhaus, K.H. Assembly of the 30S subunit from *Escherichia coli* ribosomes occurs via two assembly domains which are initiated by S4 and S7. *Biochemistry* **1988**, *27*, 7051–7055. [CrossRef]
8. Ridgeway, W.K.; Millar, D.P.; Williamson, J.R. Quantitation of ten 30S ribosomal assembly intermediates using fluorescence triple correlation spectroscopy. *Proc. Natl. Acad. Sci. USA* **2012**, *109*, 13614–13619. [CrossRef]
9. Davis, J.H.; Tan, Y.Z.; Carragher, B.; Potter, C.S.; Lyumkis, D.; Williamson, J.R. Modular Assembly of the Bacterial Large Ribosomal Subunit. *Cell* **2016**, *167*, 1610–1622.e15. [CrossRef]
10. French, S.L.; Miller, O.L. Transcription mapping of the *Escherichia coli* chromosome by electron microscopy. *J. Bacteriol.* **1989**, *171*, 4207–4216. [CrossRef]
11. Powers, T.; Daubresse, G.; Noller, H.F. Dynamics of In Vitro Assembly of 16 S rRNA into 30 S Ribosomal Subunits. *J. Mol. Biol.* **1993**, *232*, 362–374. [CrossRef] [PubMed]
12. Ramaswamy, P.; Woodson, S.A. Global Stabilization of rRNA Structure by Ribosomal Proteins S4, S17, and S20. *J. Mol. Biol.* **2009**, *392*, 666–677. [CrossRef]
13. Gerstner, R.B.; Pak, Y.; Draper, D.E. Recognition of 16S rRNA by Ribosomal Protein S4 from *Bacillus stearothermophilus*. *Biochemistry* **2001**, *40*, 7165–7173. [CrossRef] [PubMed]
14. Sapag, A.; Vartikar, J.V.; Draper, D.E. Dissection of the 16S rRNA binding site for ribosomal protein S4. *Biochim. Biophys. Acta BBA-Gene Struct. Expr.* **1990**, *1050*, 34–37. [CrossRef]
15. Vartikar, J.V.; Draper, D.E. S4-16 S ribosomal RNA complex: Binding constant measurements and specific recognition of a 460-nucleotide region. *J. Mol. Biol.* **1989**, *209*, 221–234. [CrossRef]
16. Powers, T.; Noller, H.F. A Temperature-dependent Conformational Rearrangement in the Ribosomal Protein S4 16 S rRNA Complex. *J. Biol. Chem.* **1995**, *270*, 1238–1242. [CrossRef]
17. Mayerle, M.; Bellur, D.L.; Woodson, S.A. Slow Formation of Stable Complexes during Coincubation of Minimal rRNA and Ribosomal Protein S4. *J. Mol. Biol.* **2011**, *412*, 453–465. [CrossRef]
18. Rodgers, M.L.; Woodson, S.A. Transcription Increases the Cooperativity of Ribonucleoprotein Assembly. *Cell* **2019**, *179*, 1370–1381.e12. [CrossRef]
19. Bellur, D.L.; Woodson, S.A. A Minimized rRNA-Binding Site for Ribosomal Protein S4 and Its Implications for 30S Assembly. *Proc. Natl. Acad. Sci. USA* **2009**, *37*, 1886–1896. [CrossRef]
20. Duss, O.; Stepanyuk, G.A.; Puglisi, J.D.; Williamson, J.R. Transient Protein-RNA Interactions Guide Nascent Ribosomal RNA Folding. *Cell* **2019**, *179*, 1357–1369.e16. [CrossRef]
21. Williamson, J.R. Induced fit in RNA-protein recognition. *Nat. Struct. Mol. Biol.* **2000**, *7*, 834–837. [CrossRef] [PubMed]
22. Coetzee, T.; Herschlag, D.; Belfort, M. *Escherichia coli* proteins, including ribosomal protein S12, facilitate in vitro splicing of phage T4 introns by acting as RNA chaperones. *Genes Dev.* **1994**, *8*, 1575–1588. [CrossRef]
23. Culver, G.M.; Noller, H.F. Efficient reconstitution of functional *Escherichia coli* 30S ribosomal subunits from a complete set of recombinant small subunit ribosomal proteins. *RNA* **1999**, *5*, 832–843. [CrossRef] [PubMed]
24. Culver, G.M.; Noller, H.F. In vitro reconstitution of 30S ribosomal subunits using complete set of recombinant proteins. *Methods Enzym.* **2000**, *318*, 446–460. [CrossRef]
25. Davanloo, P.; Rosenberg, A.H.; Dunn, J.J.; Studier, F.W. Cloning and expression of the gene for bacteriophage T7 RNA polymerase. *Proc. Natl. Acad. Sci. USA* **1984**, *81*, 2035–2039. [CrossRef]
26. Butler, E.T.; Chamberlin, M.J. Bacteriophage SP6-specific RNA polymerase. I. Isolation and characterization of the enzyme. *J. Biol. Chem.* **1982**, *257*, 5772–5778. [CrossRef]
27. Hickerson, R.; Majumdar, Z.K.; Baucom, A.; Clegg, R.M.; Noller, H.F. Measurement of Internal Movements within the 30S Ribosomal Subunit Using Förster Resonance Energy Transfer. *J. Mol. Biol.* **2005**, *354*, 459–472. [CrossRef]
28. Hua, B.; Han, K.Y.; Zhou, R.; Kim, H.; Shi, X.; Abeysirigunawardena, S.C.; Jain, A.; Singh, D.; Aggarwal, V.; Woodson, S.A.; et al. An improved surface passivation method for single-molecule studies. *Nat. Methods* **2014**, *11*, 1233–1236. [CrossRef]
29. Friedman, L.J.; Gelles, J. Multi-wavelength single-molecule fluorescence analysis of transcription mechanisms. *Methods* **2015**, *86*, 27–36. [CrossRef]
30. Lindahl, L. Intermediates and time kinetics of the in vivo assembly of *Escherichia coli* ribosomes. *J. Mol. Biol.* **1975**, *92*, 15–37. [CrossRef]
31. Chen, S.S.; Sperling, E.; Silverman, J.M.; Davis, J.H.; Williamson, J.R. Measuring the dynamics of *E. coli* ribosome biogenesis using pulse-labeling and quantitative mass spectrometry. *Mol. Biosyst.* **2012**, *8*, 3325. [CrossRef] [PubMed]
32. Held, W.A.; Nomura, M. Rate-determining step in the reconstitution of *Escherichia coli* 30S ribosomal subunits. *Biochemistry* **1973**, *12*, 3273–3281. [CrossRef]
33. Woodson, S.A.; Panja, S.; Santiago-Frangos, A. Proteins That Chaperone RNA Regulation. *Microbiol. Spectr.* **2018**, *6*, 385–397. [CrossRef] [PubMed]
34. Tompa, P.; Csermely, P. The role of structural disorder in the function of RNA and protein chaperones. *FASEB J.* **2004**, *18*, 1169–1175. [CrossRef] [PubMed]
35. Bear, D.G.; Ng, R.; Van Derveer, D.; Johnson, N.P.; Thomas, G.; Schleich, T.; Noller, H.F. Alteration of polynucleotide secondary structure by ribosomal protein S1. *Proc. Natl. Acad. Sci. USA* **1976**, *73*, 1824–1828. [CrossRef] [PubMed]

36. Kolb, A.; Hermoso, J.M.; Thomas, J.O.; Szer, W. Nucleic acid helix-unwinding properties of ribosomal protein S1 and the role of S1 in mRNA binding to ribosomes. *Proc. Natl. Acad. Sci. USA* **1977**, *74*, 2379–2383. [CrossRef] [PubMed]
37. Hajnsdorf, E.; Boni, I.V. Multiple activities of RNA-binding proteins S1 and Hfq. *Biochimie* **2012**, *94*, 1544–1553. [CrossRef]
38. Lund, P.E.; Chatterjee, S.; Daher, M.; Walter, N.G. Protein unties the pseudoknot: S1-mediated unfolding of RNA higher order structure. *Nucleic Acids Res.* **2020**, *48*, 2107–2125. [CrossRef]
39. Andrade, J.M.; dos Santos, R.F.; Chelysheva, I.; Ignatova, Z.; Arraiano, C.M. The RNA-binding protein Hfq is important for ribosome biogenesis and affects translation fidelity. *EMBO J.* **2018**, *37*, e97631. [CrossRef]
40. Brodersen, D.; Clemons, W.; Carter, A.; Wimberly, B.T.; Ramakrishnan, V. Crystal structure of the 30 s ribosomal subunit from Thermus thermophilus: Structure of the proteins and their interactions with 16 s RNA. *J. Mol. Biol.* **2002**, *316*, 725–768. [CrossRef]

Disclaimer/Publisher's Note: The statements, opinions and data contained in all publications are solely those of the individual author(s) and contributor(s) and not of MDPI and/or the editor(s). MDPI and/or the editor(s) disclaim responsibility for any injury to people or property resulting from any ideas, methods, instructions or products referred to in the content.

Review

Putting It All Together: The Roles of Ribosomal Proteins in Nucleolar Stages of 60S Ribosomal Assembly in the Yeast *Saccharomyces cerevisiae*

Taylor N. Ayers and John L. Woolford *

Department of Biological Sciences, Carnegie Mellon University, Pittsburgh, PA 15213, USA
* Correspondence: jw17@andrew.cmu.edu

Abstract: Here we review the functions of ribosomal proteins (RPs) in the nucleolar stages of large ribosomal subunit assembly in the yeast *Saccharomyces cerevisiae*. We summarize the effects of depleting RPs on pre-rRNA processing and turnover, on the assembly of other RPs, and on the entry and exit of assembly factors (AFs). These results are interpreted in light of recent near-atomic-resolution cryo-EM structures of multiple assembly intermediates. Results are discussed with respect to each neighborhood of RPs and rRNA. We identify several key mechanisms related to RP behavior. Neighborhoods of RPs can assemble in one or more than one step. Entry of RPs can be triggered by molecular switches, in which an AF is replaced by an RP binding to the same site. To drive assembly forward, rRNA structure can be stabilized by RPs, including clamping rRNA structures or forming bridges between rRNA domains.

Keywords: ribosome assembly; protein–rRNA interactions; rRNA folding; ribosome assembly intermediates; ribosomal proteins; assembly factors; large ribosomal subunit; pre-rRNA processing; cryo-electron microscopy

Citation: Ayers, T.N.; Woolford, J.L. Putting It All Together: The Roles of Ribosomal Proteins in Nucleolar Stages of 60S Ribosomal Assembly in the Yeast *Saccharomyces cerevisiae*. *Biomolecules* **2024**, *14*, 975. https://doi.org/10.3390/biom14080975

Academic Editors: Brigitte Pertschy and Ingrid Zierler

Received: 17 July 2024
Revised: 5 August 2024
Accepted: 7 August 2024
Published: 9 August 2024

Copyright: © 2024 by the authors. Licensee MDPI, Basel, Switzerland. This article is an open access article distributed under the terms and conditions of the Creative Commons Attribution (CC BY) license (https://creativecommons.org/licenses/by/4.0/).

1. Introduction

Ribosomes are ribonucleoprotein nanomachines that translate the genetic code in mRNA and catalyze peptide bond formation. The small ribosomal subunit (SSU) houses the mRNA decoding center, while the large ribosomal subunit (LSU) contains the peptidyl-transferase center (PTC). In yeast, the SSU is comprised of the 18S rRNA (1800 nucleotides (nts)) and 33 different ribosomal proteins (RPs). The LSU contains the 5S rRNA (121 nts), 5.8S rRNA (158 nts), 25S rRNA (3396 nts), and 46 different RPs.

The LSU 25S rRNA is organized into six phylogenetically conserved domains of secondary structure (domains I–VI, 5′ to 3′) that intertwine with each other (Figure S1A,B) [1,2]. Each rRNA domain originates from a root helix created by base-pairing between the first and last nucleotide sequences of that domain, defining the domain "boundaries" (Figure S1A,C). The 5.8S rRNA base-pairs with the 5′ end of rRNA domain I to form the proximal stem (Figure S1A), while the 5S rRNA docks onto rRNA domain V to form the central protuberance, a bridge-like structure that links the two subunits during protein synthesis. RPs decorate the outer surface of rRNA and burrow into the rRNA core to help knit together the multiple rRNA domains (Figure S1D).

Assembly of the ribosomal subunits begins in the nucleolus, with transcription of the 35S pre-rRNA by RNA polymerase I and the 5S rRNA by RNA polymerase III (Figure S2A). The 35S primary rRNA transcript contains the sequences for the mature 18S, 5.8S, and 25S rRNAs, which are separated and flanked by internal and external transcribed spacers (Figure S2B). During, and immediately after its synthesis, rRNA undergoes folding, modification, and removal of some of the spacer sequences, accompanied by binding of a subset of the RPs (Figure S2A,B) [3–6]. The 5S rRNA is separately packaged into a 5S ribonucleoprotein complex (RNP) before joining nascent 60S subunits [7–9]. Small subunit precursor particles are released from the transcription complex upon endonucleolytic cleavage of rRNA at the A_2 site, leading to rapid transit of pre-40S particles out of the nucleolus (Figure S2A). In contrast, free pre-60S particles are only generated upon termination of rRNA transcription and then undergo multiple remodeling steps before moving into the nucleoplasm (Figure S2A). Additional restructuring of precursors to both subunits occurs in the nucleoplasm, followed by final stages of maturation and quality control in the cytoplasm [10–13]. In rapidly growing yeast cells, 2000 ribosomes are produced each minute [14]. The accuracy and efficiency of this complex pathway of ribosome biogenesis are enabled at different steps by more than 200 AFs in yeast (Figure S3) [3,5,6]. Of these AFs, approximately 90 participate in LSU biogenesis.

2. rRNA Folding Is a Critical Component of Ribosome Assembly

Arguably, the most challenging task of assembling ribosomes is the folding of their rRNA [15]. Because functional centers of the ribosome are largely comprised of rRNA, e.g., the peptidyltransferase center and the polypeptide exit tunnel in the LSU, proper formation of the three-dimensional structure of rRNA is especially critical for ribosome function [16].

RNA structure probing and near-atomic resolution cryo-electron microscopy (cryo-EM) structures of ribosome assembly intermediates have provided an initial glimpse into the rRNA folding pathway [5,6,17–20]. As rRNA is transcribed, it begins to fold into its secondary structure and bind to RPs and AFs (Figure S2A). Yet the path to do so is not always straightforward. The complementary strands of some rRNA helices in mature ribosomes, including the root helices at the base of each domain, are far apart in the rRNA primary sequence (Figure S1A). Therefore, alternative rRNA helices can form initially via base-pairing with other sequences 5′ of the mature partner before the second strand of that root helix is even transcribed. These initial helices then must be remodeled to promote productive pairing between downstream rRNA partners [21–25]. Once rRNA secondary structures are established, the resulting helices then reorient to form the complex tertiary structures necessary for rRNA function.

One challenge to defining the rRNA folding pathway (for both cells and experimentalists!) is that a significant fraction of the rRNA is structurally heterogeneous during certain stages and thus cannot be resolved via cryo-EM. For example, newly synthesized rRNA is highly flexible in the earliest assembly intermediate, with less than one fourth of the LSU rRNA visible in the structure of the Noc1-Noc2 particle [26,27]. Likewise, most of rRNA domains III, IV, and V are flexible and do not stably compact onto pre-ribosomes until late nucleolar stages, after the installation of domains I, II, and VI (Figure S3) [18,20,28]. Thus, it remains unclear precisely how and when these initially flexible RNAs fold into secondary and tertiary structures and bind to RPs and AFs.

3. How to Determine Functions of Ribosomal Proteins

The primary function of RPs and many AFs is to enable the proper folding of rRNA. This facilitates the recruitment of other RPs and AFs to promote further remodeling of the pre-rRNP structure to drive assembly forward [6,29]. Before reviewing the roles of RPs in LSU assembly, we will first describe the experimental approaches used to assay RP functions.

3.1. Investigating When Each RP Assembles into Pre-Ribosomes

Determining the order of RP assembly into pre-ribosomes provides useful hints for delineating the specific role of RPs in subunit maturation. Thirty-nine of the forty-six RPs in the large subunit associate with pre-ribosomes in the nucleus, while the remaining seven join nascent 60S subunits in the cytoplasm (Figure S2C) [30]. To more precisely determine when each RP assembles, the consecutive pre-rRNA processing intermediates (Figure S2B) and pre-ribosome assembly intermediates (Figure S2C) with which each RP co-immunopurifies have been identified [31–36]. However, these approaches provide only rough estimates of RP entry times since there are only a handful of different pre-rRNA processing intermediates or purified pre-ribosome assembly intermediates that one can distinguish (Figure S2B,C). Pre-ribosomes are captured using epitope-tagged AFs that copurify with multiple different assembly intermediates that span early, middle, and/or late intervals. Thus, affinity-purified pre-ribosomes usually comprise a heterogeneous population. In this review, we indicate the point of entry for each RP based upon when it is first resolved in pre-ribosome intermediates via cryo-EM. Note that biochemical assays and cryo-EM structures of pre-ribosomes do not always agree. Failure to visualize a protein in a particle may result from the structural heterogeneity and/or flexibility of that protein.

3.2. Depleting Ribosomal Proteins

Of the 46 RPs in the yeast LSU, 35 are essential for cell growth, while 11 are not essential under laboratory conditions [30,37]. However, deletion of the "nonessential" RP genes usually results in slow growth at 30 °C or an inability to grow at lower temperatures (cold sensitivity) [37,38]. Conditional expression has been utilized to deplete RPs whose absence causes a lethal or severe growth defect at 30 °C. To deplete an RP, a glucose-repressible *GAL* promoter is fused to the RP gene of interest. Cell cultures shifted from galactose- to glucose-containing medium halt transcription of the RP mRNA [39]. After waiting for the pre-existing RP mRNA to decay, and for the pool of unassembled RPs to be depleted by turnover or by assembly into pre-ribosomes, the depletion phenotype can be assessed. Phenotypes of cold-sensitive knockouts of nonessential RP genes can be assayed by shifting cultures to low temperatures.

To date, strains conditional for the expression of 42 of the 46 yeast LSU RPs have been constructed. Most of these mutants exhibit a shortage of 60S subunits relative to 40S subunits and are blocked at a particular stage of pre-rRNA processing, although some arrest assembly occurs only after all steps of rRNA processing have occurred [30]. The effects of the depletions on subunit maturation have been examined in more detail for 28 of the RPs. To do so, semi-quantitative mass spectrometry was used to assay changes in protein constituents of affinity-purified pre-ribosomes [40].

Interpreting RP depletion phenotypes is not necessarily straightforward. Depletion of an RP using conditional promoter systems can take several hours relative to the 15–20 min required for subunit assembly [40–42]. Thus, one must be aware of indirect effects. Defects in early steps of assembly can result when a late step is blocked by RP depletion due to improper recycling of early-acting AFs from the abortive late intermediates [42,43]. RPs could also impact assembly even before they form stable complexes with their rRNA ligands by first chaperoning the binding of other RPs [44]. Conversely, an assembly defect might only be observed downstream of the stable installation of an RP with an rRNA domain if that entire domain is initially flexible and then stably incorporated into pre-ribosomes at a later step [7]. Finally, an RP may be important for multiple stages of subunit maturation, yet its depletion may only reveal the earliest defect. For example, an RP may stably contact multiple neighborhoods of rRNA, with the potential to affect the assembly of more than one rRNA domain at different stages of biogenesis [45].

4. Roles of RPs in LSU Assembly

Here, we review the roles of RPs in the nucleolar stages of LSU biogenesis revealed by biochemical and genetic approaches in yeast and speculate how recently obtained cryo-EM structures of LSU assembly intermediates might help us better imagine the mechanistic principles underlying ribosome assembly. We discuss RPs whose association with pre-ribosomes has been detected in the nucleolus, including three RPs that are not visualized in particles by cryo-EM until the nucleoplasm (L5, L11, and L39). We assign each RP to one RNP domain in the LSU based on the rRNA with which that RP establishes the most contacts (Figure 1). However, we note that in many cases, RPs have significant contacts with more than one rRNA domain. The construction of each "neighborhood" in the LSU is discussed based on the order of their assembly: rRNA domain I, rRNA domain II, the proximal stem, rRNA domain VI, rRNA domain III, rRNA domain IV, rRNA domain V, the polypeptide exit tunnel, and the central protuberance. This order is not strictly 5' to 3' with respect to the rRNA sequence; domains at the two ends of the pre-rRNA (domains I, II, and VI) destined for the LSU assemble before those in the middle portion (domains III, IV, and V) (Figure 1, Figure S2 and S3) [18,46].

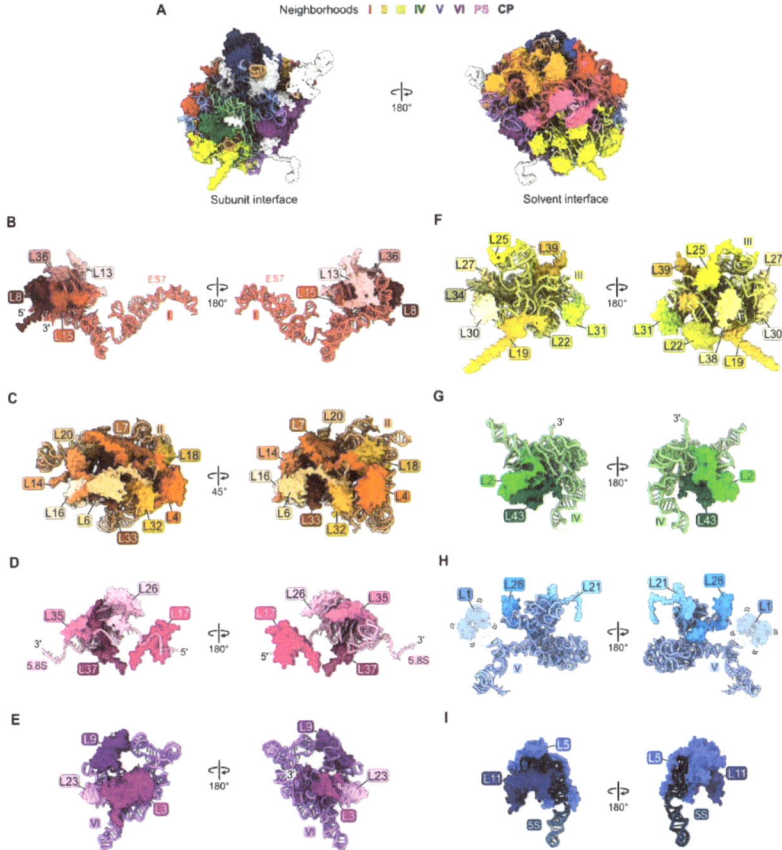

Figure 1. Each neighborhood of rRNA and RPs in the crystal structure of the yeast mature large ribosomal subunit [47], in the order with which they assemble: (**A**) The entire large subunit. Ribosomal proteins are color coded according to the neighborhood in which they are located. Ribosomal proteins colored in white assemble after nucleolar stages and thus are not discussed in this review; (**B**) domain I (red); (**C**) domain II (orange); (**D**) proximal stem, PS (pink); (**E**) domain VI (purple); (**F**) domain III (yellow); (**G**) domain IV (green); (**H**) domain V (blue); (**I**) central protuberance, CP (navy blue). Structures are shown in the subunit interface view on the left and the solvent-exposed surface on the right, except for domain II. In this case, orientations are shown to optimize the visualization of each RP. Note, though ribosomal protein L1 is present, it is not resolved in the mature LSU crystal structure (PDB: 4V88).

4.1. rRNA Domain I: L8, L13, L15, L36

In mature ribosomes, RPs L8, L13, L15, and L36 primarily contact rRNA domain I (nucleotides 1–651) (Figures S1A and 1B), but they also form bridges with neighboring domains II, V, and 5.8S rRNA (Figure S3). Most likely, these four RPs assemble co-transcriptionally, as they are first visible in the cryo-EM structure of the co-transcriptional Noc1-Noc2 particle (Figure 2A) [26]. L8, L13, and L15 associate with rRNA domain I when only this fragment of rRNA is expressed, while L36 only assembles once a construct containing both domains I and II is expressed [48]. The so-called "A_3" AFs Brx1, Cic1, Has1, Ebp2, Erb1, Nop7, Nop15, and Rlp7 form a shell surrounding the RPs bound to domain I rRNA, and also assemble early, in the Noc1-Noc2 particle (Figure 2A).

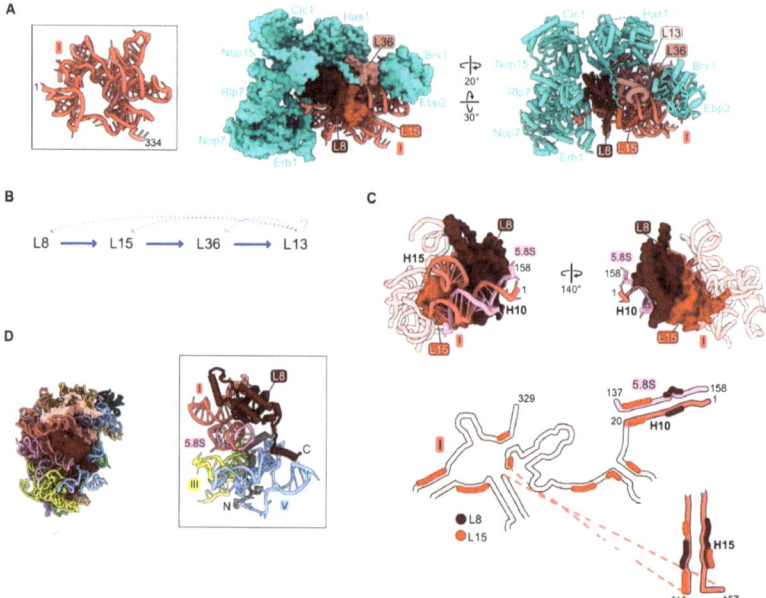

Figure 2. Domain I (RPs L8, L13, L15, L36): (**A**) Eight assembly factors (AFs) (cyan) bind to the exterior of the domain I RNP in the Noc1-Noc2 particle during early stages of assembly (PDB: 8E5T) [26]; (**B**) the assembly hierarchy implied by the effects of depleting RPs L8, L15, or L36; (**C**) binding of L8 and L15 to domain I rRNA and to each other in Noc1-Noc2 pre-ribosomes (PDB: 8E5T). Flexible rRNA is depicted as transparent in the secondary structure. Regions where L8 and L15 contact rRNA are indicated; (**D**) the globular domain of L8 (red) binds to rRNA domain I, while the 70-amino acid long N-terminal domain (grey) binds to rRNA domain V (PDB: 4V88).

The effects of depleting L8, L15, and L36 have been investigated, but not yet for L13 [32,40,46,49]. Primer extension assays have revealed the accumulation of modest amounts of pre-rRNAs with 5' ends at the A_2, A_3, and B_L sites. Northern blotting showed increased levels of 35S and 23S pre-rRNAs, while amounts of other downstream processing intermediates and mature rRNAs were significantly decreased. Consistent with these steady-state assays, pulse-chase experiments showed that the 27S pre-rRNAs are rapidly turned over. Thus, blocking assembly of domain I results in disassembly and degradation of early assembly intermediates.

Analysis of pre-ribosome protein constituents following domain I RP depletions suggested that assembly of the four domain I RPs is largely, but not completely, hierarchical (Figure 2B). L13, L15, and L36 are the RPs most strongly diminished in the absence of L8, while L13 is affected by the depletion of L15 or L36, but L8 is much less so. However, this pattern of linear dependence does not appear to be consistent with the structure of domain I in the mature LSU. L15 is located on the interior of domain I embedded within domain I rRNA, while L8 is located on the outer surface bound to L15 (Figure 2C). Both L8 and L15 contact distant rRNA helices 10 and 15, appearing to pinch them together. In this configuration, we imagine that L8 would depend upon L15 to assemble, as well as vice versa. For example, we suspect that because L15 associates with numerous domain I rRNA regions, L15 could help the rRNA fold co-transcriptionally. This chaperoning could facilitate L8 binding, which requires the coming together of distal secondary structure elements, ultimately stabilizing the proximity of helices 10 and 15 in three dimensions. Thus, the seeming contradiction between molecular assays and existing structural snapshots suggests that the pathway of assembly may be more complex than initially evident from depletion experiments. This example signifies the need for more sophisticated real time assays of subunit maturation (see below in Section 6).

Depletion of L8, L15, or L36 affects steps upstream and downstream of their assembly. Amounts of the very early scaffolding AFs Nop4, Rrp5, Noc1, and Noc2 increase in the population of purified pre-ribosomes, suggesting that early stages of assembly cannot proceed efficiently when domain I cannot properly form. Alternatively, a pool of pre-existing earlier intermediates containing these scaffolding factors might remain in the depletion strains and could be more stable than later particles in which assembly of domain I is aborted. Meanwhile, there are decreased amounts of the A_3 AFs surrounding the RPs bound to domain I (Figure 2A). Interestingly, the reverse is not true. Depletion of A_3 factors does not affect the assembly of domain I RPs, consistent with these AFs coating the exterior surface of domain I RNP in pre-ribosomes [50,51]. The absence of L8, L15, and L36 also impairs the assembly of other RPs. Amounts of the proximal stem and domain III RPs L17, L26, L19, and L31 that join in downstream steps of subunit maturation strongly decrease, yet the effects on assembly of other RPs with rRNA domains II or III are much weaker.

Strikingly, many RPs contain N- or C-terminal domains or internal loops that extend from their globular core. These extensions thread through neighboring rRNA domains across the surface of the ribosome or penetrate into the ribosome core [52]. Domain I RP L8 is one example of such a protein that contacts multiple rRNP domains. While the globular domain of L8 binds to rRNA domain I, the N-terminal extension bridges rRNA domain I with domain V and the proximal stem. Greater than half of the L8 contacts with rRNA occur via its N-terminal extension (Figure 2D).

Deletion of the L8 extension revealed that this portion of L8 participates in late stages of assembly that involve domain V [45]. A mutant RP L8 lacking the 70 N-terminal amino acids can assemble into pre-ribosomes, but it causes a dominant lethal growth defect. While depletion of L8 aborts early steps of pre-rRNA processing and subunit assembly, the *rpl8 Δ1-70* mutation blocks later steps: processing of the 7S pre-rRNA, a precursor of 5.8S rRNA, and assembly of AFs required for remodeling of domain V and the 5S RNP. Thus, binding of the globular domain of L8 with domain I rRNA is important for the construction of domain I in the early steps of LSU biogenesis, while insertion of its N-terminal extension into domain V plays a role in later steps of subunit maturation that are coupled with the compaction of domain V and 7S pre-rRNA processing.

4.2. rRNA Domain II: L4, L6, L7, L14, L16, L18, L20, L32, L33

Nine RPs have extensive contacts with rRNA domain II (nucleotides 652–1455) in mature LSUs (Figure S1 and Figure 1C). Among these proteins, L4, L7, L18, and L32 form one "node" that bridges domain II with rRNA domain I, while L6, L14, L16, L20, and L33 form a separate node that contacts both rRNA domains II and VI (Figures 3A and S3).

Domain II (as well as domain I) adopts its mature conformation very early; both domains are organized in the Noc1-Noc2 particle in a fashion that is very similar to that in mature LSUs [26,53]. During the transition from the Noc1-Noc2 particle to the State 2 intermediate, a molecular switch draws together domains I and II (Figure 3B). Rrp1 and Mak16 bind primarily to domain II in the Noc1-Noc2 particle, but Nsa1 and Rpf1 are only loosely tethered to this complex. Upon release of Noc1 and Noc2, Nsa1 and Rpf1 become stably accommodated into State 2 particles, working with Rrp1 and Mak16 to form a bridge connecting domains I and II.

Anchoring of domain VI onto domain II, mediated by RP bridges, may function as a "lid" to protect the 5' end of 5.8S rRNA from degradation by nucleases [34]. Node II RPs L14 and L16 bound to domain II RNA may function in this compaction of domain VI by acting as clamps to stabilize the docking of domains II and VI. The globular portions of L14 and L16 are adjacent to one another, as are contact domains II and VI. The C-terminal extensions of these RPs contact one another in the Noc1-Noc2 particle, similar to the mature structure, while the C-terminal domain of L16 also cradles domain VI rRNA in mature subunits (Figure 3C). Truncating the C-terminal portion of either of these proteins (*rpL14-Δ109*, *rpl16-Δ171*) results in lethality [34,54,55]. The truncated L16 Δ171 protein assembles into pre-ribosomes, but pre-rRNA processing is affected similarly to a depletion of L16. Thus, analogous to the behavior of L8 in domain I, the globular domain of L16 is sufficient for assembly of this RP, but the C-terminal extension plays an important role in LSU maturation.

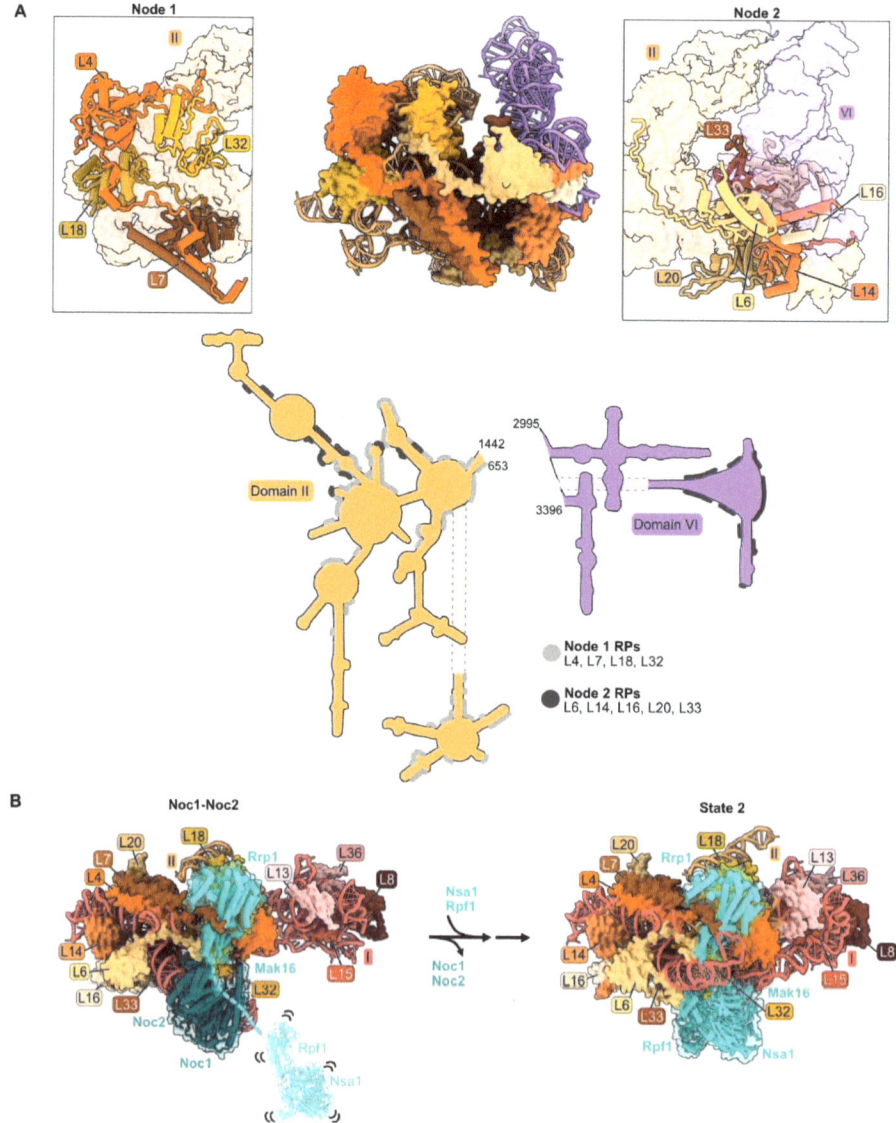

Figure 3. *Cont.*

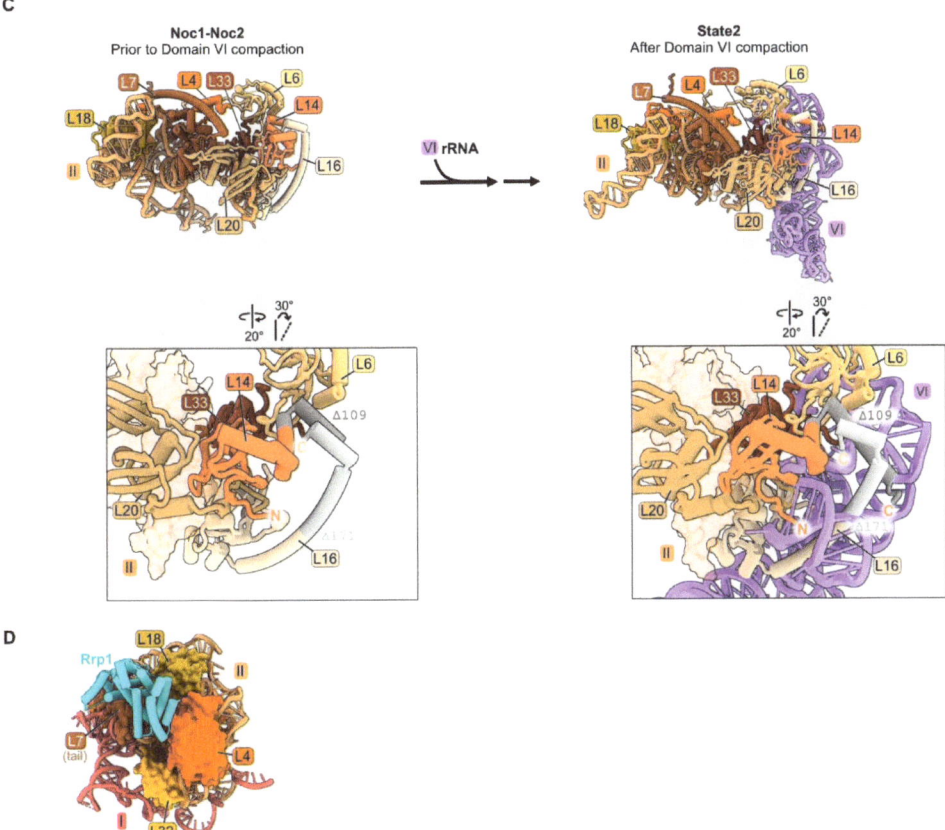

Figure 3. Domain II: (**A**) Structure of node 1 containing RPs L4, L7, L18, and L32 (left), and of node 2 containing RPs L6, L14, L16, L20, and L33 (right) (PDB: 4V88). Domain VI (purple) is shown docked onto domain II rRNA. The bottom panel depicts the contact points of each protein with the rRNA; (**B**) a molecular switch between the co-transcriptional Noc1-Noc2 particle and the State 2 intermediate (PDB: 8E5T, 6C0F) [20,26]. State A (lower resolution) precedes State 2 and is not shown; (**C**) docking of domain VI rRNA onto domain II rRNA during the transition from the co-transcriptional Noc1-Noc2 particle to the State 2 intermediate. Panels below show extensions of L14 and L16 binding to domain VI after compaction. Truncations of L14 and L16 C-termini (grey) result in lethality (PDB: 8E5T, 6C0F); (**D**) binding of AF Rrp1 to node 1 domain II RPs (PDB: 8E5T).

The effects of depleting each of the domain II RPs have been examined, except for L6 [32,34,40,54,55]. In each case, as observed for depletion of RPs present in domain I, amounts of $27SA_2$ and $27SA_3$ pre-rRNAs increase slightly, then are rapidly degraded. When domain II RPs are depleted, the association of RPs with domain I is less affected than for domain II RPs, and vice versa. Thus, the assembly of rRNA domains I and II is somewhat independent of each other, consistent with their initial physical separation from each other in Noc1-Noc2 particles (Figure 3B). However, in each depletion, abortive pre-ribosomes rapidly disassemble, and the pre-rRNA is degraded.

When domain II RPs are depleted, the amounts of RPs and AFs in pre-ribosomes that assemble downstream of domain II are diminished. This likely results from the turnover of particles containing 27SA pre-rRNAs, preventing the construction of later intermediates containing 27SB pre-rRNA. However, as observed for depletion of domain I RPs, a subset of RPs in the proximal stem (L17 and L26) and in domain III (L19 and L31) are consistently affected more than others (see discussion in Section 4.5 below).

Assembly of AF Nsa1 is also decreased upon depletion of domain II RPs. This probably results from preventing the transition of Nsa1 from a flexible state, loosely associated with pre-ribosomes, to being stably installed into State 2 particles following the exit of AFs Noc1 and Noc2 (Figure 3B) [26].

Binding of the AF Rrp1 to pre-ribosomes is also strongly affected by depletion of L4, L18, or L32. L18 and the globular body of L4 appear to form the binding site for Rrp1 in the Noc1-Noc2 particle (Figure 3D). Thus, Rrp1 relies on the accurate incorporation of L4 and L18 into pre-ribosomes. However, Rrp1 does not appear to make direct contact with L32. We speculate that the absence of L32 disrupts the local structure of this neighborhood, impairing the binding site for Rrp1 [26].

4.3. Proximal Stem: L17, L26, L35, L37

In mature 60S subunits, the proximal stem is formed by base-pairing portions of 5.8S rRNA with nucleotides in domain I of 25S rRNA (Figures S1A, 1D and 4A,B). RPs L17, L26, L35, and L37 bind to rRNA in the proximal stem but also contact multiple other domains of rRNA (Figures 1D, 4A and S3). L17 binds to helix 2, formed by base-pairing the very 5' end of mature 5.8S rRNA with 25S rRNA, while L26 is positioned downstream bound to helix 4. The globular domain of L35 contacts folded 5.8S rRNA, while its C-terminal extension snakes into domain I. Similarly, L37 binds 5.8S rRNA but also bridges domains I and III.

The proximal stem begins to form during the early stages of LSU assembly. Nucleotides in 5.8S rRNA base-pair with those in 25S rRNA co-transcriptionally, visualized in Noc1-Noc2 particles [26]. Formation of these base-pairs creates the pre-ribosome "foot structure" containing the ITS2 spacer RNA sequences that lie between 5.8S rRNA and domain I in the 27S pre-rRNA (Figure 4B). Subsequent nucleolytic removal of ITS2, at much later steps of LSU assembly in the nucleoplasm, creates the 3' end of mature 5.8S rRNA and the 5' end of mature 25S rRNA, whose proximal sequences base-pair with each other to form helix 10 (Figure 4B).

While the proximal stem rRNA structure is formed early, all four proximal stem RPs are visualized in pre-ribosomes only after assembly of rRNA domains I and II, upon conversion of the Noc1-Noc2 particle to the State A intermediate (Figure 4C). Their incorporation during this transition is enabled by a molecular switch (Figure 4D) [26]. In the Noc1-Noc2 particle, Noc1 encircles helix 2. Upon release of Noc1, L17 binds to helix 2. Likewise, the exit of Noc2 bound to helix 25 enables the incorporation of L26 as well as AFs Nsa1 and Rpf1.

Although ribosomal protein L26 is evolutionarily conserved, it is one of 11 yeast RPs in the LSU that is not essential for cell growth [37]. Deletion of both *RPL26* genes has only mild effects on pre-rRNA processing, the composition of pre-ribosomes, or the transit of pre-ribosomes from the nucleus to the cytoplasm [56]. In contrast, L17, L35, and L37 are essential RPs. Their depletion strongly blocks processing of 27SB pre-rRNA, leading to turnover of the rRNAs, although more slowly than occurs for $27SA_2$ and $27SA_3$ pre-rRNA when domain I or II RPs are depleted [32,46,57,58].

The proximal stem RPs do not appear to depend on each other to stably associate with pre-ribosomes. Depletion of any one of them does not strongly affect the assembly of the other three. In contrast, downstream assembly of other RPs and AFs is diminished, most likely as a result of the turnover of the abortive earlier intermediates. Again, as with domain I and II RP depletions, the assembly of L19 and L31 is consistently affected significantly more than that of the other six RPs that contact domain III (see discussion below in Section 4.5).

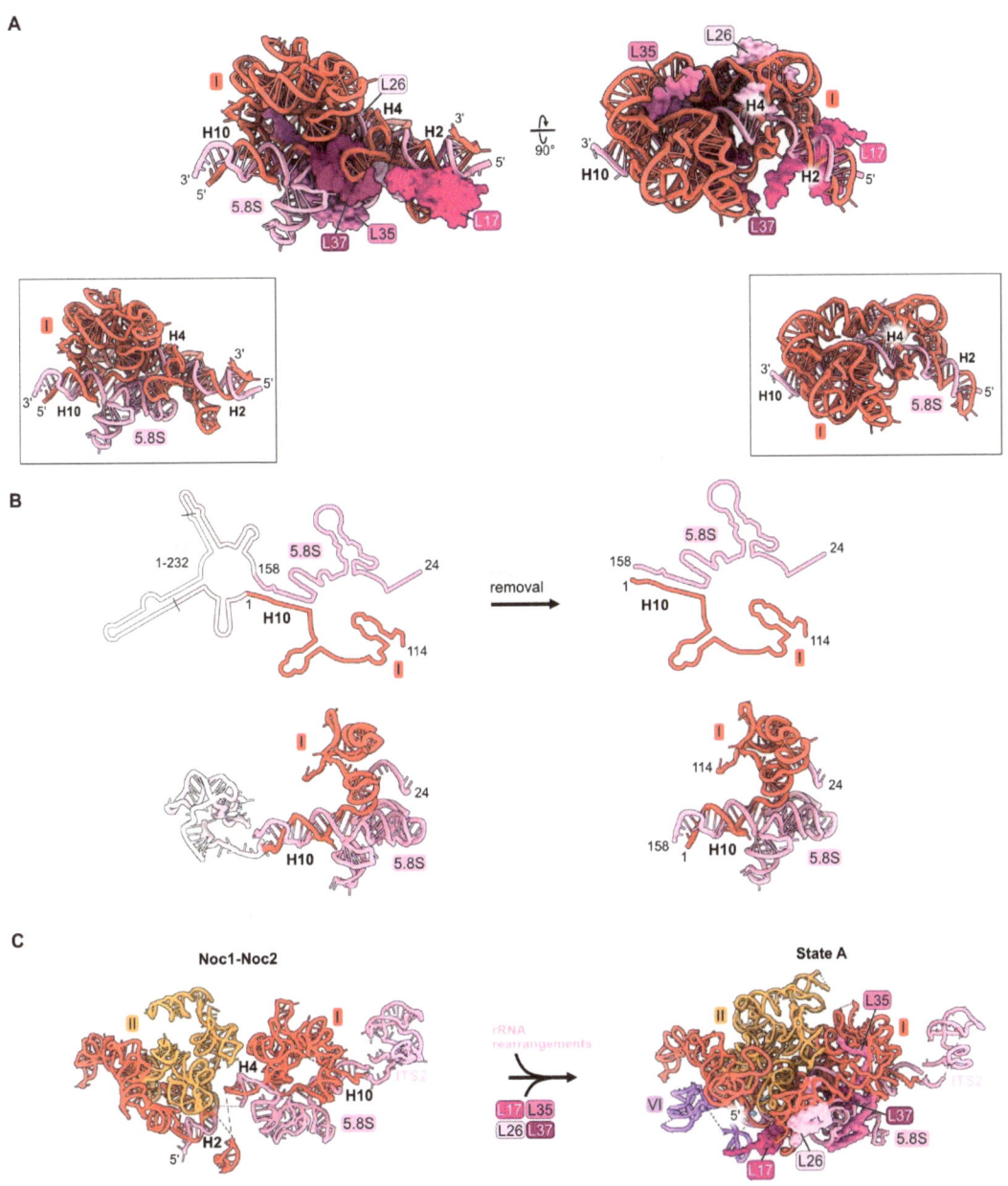

Figure 4. *Cont.*

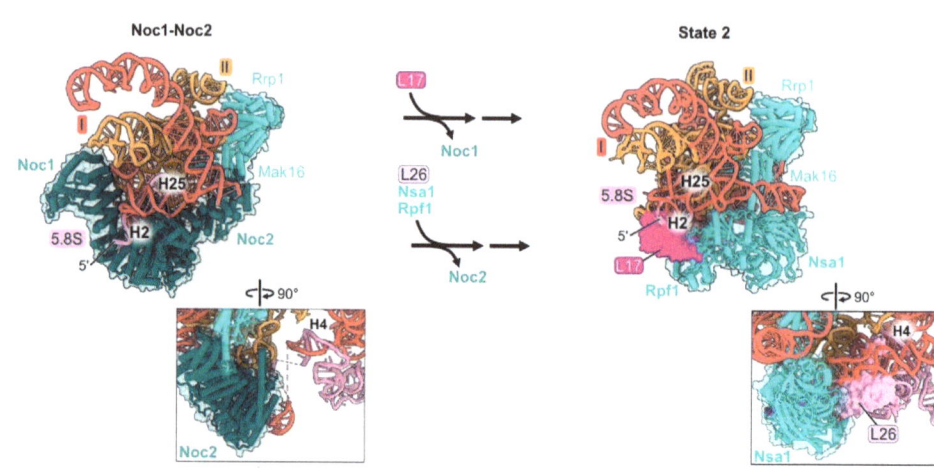

Figure 4. Proximal stem (RPs L17, L26, L35, and L37): (**A**) The structure of the mature proximal stem RNP (PDB: 4V88); (**B**) processing of the ITS2 spacer rRNA to create the mature proximal stem (PDB: 6C0F, 4V88); (**C**) Maturation of the proximal stem during the transition from the Noc1-Noc2 particle to the State A intermediate (PDB: 8E5T, 6EM3) [18,26]; (**D**) a molecular switch resulting in the replacement of AF Noc1 with RP L17 and AF Noc2 with L26, Nsa1, and Rpf1 during the transition from Noc1-Noc2 to State 2 (PDB: 8E5T, 6C0F).

4.4. rRNA Domain VI: L3, L9, L23

rRNA domain VI (nucleotides 2995–3396), (Figures S1A, 1E and 5A) is stably accommodated into pre-ribosomes in a stepwise fashion during the early stages of large subunit maturation [18,20]. Approximately 22 percent (88/401 nucleotides) of the rRNA in domain VI becomes visible in State A particles, while the remainder of domain VI can first be seen by cryo-EM in State 2 particles (Figures S3 and 5A). RPs L3 and L23 primarily contact domain VI rRNA and are first visible in State 2 particles (Figures 1E and 5A). L9 interacts with rRNA in both domains IV and V, as well as domain VI, and is first visible in State C assembly intermediates (Figures S3, 1E and 5A).

Pulse-chase and steady-state assays of L3-depleted cells reveal only a minor effect on amounts of 27SA$_2$ pre-rRNA compared to wild-type cells. In contrast, levels of 27SB pre-rRNA are strongly reduced [32,46,59]. Consistent with this early block in subunit maturation, the amounts of RPs and AFs recovered in affinity-purified pre-ribosomes are decreased in all but the earliest assembly intermediates. These strong effects upon L3 depletion might result from the fact that L3 functions early in assembly, together with the Npa1 complex of AFs, to bind to and consolidate rRNA from multiple domains [60]. In mature subunits, L3 is proximal to the clustered root helices of all six rRNA domains, as well as the 5′ end of 5.8S rRNA and the 3′ end of 25S rRNA (Figure 5B).

Depletion of either L9 or L23 blocks a later step in subunit maturation than when L3 is depleted and results in slower turnover [32,46]. However, L23 and L9 are important for different steps of assembly. Upon depletion of L23, the formation of State 2 particles is blocked, resulting in decreased amounts of AFs Tif6, Rlp24, and Nog1 that first enter these State 2 particles (Figure S3) and in decreased amounts of proteins that associate with downstream intermediates. Tif6 binds directly to L23, while Rlp24 and Nog1 are proximal to L23 but do not contact it directly. Without proper incorporation of L23, it is possible that local RNA structure is perturbed, impacting binding of Rlp24 and Nog1 (Figure 5C). Interestingly, during the remodeling of helices 90–92 in the PTC by the RNA helicase Dbp10, L23 stabilizes both the initial alternate helix 92 as well as the subsequent mature helix 92 (Figure 5D) [23]. In both structures, this function may be enabled by the interaction of L23 with nearby domain VI rRNA.

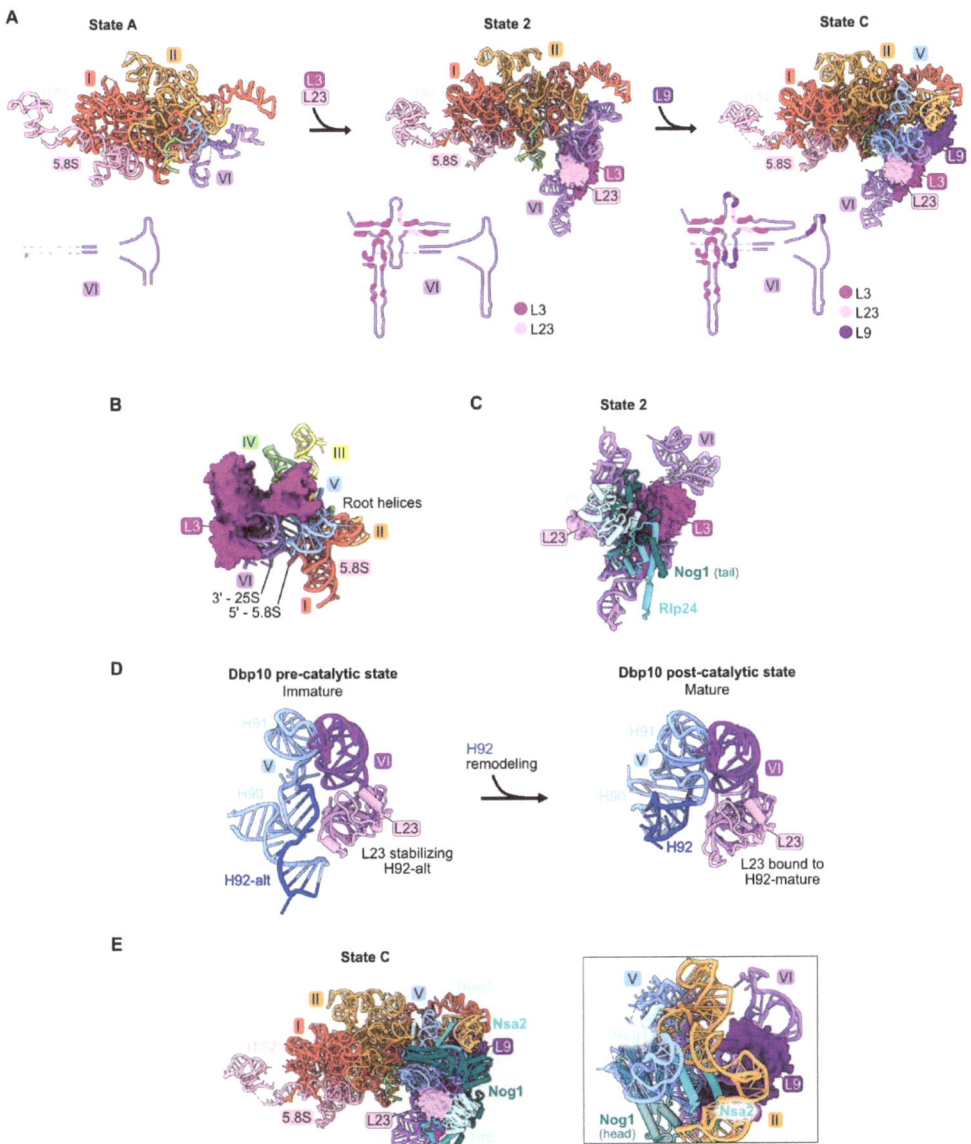

Figure 5. Domain VI (RPs L3, L9, and L23): (**A**) The sequential assembly of domain VI from State A to State C intermediates (PDB: 6EM3, 6C0F, 6EM1). Bottom panel: secondary structure diagrams indicating the successive stabilization of domain VI rRNA, including contacts with the domain VI RPs; (**B**) L3 contacts all six root helices, and is proximal to the 5′ end of 5.8S rRNA and the 3′ end of 25S rRNA in the mature LSU (PDB: 4V88); (**C**) the assembly of AFs Nog1, Rlp24, and Tif6 proximal to L23 is most affected by depletion of L23 (PDB: 6C0F); (**D**) the assembly of AFs Nug1, Nsa2, and Dbp10 adjacent to L9 is most affected by its depletion (PDB: 6EM1). Note that Dbp10 is invisible in these particles; (**E**) L23 stabilizes both conformations of H92 (H92-alt, then H92 in its mature form) (PDB: 8V83, 8V87) [23].

Depletion of L9 blocks the stable construction of State C particles. This is evident by the decreased amounts of AFs Nug1, Nsa2, and Dbp10 that enter pre-ribosomes during the transition of State 2 to State C (Figure 5E). Although the head domain of Nog1 lies near L9, assembly of this factor is not affected by L9 depletion, likely because the middle domain of Nog1 is anchored onto particles before entry of L9 [18].

4.5. rRNA Domain III: L19, L22, L25, L27, L30, L31, L34, L38

The nine RPs L19, L22, L25, L27, L30, L31, L34, L38, and L39 primarily contact rRNA domain III in mature 60S subunits (nucleotides 1456–1877, Figure S1, 1F and 6A). Because L39 seems to have completed its assembly and function after all of the other domain III RPs, it will be discussed in a later section (see below Section 4.8). L22, L25, L27, L30, and L38 bind to the exterior surface of domain III, while L19, L31, and L34 are embedded within domain III rRNA. The long C-terminal alpha-helix of L19 extends outward from the ribosome to form an inter-subunit bridge with 40S subunits during translation [47,61]. L31 bridges the root helix of domain III with domain VI rRNA, while the N-terminal extension of L25 threads from domain III into domain I. Deleting 61 amino acids from the N-terminus of L25 does not impact its ability to recognize its binding site in a fragment of 25S rRNA in vitro, suggesting that the C-terminal globular domain could be responsible for initial assembly onto the pre-ribosome [62].

Domain III rRNA is remodeled from an initially flexible state to stably compact onto the body of State D pre-ribosomes, where it is first visible (Figure S3) [18]. It is not yet completely clear precisely when RPs assemble with domain III rRNA, because this rRNA and any proteins bound to it are not visible by cryo-EM until State D particles (but see below).

Interestingly, of the 11 nonessential RPs in the LSU, L22, L30, L38, and L39 are bound to domain III rRNA. The *rpl22* knockout strain grows slowly at 30 °C and is unable to grow at low temperatures [37]. Depletion of the essential RPs L19, L25, L27, L31, or L34 has a strong pre-rRNA processing defect with the accumulation of 27SB pre-rRNA followed by its slow turnover [32,33,46,61,63].

Effects on pre-ribosome maturation have been examined upon depletion of L19 and L25, but not yet for other domain III RPs [33,63]. When either of these RPs is depleted, most of the proteins that assemble after the compaction of domain III fail to associate with pre-ribosomes. However, incorporation of many of the RPs that bind to domain III rRNA is not significantly affected, except for L19, L31, and L39. Similarly, assembly of L19, L31, and L39, but not that of the other domain III RPs, is also strongly diminished when maturation of domains I, II, VI, or the proximal stem is blocked during earlier stages of subunit maturation, before domain III is stably compacted and visible [34,46]. Taken together, these results suggest that L22, L25, L27, L30, L34, and L38 might associate with flexible domain III rRNA before it undergoes remodeling to compact into/onto the pre-ribosome. Their assembly may be independent of that of domains I, II, and VI during the early stages of LSU assembly and also might not depend on the completion of domain III construction at later stages of subunit maturation. Because assembly of L19 and L31 is coincident with compaction of domain III, and both proteins are juxtaposed with the root helices of domain III (helices 47, 48, and 60) (Figure 6A), it is tempting to speculate that compaction of domain III is driven by remodeling of its root helices coupled with insertion of L19 and L31. Alternatively, the formation of the mature root helix for domain III may simply create binding sites for L19 and L31.

Interestingly, amounts of RPs bound to the proximal stem decrease when domain III RPs L19 and L25 are depleted, even though this neighborhood assembles upstream of domain III compaction (Figure S2C). This apparent anomaly might be explained by the fact that the proximal stem RPs L17, L26, L35, and L37, together with domain III RPs L19, L25, and L31, are components of the exit platform of the polypeptide exit tunnel (Figure 6B). We speculate that failure to complete the tunnel exit platform by incorporation of L19, L25, and L31, and/or failure to compact domain III rRNP to complete the wall of the tunnel, might destabilize the platform and cause the RPs that had previously joined the platform to disassemble.

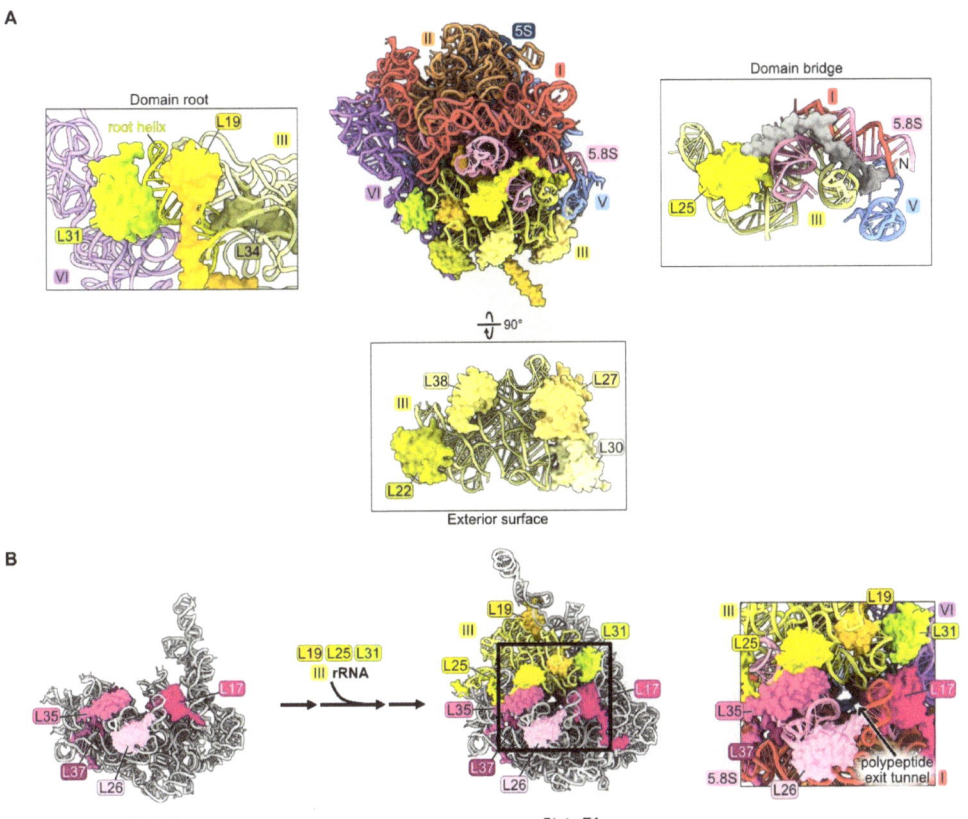

Figure 6. Domain III (RPs L19, L22, L25, L27, L30, L31, L34, and L38): (**A**) The locations of domain III RPs in mature LSU based on structural features. L19, L31, and L34 are shown embedded on the domain III root helix. L22, L27, L30, L38 are bound to the exterior surface of the domain III rRNA. The L25 globular domain (yellow) binds domain III rRNA, and the N-terminal extension of L25 containing amino acids 1–61 (grey) contacts neighboring rRNA domains (PDB: 4V88); (**B**) the completion of exit platform construction by assembly of RPs L19, L25, and L31 (PDB: 6C0F, 7NAC) [20,21].

4.6. rRNA Domain IV: L2, L43

Ribosomal proteins L2 and L43 bind to each other, sandwiched between rRNA domains II, III, IV, and V in mature ribosomes (Figures 1G and 7A). Consistent with the extensive contact between L2 and L43, their assembly is mutually interdependent. These two RPs are first visible by cryo-EM in pre-ribosome State NE1, coincident with the transition from State E2 to NE1 (Figure 7B, left and center panels) [21,28]. The previously flexible portion of rRNA domain IV can now be visualized compacted onto rRNA domains II and III, stabilized by contacts of L2 and L43 with all three of these domains (Figure 7B, center panel). In the subsequent State NE2, the L1 stalk of rRNA domain V now becomes visible, compacted onto the body of the particle to contact L2 and L43 (Figure 7B, right panel).

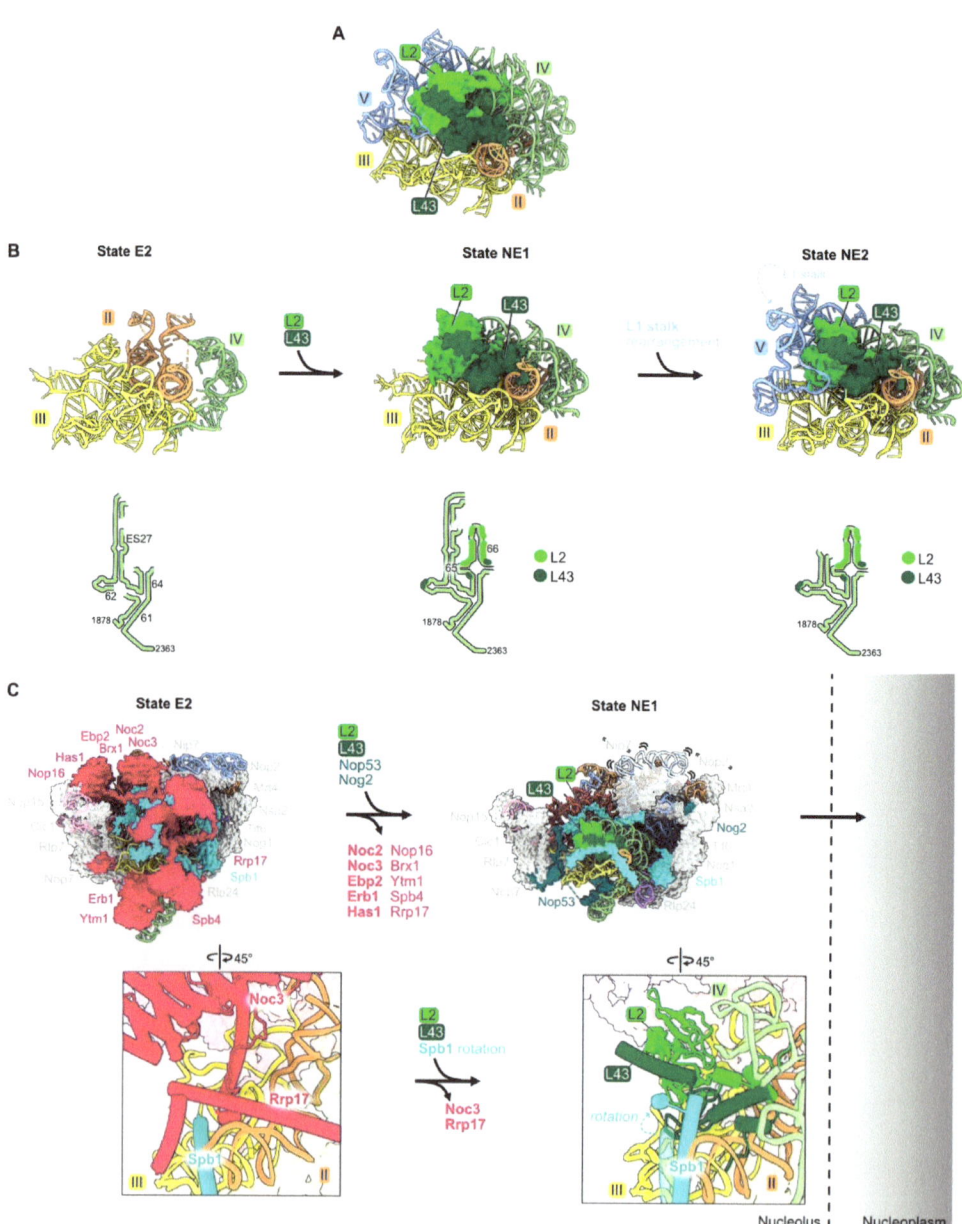

Figure 7. Domain IV (RPs L2 and L43): (**A**) L2 and L43 are positioned between rRNA domains II, III, IV, and V in the mature LSU (PDB: 4V88); (**B**) the entry of L2 and L43 enables compaction of rRNA domain IV and stabilizes the rotation of the L1 stalk (PDB: 7NAC, 7U0H, 6YLY) [21,28]. The bottom panel depicts changes in the stabilization of domain IV rRNA during this interval; (**C**) significant particle remodeling occurs during the entry of L2 and L43 just prior to the exit of pre-60S particles from the nucleolus to the nucleoplasm. AFs that contain intrinsically disordered domains and that exit at this interval are shown in bold. Note that Nip7, Nop2, and the L1 stalk are invisible in State NE1. (PDB: 7NAC, 7U0H).

This cascade of rRNA compaction is triggered by the release of ten AFs from the E2 particle by the Rea1 AAA-ATPase (Figure 7C) [21,64]. This remodeling includes yet another molecular switch. When Noc3 and Rrp17 are released from the E2 particle, L2 and L43 can then bind to the same sites in the NE1 intermediate. The subsequent rotation of an Spb1 alpha-helix located above L43 appears to "lock down" L43 onto the particle, driving assembly forward by preventing re-entry of Rrp17 (Figure 7C).

Both L2 and L43 co-immunoprecipitate significant amounts of 7S pre-rRNA and small amounts of 27SB pre-rRNA, suggesting that they assemble just before cleavage and removal of the ITS2 spacer RNA in 27SB pre-rRNA to produce 7S pre-rRNA [33]. Consistent with this timing of their entry into pre-ribosomes, depletion of L2 or L43 blocks this cleavage event, resulting in the accumulation of pre-ribosomes containing the three AFs that bind to ITS2 (Nop15, Rlp7, and Cic1) [32,65].

Importantly, L2 and L43 are the last RPs to associate with pre-60S particles before they transit from the nucleolus into the nucleoplasm. Assembly of these RPs was hypothesized to facilitate this exit of pre-ribosomes from the nucleolus (Figure 7C) [66]. Installation of L2 and L43 is coincident with the release of AFs Noc3, Ebp2, Erb1, and Nop16 that contain intrinsically disordered regions (IDRs), and with the compaction of flexible rRNA domain IV. The IDRs in these AFs and this flexible rRNA are hypothesized to form an interaction network between pre-ribosomes to create the nucleolar condensate [66]. Thus, these remodeling events, including the entry of L2 and L43, could decrease the potential of pre-ribosomes to interact with each other and therefore favor the release of pre-ribosomes from the nucleolar condensate. Indeed, depletion of L2 or L43 prevents the nucleolar exit of pre-ribosomes, whereas blocking the next step, maturation of NE2 particles to Nog2 particles, does not [65].

4.7. rRNA Domain V: L1, L21, L28

Both L21 and L28 primarily contact rRNA domains II and V on the top surface of the mature large subunit, directly underneath the central protuberance (Figures 1H and 8A). L21 spans the solvent-exposed and inter-subunit surfaces, while L28 binds to the solvent-exposed surface. A helical portion of L21 is initially resolved in State D. Then, in State E1, the C-terminal domain of L21 is visible contacting domain II RPs L7 and L20 (Figure 8B) [21]. The C-terminal globular portion of L28 is first seen in State NE1, whereas the N-terminal domains of both L21 and L28 are not visible until the cytoplasmic stages of maturation (Figure 8C) [19]. The trigger for entry of L28 into pre-ribosomes is another example of a molecular switch, where the AF Brx1 is removed by the AAA-ATPase Rea1 to free up the binding site for the C-terminal portion of L28. The C-terminal domain of RP L13 then clamps down on L28 in state NE1 to stabilize its binding (Figure 8C).

Depletion of L21 and L28 appears to primarily affect steps downstream in the nucleoplasm. Small amounts of 27SB pre-rRNA and much larger amounts of 7S pre-rRNA accumulate [32,33,46]. In addition, depletion of L21 results in modest accumulation of 25.5S and 6S pre-rRNAs. Levels of AFs released from pre-ribosomes in the nucleoplasm moderately increase in L21-depleted cells. The release of AFs Rsa4, Nog2, Rpf2, and Rrs1 that are proximal to L21 appears to be slightly delayed, suggesting a partial effect on the remodeling of domain V, including the rotation of the 5S RNP that sits above L21 (Figure 10). Upon depletion of L28, the binding of nuclear export factors Bud20 and Nmd3 to pre-ribosomes is slightly diminished, consistent with their entry after L28. L10, L24, L29, and L40 are the only RPs markedly affected by the absence of L21 and L28 (Figure 8D) [33,46]. Of these, L10, L29, and L40 lie proximal to L21 and L28. L10 and L24 assemble in the cytoplasm, and therefore their assembly may be precluded due to a block in nuclear export.

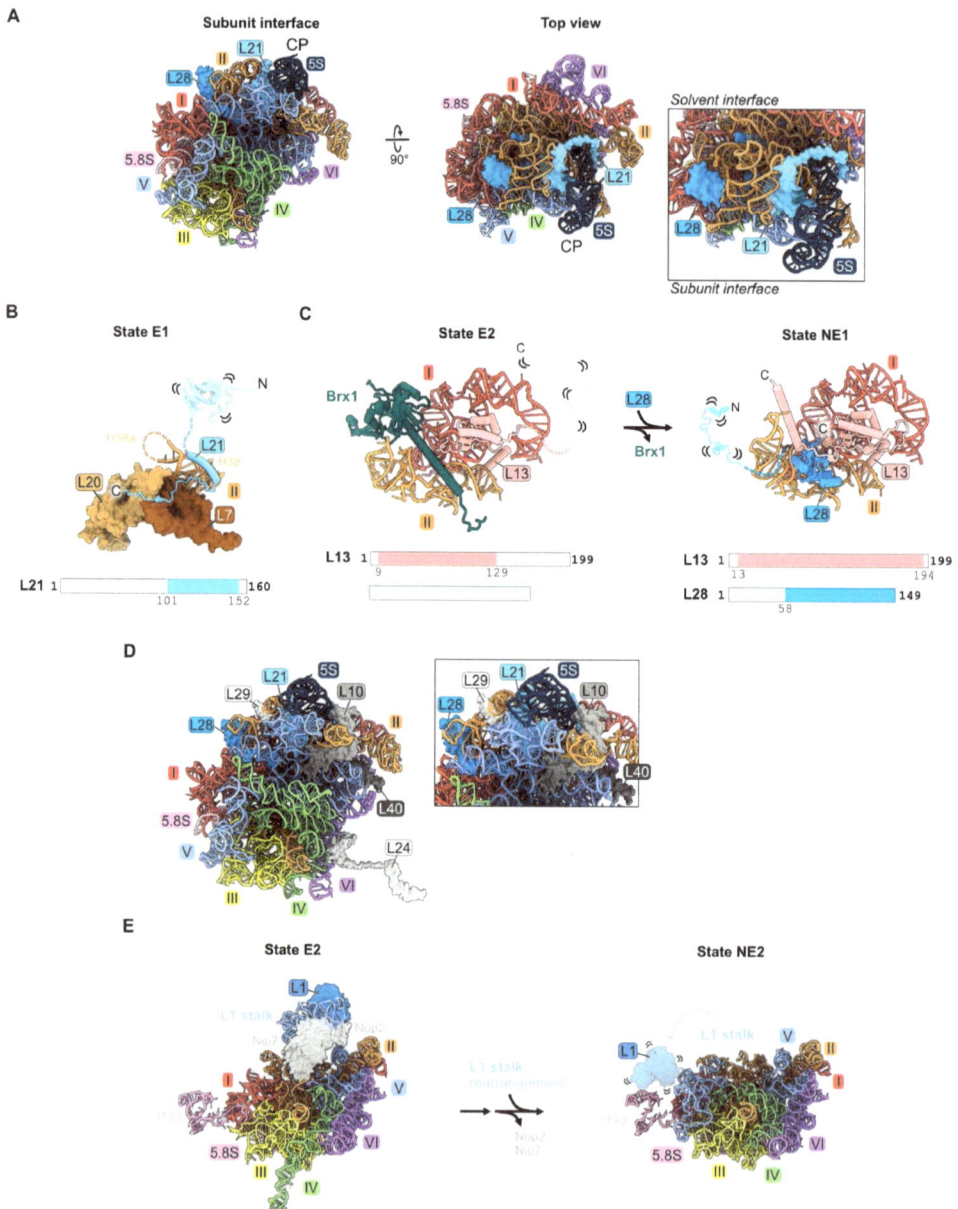

Figure 8. Domain V (RPs L1, L21, and L28): (**A**) The location of L21 and L28 in domain V beneath the central protuberance (CP) in the mature LSU. Note that L1 is not resolved in this structure (PDB: 4V88); (**B**) the C-terminal domain of L21 is stabilized on domain II RPs in State E1 (PDB: 7R7A); (**C**) a molecular switch: the exit of Brx1 enables the entry of the C-terminal globular domain of L28 (7NAC, 7U0H); (**D**) the assembly of L10, L24, L29, and L40 during nucleoplasmic and cytoplasmic stages of LSU construction is most affected by the absence of L21 or L28 (PDB: 4V88); (**E**) the rotation of the L1 stalk during the transition of the state E1 intermediate to state NE2. Note that L1 and H77-78 are not resolved following L1 stalk rotation in the nucleus. L1 and H77-78 are overlayed onto the NE2 structure (PDB: 7NAC, 6YLY).

Ribosomal protein L1 binds the loop formed by helix 77 in rRNA domain V to form the L1 stalk, located adjacent to the E site in mature 60S subunits. There, L1 participates in translation elongation by facilitating the release of E site tRNAs and translation factors [67]. L1 is first visualized in state E1/E2 particles prior to the rotation of the L1 stalk (Figure 8E) [21]. In the NE2 particle, the L1 stalk swivels to its mature position, yet L1 is not visible in this conformation (Figure 8E).

Depletion of L1 results in a modest increase in 7S pre-rRNA [32]. Consistent with this late phenotype, L1 co-immunopurifies with more 7S pre-rRNA than 27S pre-rRNA.

A portion of the essential ribosome export factor Nmd3 binds to L1 and holds the L1 stalk in a closed conformation in late nuclear assembly intermediates [19,68]. Restraining this hydrophilic stalk onto the pre-ribosome body might be important for particles to efficiently travel through the constrained, hydrophobic environment of the nuclear pore complex. Nmd3 as well as the export factors Bud20 and Arx1 are efficiently recruited to pre-ribosomes lacking L1 [69], but the export factor Mex67/Mtr2 does not assemble. Exactly how L1 affects the assembly of this factor remains unclear. Nevertheless, the absence of L1 or truncation of the rRNA that binds to L1 results in inefficient nuclear export of pre-60S particles. Some fraction of mutant particles enter the cytoplasm and associate with polysomes. Yet, their enrichment with relatively smaller polyribosomes suggests a defect in translation elongation, consistent with the role of L1 in this process. The effect of depleting L1 on the assembly of other RPS or AFs has not been reported.

4.8. The Polypeptide Exit Tunnel: L39

Ribosomal protein L39 is one of three RPs situated in the nascent polypeptide exit tunnel (NPET) of the 60S ribosomal subunit (Figures 1F and 9A). L39 is embedded within helices 49–51 of domain III rRNA in the wall of the tunnel, near the tunnel exit site (Figure 9A). In eubacteria, the C-terminal extension of RP uL23 (a homologue of eukaryotic RP L25) is present adjacent to the NPET, but it is replaced by L39 in eukaryotes. Internal loops of two other RPs, L4 and L17, extend into the tunnel proximal to the PTC to form the so-called "constriction site". L39 reduces the diameter of the NPET right at the exit of the tunnel, which could affect protein synthesis (Figure 9B). Indeed, L39 is important for the folding of nascent polypeptides to minimize their aggregation [38].

L39 is first visualized fully accommodated into the NPET of pre-ribosomes in Nog2 particles at the late nuclear stages of subunit maturation (Figure 9C) [70]. However, L39 appears to initially begin loading onto pre-ribosomes at State C or earlier. L39 copurifies with pre-ribosomes containing the AF Ssf1 that exits before State D particles, and partial densities of L39 are present in cryo-EM structures of E and NE1 particles [38]. The inability to visualize all of L39 in pre-ribosomes at these early stages of subunit assembly may result from its incomplete accommodation into the particle, leading to a flexible conformation. Thus, L39 may begin to assemble during or soon after the compaction of domain III, but it is not completely accommodated into pre-ribosomes until later.

While L39 is not essential under normal growth conditions, *rpl39Δ* mutants grow very slowly in the cold and are defective in 60S subunit assembly [38]. In the absence of L39, small amounts of 27SB pre-rRNA and much larger amounts of 7S pre-rRNA accumulate, then both undergo slow turnover. This effect on processing of both 27SB and 7S pre-rRNAs is observed in other mutants blocked during late nucleolar stages, e.g., upon depletion of L21, L28, Nog2, or Rsa4 [32,71,72].

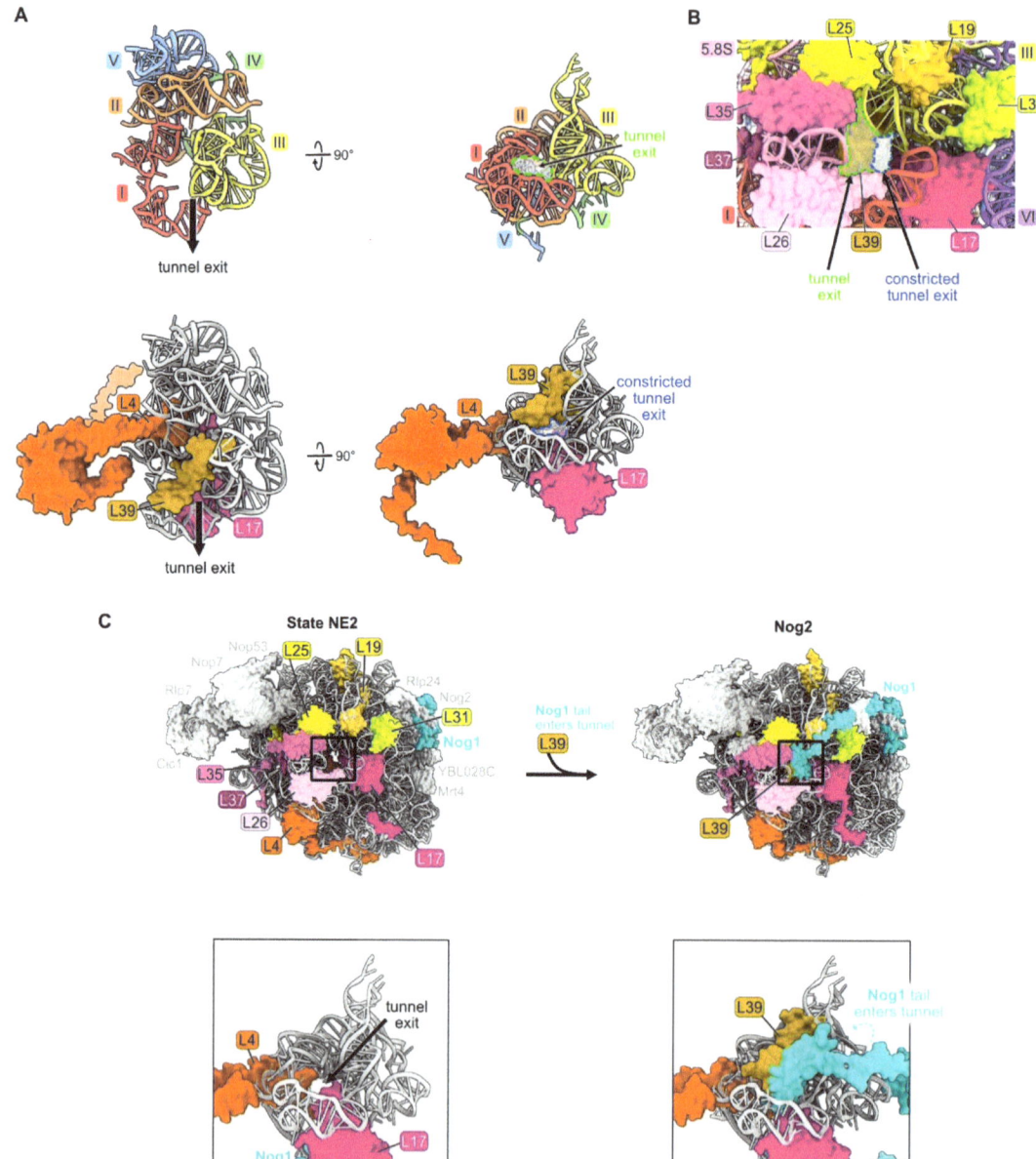

Figure 9. The polypeptide exit tunnel (RP L39): (**A**) The polypeptide exit tunnel is comprised of portions of rRNA domains I, II, III, IV, and V. Segments of L4 and L17 are inserted into the tunnel and L39 is embedded near the tunnel exit (PDB: 4V88); (**B**) the presence of L39 reduces the diameter of the tunnel exit. The space outlined in green highlights the size of the tunnel exit in the absence of L39. In contrast, the space outlined in blue indicates the reduced dimensions of the tunnel exit when L39 is present (PDB: 4V88); (**C**) particles before and after L39 entry, including RPs surrounding the tunnel exit. The C-terminal tail of Nog1 enters the tunnel coincident with L39 in the Nog2 assembly intermediate (PDB: 6YLY, 3JCT) [28,70].

Despite the fact that L39 binds to domain III rRNA, the absence of L39 does not appear to impact the compaction of domain III, likely because the protein does not stably associate with pre-ribosomes until after this compaction has occurred (Figure 9C) [70]. Consistent with this idea, the stable installation of L39 is necessary for later nucleoplasmic steps of subunit assembly. In *rpl39Δ* knockout strains, AFs that are required for rotation of the 5S RNP and for nuclear export of pre-60S particles assemble much less efficiently. Interestingly, this same phenotype is observed in *rpl4Δ63-87* mutants where the RP L4 tunnel constriction site is deleted. L39 fails to assemble in this *rpl4Δ63-87* mutant, but assembly of L4 is not perturbed in the *rpl39Δ* mutant [38,73]. Thus, this failure of L39 assembly may account for the *rpl4* mutant phenotype.

4.9. Central Protuberance: L5, L11

RPs L5 and L11 bind to 5S rRNA to form the central protuberance (CP). This conserved structural feature of LSUs includes helices 80 and 82–88 of rRNA domain V situated on top of mature large ribosomal subunits, just above the peptidyl-transferase center (Figures S1B,D and 1I).

As stated earlier, 5S rRNA is transcribed separately from the 35S precursor of the mature 18S, 5.8S, and 25S rRNAs. The 5S rRNA is assembled together with RPs L5 and L11 onto pre-ribosomes at an early, nucleolar stage of subunit maturation in particles containing 27SA$_2$ pre-rRNA [7]. First, the dedicated chaperone Syo1 binds L5 and L11 in the cytoplasm and facilitates their import into the nucleus [74]. L5 and L11 join 5S rRNA and AFs Rpf2 and Rrs1 to form the 5S rRNP, which incorporates together into pre-ribosomes [7,8]. These four proteins and 5S rRNA, as well as helices 80 and 82–88 of domain V, are present but not visible by cryo-EM in the early assembly intermediates. Thus, the nascent CP, including portions of domain V, is initially flexible (Figure 10). Surprisingly, when the CP is first visible in Nog2 particles by cryo-EM, it is rotated almost 180 degrees backwards from its mature conformation, linked to domain V by Rpf2 and Rrs1. Release of Rpf2 and Rrs1 then destabilizes the pre-rotation state, helices 80 and 82–88 undergo remodeling, and additional AFs, including the AAA-ATPase Rea1, join the particle to stabilize the rotated state of the CP [75,76]. Subsequent ATP hydrolysis by Rea1 activates the release of Nog2 and other AFs to enable binding of the nuclear export adaptor Nmd3 and trigger the export of pre-60S subunits from the nucleus to the cytoplasm.

Depletion of L5, L11, Rpf2, or Rrs1 prevents association of the 5S rRNP with pre-ribosomes [7]. Specifically, the C-terminal extensions of Rpf2 and Rrs1 are required for initial docking by tethering the RNP to early assembly intermediates [8]. Interestingly, preventing entry of the 5S RNP into pre-ribosomes does not block early steps of large subunit maturation, probably because the tethered, flexible 5S RNP does not physically impact other neighborhoods of the pre-60S particles [7]. Likewise, in mutants where the assembly of other rRNA domains is compromised during the early stages of subunit maturation, the 5S rRNP can still assemble into the mutant particles, before they are turned over [46]. Thus, the initial docking of the 5S RNP is independent of the assembly of other domains during the early stages of subunit maturation. The presence of the 5S RNP in pre-ribosomes is only important at later steps, when it is stably accommodated into Nog2 particles together with rRNA domain V (Figure 10). When L5, L11, Rpf2, or Rrs1 are depleted, 27SB and 7S pre-rRNAs accumulate, and the AF Nog2 fails to efficiently assemble.

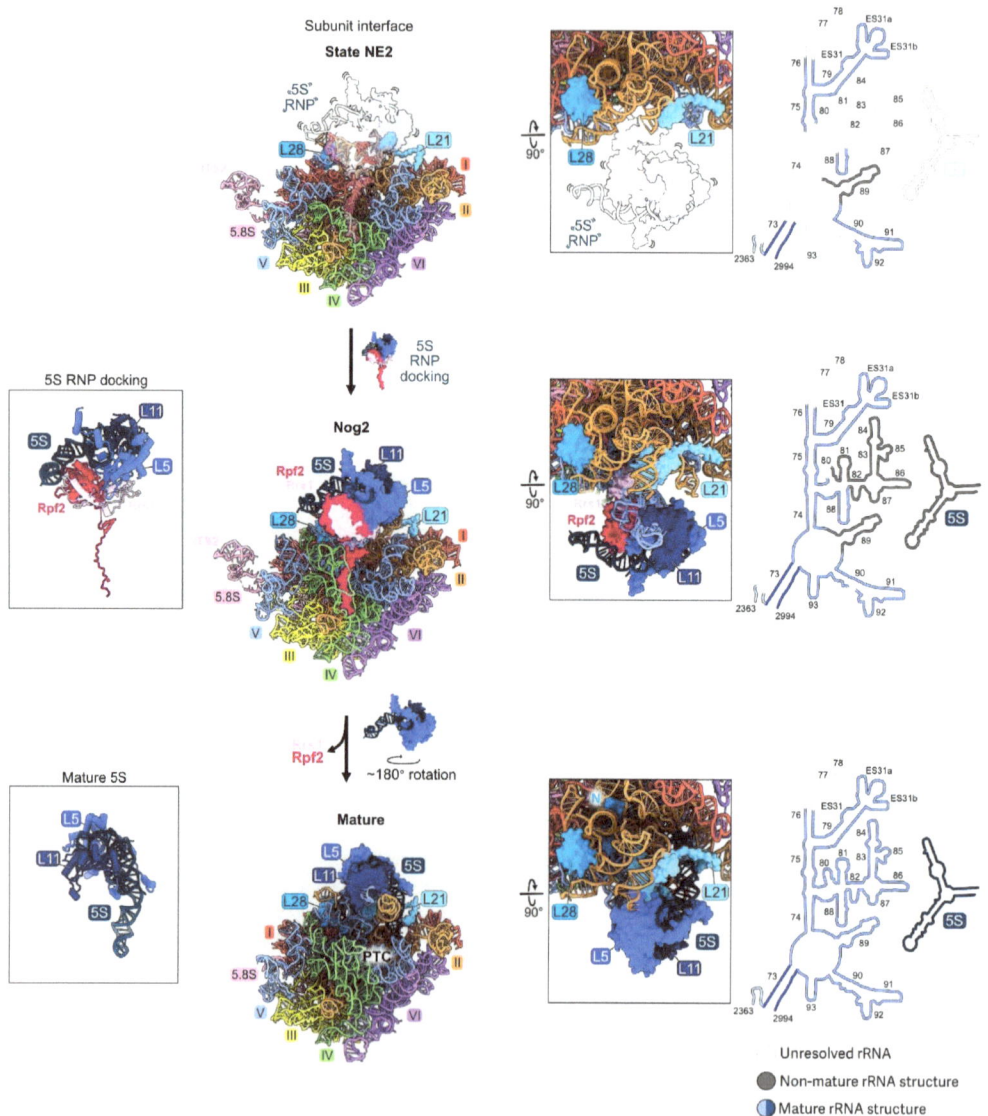

Figure 10. The central protuberance (RPs L5, L11): RPs L5 and L11 bind to the separately transcribed 5S rRNA and dock onto domain V on top of the LSU to form the central protuberance. The 5S RNP assembles with pre-ribosomes early, but is not visible by cryo-EM until the transition from state NE2 (top) to the Nog2 particle (middle). The 5S RNP undergoes ~180° rotation to form the mature structure (bottom) (PDB: 6YLY, 3JCT, 4V88) [28,47,70]. Abbreviation: peptidyltransferase center, PTC. Adapted from [28].

5. Discussion: Lessons Learned from RPs

5.1. An Ordered Pathway?

Unexpectedly, mapping the molecular phenotypes of RP depletions onto the crystal structure of yeast 60S subunits revealed that the 5′ and 3′ domains of rRNA (domains I, II, and VI) are assembled with RPs before middle domains III, IV, and V assemble into stable neighborhoods [46]. This was later confirmed by cryo-EM [18]. This pathway makes sense from the perspective of the 60S subunit structure: the solvent-exposed surface is built prior to the inter-subunit interface that contains the functional centers. However, the extent of alternate assembly pathways, as observed for bacterial ribosome subunit reconstitution in vitro [77], is not yet clear.

5.2. Individual Neighborhoods May Assemble in Discrete Stages

Portions of rRNA domains IV, V, or VI, including the proteins bound to them, are visualized by cryo-EM in a stepwise fashion (Figures 5A, 7B and 10). Similarly, domain III appears to assemble in several different steps. Even though domain III is largely invisible to cryo-EM in early stages, a subset of domain III RPs might bind to flexible domain III before the stable installation of the domain onto the body of the pre-ribosome. The subsequent compaction of domain III might be driven in part by the binding of the remaining RPs to the root helix of domain III (Figure 6).

Functional centers also assemble in multiple steps. For example, the exit platform for the polypeptide exit tunnel is constructed in two separate stages. L17, L26, L35, and L37 bind to the proximal stem portion of the platform several steps before L19, L25, and L31 bind to the section comprised of a portion of domain III (Figures 6B and S2C).

5.3. RPs Enable rRNA Folding

Extended domains of RPs may function as bridges between rRNA domains. Globular domains of some RPs bind to one rRNA domain, while an N- or C-terminal extension of the same RP threads into an adjacent domain (e.g., L8, L16, and L25) (Figures 2D, 3C and 6A). Bridges between rRNA domains are also mediated by different portions of the globular domains of RPs (L2 and L43 with domains III, IV, and V, or L31 with domains III and VI) (Figures 6A and 7A). RPs might also function as clamps to stabilize the binding of another RP, e.g., L13 with L28, and thus to drive assembly forward (Figure 8C).

RPs may not establish contact with all of their rRNA ligands in one step. In several cases, portions of an RP are not visible in the cryo-EM structure of assembly intermediates (e.g., L39). Yet, RPs can be detected by co-immunoprecipitation assays, although weakly associated with the rRNA. Such an RP may initially form "encounter complexes", in which it searches multiple RNA conformations before stably binding its substrate [44]. Thus, depletion of such an RP can affect a stage of assembly before it is visible by cryo-EM.

5.4. Molecular Switches

The entry of an RP may be dictated by the release of an AF that occupies the same binding site, e.g., Noc1 and L17, Noc2 and L26, or Rrp17 and Noc3 with L2 and L43 (Figures 4D and 7C).

6. Going Forward

Despite the recent near-atomic structures of assembly intermediates, we have only pre-ribosome snapshots, not movies. We lack a detailed understanding of the dynamics of ribosome assembly. Better approaches are needed to interrogate rRNA structure and rRNA folding. Exactly how does each RP establish all of its contacts with rRNA? How do the structures of the proteins and rRNAs change and affect each other during these encounters? What roles do RPs play in the compaction of flexible domains of rRNA? How are structural changes transmitted to proximal and distal rRNP neighborhoods in the pre-ribosome as assembly progresses? Single-molecule assays of RP binding and rRNA folding in real time, such as those carried out by the Woodson and Williamson labs for the assembly of individual domains of bacterial ribosomes, could shed light on these questions [44,78].

Supplementary Materials: The following supporting information can be downloaded at: https://www.mdpi.com/article/10.3390/biom14080975/s1, Figure S1: (A) Phylogenetically conserved secondary structure of LSU rRNA. The 5S rRNA, 5.8S rRNA, and six domains of 25S rRNA are color coded. The six root helices are boxed and highlighted [1,2]; (B) The tertiary structure of rRNA in the LSU, color coded as in (A); (C) The tertiary structure of the six LSU root helices, color coded as in (A);

(D) The RPs in the mature LSU (color coded as in (A)). The RPs colored in white are not discussed in this review. Abbreviations: proximal stem (PS), central protuberance (CP) (PDB: 4V88) [47]; Figure S2: (A) Cartoon depicting transcription of rRNA and early stages of small and large subunit assembly. Co-transcriptional assembly of some RPs is indicated. In co-transcriptional assembly, endonucleolytic cleavage occurs in the nascent rRNA at site A2 to separate the pathway of assembly for the two subunits (pre-40S, pre-60S). In post-transcriptional assembly, cleavage occurs at the A3 site only after transcription has completed; (B) Pathway of endo- and exonucleolytic processing of pre-rRNA. The co- and post-transcriptional pathways are shown; (C) Pathway of RP assembly into the LSU. The order is based upon when each RP is first visualized by cryo-EM. RPs in each neighborhood are color coded as in Figure S1. RPs shown in grey are not discussed in this review. Note that L41 is visualized in a few fungal LSUs, but otherwise is only observed in the SSU of archaea and eukaryotes, and thus is designated as eS32 in these organisms (M. Leibundgut and N. Ban, personal communication). PDB files for each structure are shown. Abbreviations: proximal stem (PS), central protuberance (CP); Figure S3: Assembly pathway of the LSU showing consecutive nuclear intermediates discussed in this review. This schematic details the association of RPs and the entry and exit of AFs as visualized by cryo-EM. Note that some proteins may be present at other points in assembly but may not be resolved in a structural intermediate. PDB files are indicated for each structure. RPs are color coded according to their neighborhood (Figure S1), whereas all AFs are colored in cyan (first row). Consecutive assembly of RPs (second row). Entry and exit of AFs (third row). Visualization of rRNA tertiary structure as it undergoes stabilization (fourth row). Secondary structure of resolved rRNA. Segments not resolved are shown as transparent (fifth row). Adapted from [18].

Author Contributions: Conceptualization, T.N.A. and J.L.W.; writing (original draft preparation, review, and editing), T.N.A. and J.L.W.; visualization, T.N.A. and J.L.W.; supervision, T.N.A.; funding acquisition, T.N.A. and J.L.W. All authors have read and agreed to the published version of the manuscript.

Funding: This research was funded by the National Institutes of Health Grant 5R01GM028301-43 to J.L.W. and by the National Science Foundation Grant DGE2140739 to T.N.A.

Institutional Review Board Statement: Not applicable.

Informed Consent Statement: Not applicable.

Data Availability Statement: Not applicable.

Acknowledgments: We thank the Woolford Lab, specifically Stefanie Hedayati and Collin Bachert, for their helpful discussion and feedback.

Conflicts of Interest: The authors declare no conflicts of interest.

References

1. Petrov, A.S.; Bernier, C.R.; Hershkovits, E.; Xue, Y.; Waterbury, C.C.; Hsiao, C.; Stepanov, V.G.; Gaucher, E.A.; Grover, M.A.; Harvey, S.C.; et al. Secondary Structure and Domain Architecture of the 23S and 5S rRNAs. *Nucleic Acids Res.* **2013**, *41*, 7522–7535. [CrossRef] [PubMed]
2. Petrov, A.S.; Bernier, C.R.; Gulen, B.; Waterbury, C.C.; Hershkovits, E.; Hsiao, C.; Harvey, S.C.; Hud, N.V.; Fox, G.E.; Wartell, R.M.; et al. Secondary Structures of rRNAs from All Three Domains of Life. *PLoS ONE* **2014**, *9*, e88222. [CrossRef] [PubMed]
3. Baßler, J.; Hurt, E. Eukaryotic Ribosome Assembly. *Annu. Rev. Biochem.* **2019**, *88*, 281–306. [CrossRef] [PubMed]
4. Fernández-Pevida, A.; Kressler, D.; De La Cruz, J. Processing of Preribosomal RNA in Saccharomyces Cerevisiae: Yeast Pre-rRNA Processing. *WIREs RNA* **2015**, *6*, 191–209. [CrossRef] [PubMed]
5. Vanden Broeck, A.; Klinge, S. Eukaryotic Ribosome Assembly. *Annu. Rev. Biochem.* **2024**, *93*, 189–210. [CrossRef] [PubMed]
6. Klinge, S.; Woolford, J.L. Ribosome Assembly Coming into Focus. *Nat. Rev. Mol. Cell Biol.* **2019**, *20*, 116–131. [CrossRef] [PubMed]
7. Zhang, J.; Harnpicharnchai, P.; Jakovljevic, J.; Tang, L.; Guo, Y.; Oeffinger, M.; Rout, M.P.; Hiley, S.L.; Hughes, T.; Woolford, J.L. Assembly Factors Rpf2 and Rrs1 Recruit 5S rRNA and Ribosomal Proteins rpL5 and rpL11 into Nascent Ribosomes. *Genes Dev.* **2007**, *21*, 2580–2592. [CrossRef] [PubMed]
8. Castillo Duque De Estrada, N.M.; Thoms, M.; Flemming, D.; Hammaren, H.M.; Buschauer, R.; Ameismeier, M.; Baßler, J.; Beck, M.; Beckmann, R.; Hurt, E. Structure of Nascent 5S RNPs at the Crossroad between Ribosome Assembly and MDM2–P53 Pathways. *Nat. Struct. Mol. Biol.* **2023**, *30*, 1119–1131. [CrossRef] [PubMed]
9. Lau, B.; Huang, Z.; Kellner, N.; Niu, S.; Berninghausen, O.; Beckmann, R.; Hurt, E.; Cheng, J. Mechanism of 5S RNP Recruitment and Helicase-surveilled rRNA Maturation during PRE-60S Biogenesis. *EMBO Rep.* **2023**, *24*, e56910. [CrossRef]

10. Lo, K.-Y.; Li, Z.; Bussiere, C.; Bresson, S.; Marcotte, E.M.; Johnson, A.W. Defining the Pathway of Cytoplasmic Maturation of the 60S Ribosomal Subunit. *Mol. Cell* **2010**, *39*, 196–208. [CrossRef]
11. Nerurkar, P.; Altvater, M.; Gerhardy, S.; Schütz, S.; Fischer, U.; Weirich, C.; Panse, V.G. Eukaryotic Ribosome Assembly and Nuclear Export. *Int. Rev. Cell Mol. Biol* **2015**, *319*, 107–140.
12. Parker, M.D.; Karbstein, K. Quality Control Ensures Fidelity in Ribosome Assembly and Cellular Health. *J. Cell Biol.* **2023**, *222*, e202209115. [CrossRef] [PubMed]
13. Yang, Y.-M.; Karbstein, K. Ribosome Assembly and Repair. *Annu. Rev. Cell Dev. Biol.* **2024**. [CrossRef] [PubMed]
14. Warner, J.R. The Economics of Ribosome Biosynthesis in Yeast. *Trends Biochem. Sci.* **1999**, *24*, 437–440. [CrossRef]
15. Rodgers, M.L.; Woodson, S.A. A Roadmap for rRNA Folding and Assembly during Transcription. *Trends Biochem. Sci.* **2021**, *46*, 889–901. [CrossRef]
16. Polacek, N.; Mankin, A.S. The Ribosomal Peptidyl Transferase Center: Structure, Function, Evolution, Inhibition. *Crit. Rev. Biochem. Mol. Biol.* **2005**, *40*, 285–311. [CrossRef]
17. Burlacu, E.; Lackmann, F.; Aguilar, L.-C.; Belikov, S.; Nues, R.V.; Trahan, C.; Hector, R.D.; Dominelli-Whiteley, N.; Cockroft, S.L.; Wieslander, L.; et al. High-Throughput RNA Structure Probing Reveals Critical Folding Events during Early 60S Ribosome Assembly in Yeast. *Nat. Commun.* **2017**, *8*, 714. [CrossRef]
18. Kater, L.; Thoms, M.; Barrio-Garcia, C.; Cheng, J.; Ismail, S.; Ahmed, Y.L.; Bange, G.; Kressler, D.; Berninghausen, O.; Sinning, I.; et al. Visualizing the Assembly Pathway of Nucleolar Pre-60S Ribosomes. *Cell* **2017**, *171*, 1599–1610. [CrossRef] [PubMed]
19. Zhou, Y.; Musalgaonkar, S.; Johnson, A.W.; Taylor, D.W. Tightly-Orchestrated Rearrangements Govern Catalytic Center Assembly of the Ribosome. *Nat. Commun.* **2019**, *10*, 958. [CrossRef]
20. Sanghai, Z.A.; Miller, L.; Molloy, K.R.; Barandun, J.; Hunziker, M.; Chaker-Margot, M.; Wang, J.; Chait, B.T.; Klinge, S. Modular Assembly of the Nucleolar Pre-60S Ribosomal Subunit. *Nat.* **2018**, *556*, 126–129. [CrossRef]
21. Cruz, V.E.; Sekulski, K.; Peddada, N.; Sailer, C.; Balasubramanian, S.; Weirich, C.S.; Stengel, F.; Erzberger, J.P. Sequence-Specific Remodeling of a Topologically Complex RNP Substrate by Spb4. *Nat. Struct. Mol. Biol.* **2022**, *29*, 1228–1238. [CrossRef] [PubMed]
22. Aquino, G.R.R.; Hackert, P.; Krogh, N.; Pan, K.-T.; Jaafar, M.; Henras, A.K.; Nielsen, H.; Urlaub, H.; Bohnsack, K.E.; Bohnsack, M.T. The RNA Helicase Dbp7 Promotes Domain V/VI Compaction and Stabilization of Inter-Domain Interactions during Early 60S Assembly. *Nat. Commun.* **2021**, *12*, 6152. [CrossRef]
23. Cruz, V.E.; Weirich, C.S.; Peddada, N.; Erzberger, J.P. The DEAD-Box ATPase Dbp10/DDX54 Initiates Peptidyl Transferase Center Formation during 60S Ribosome Biogenesis. *Nat. Commun.* **2024**, *15*, 3296. [CrossRef] [PubMed]
24. Mitterer, V.; Thoms, M.; Buschauer, R.; Berninghausen, O.; Hurt, E.; Beckmann, R. Concurrent Remodelling of Nucleolar 60S Subunit Precursors by the Rea1 ATPase and Spb4 RNA Helicase. *Elife* **2023**, *12*, e84877. [CrossRef] [PubMed]
25. Mitterer, V.; Hamze, H.; Kunowska, N.; Stelzl, U.; Henras, A.K.; Hurt, E. The RNA Helicase Dbp10 Coordinates Assembly Factor Association with PTC Maturation during Ribosome Biogenesis. *Nucleic Acids Res.* **2024**, *52*, 1975–1987. [CrossRef] [PubMed]
26. Sanghai, Z.A.; Piwowarczyk, R.; Broeck, A.V.; Klinge, S. A Co-Transcriptional Ribosome Assembly Checkpoint Controls Nascent Large Ribosomal Subunit Maturation. *Nat. Struct. Mol. Biol.* **2023**, *30*, 594–599. [CrossRef] [PubMed]
27. Ismail, S.; Flemming, D.; Thoms, M.; Gomes-Filho, J.V.; Randau, L.; Beckmann, R.; Hurt, E. Emergence of the Primordial Pre-60S from the 90S Pre-Ribosome. *Cell Rep.* **2022**, *39*, 110640. [CrossRef] [PubMed]
28. Kater, L.; Mitterer, V.; Thoms, M.; Cheng, J.; Berninghausen, O.; Beckmann, R.; Hurt, E. Construction of the Central Protuberance and L1 Stalk during 60S Subunit Biogenesis. *Mol. Cell* **2020**, *79*, 615–628.e5. [CrossRef]
29. Woodson, S.A. Taming Free Energy Landscapes with RNA Chaperones. *RNA Biol.* **2010**, *7*, 677–686. [CrossRef]
30. De La Cruz, J.; Karbstein, K.; Woolford, J.L. Functions of Ribosomal Proteins in Assembly of Eukaryotic Ribosomes In Vivo. *Annu. Rev. Biochem.* **2015**, *84*, 93–129. [CrossRef]
31. Harnpicharnchai, P.; Jakovljevic, J.; Horsey, E.; Miles, T.; Roman, J.; Rout, M.; Meagher, D.; Imai, B.; Guo, Y.; Brame, C.J.; et al. Composition and Functional Characterization of Yeast 66S Ribosome Assembly Intermediates. *Mol. Cell* **2001**, *8*, 505–515. [CrossRef] [PubMed]
32. Pöll, G.; Braun, T.; Jakovljevic, J.; Neueder, A.; Jakob, S.; Woolford, J.L.; Tschochner, H.; Milkereit, P. rRNA Maturation in Yeast Cells Depleted of Large Ribosomal Subunit Proteins. *PLoS ONE* **2009**, *4*, e8249. [CrossRef] [PubMed]
33. Ohmayer, U.; Gamalinda, M.; Sauert, M.; Ossowski, J.; Pöll, G.; Linnemann, J.; Hierlmeier, T.; Perez-Fernandez, J.; Kumcuoglu, B.; Leger-Silvestre, I.; et al. Studies on the Assembly Characteristics of Large Subunit Ribosomal Proteins in S. Cerevisae. *PLoS ONE* **2013**, *8*, e68412. [CrossRef] [PubMed]
34. Ohmayer, U.; Gil-Hernández, Á.; Sauert, M.; Martín-Marcos, P.; Tamame, M.; Tschochner, H.; Griesenbeck, J.; Milkereit, P. Studies on the Coordination of Ribosomal Protein Assembly Events Involved in Processing and Stabilization of Yeast Early Large Ribosomal Subunit Precursors. *PLoS ONE* **2015**, *10*, e0143768. [CrossRef] [PubMed]
35. Dragon, F.; Gallagher, J.E.G.; Compagnone-Post, P.A.; Mitchell, B.M.; Porwancher, K.A.; Wehner, K.A.; Wormsley, S.; Settlage, R.E.; Shabanowitz, J.; Osheim, Y.; et al. A Large Nucleolar U3 Ribonucleoprotein Required for 18S Ribosomal RNA Biogenesis. *Nature* **2002**, *417*, 967–970. [CrossRef]
36. Baßler, J.; Grandi, P.; Gadal, O.; Leßmann, T.; Petfalski, E.; Tollervey, D.; Lechner, J.; Hurt, E. Identification of a 60S Preribosomal Particle That Is Closely Linked to Nuclear Export. *Mol. Cell* **2001**, *8*, 517–529. [CrossRef]
37. Steffen, K.K.; McCormick, M.A.; Pham, K.M.; MacKay, V.L.; Delaney, J.R.; Murakami, C.J.; Kaeberlein, M.; Kennedy, B.K. Ribosome Deficiency Protects Against ER Stress in Saccharomyces Cerevisiae. *Genet.* **2012**, *191*, 107–118. [CrossRef]

38. Micic, J.; Rodríguez-Galán, O.; Babiano, R.; Fitzgerald, F.; Fernández-Fernández, J.; Zhang, Y.; Gao, N.; Woolford, J.L.; de la Cruz, J. Ribosomal Protein eL39 Is Important for Maturation of the Nascent Polypeptide Exit Tunnel and Proper Protein Folding during Translation. *Nucleic Acids Res.* **2022**, *50*, 6453–6473. [CrossRef]
39. Moritz, M.; Paulovich, A.G.; Tsay, Y.F.; Woolford, J.L. Depletion of Yeast Ribosomal Proteins L16 or Rp59 Disrupts Ribosome Assembly. *J. Cell Biol.* **1990**, *111*, 2261–2274. [CrossRef]
40. Jakovljevic, J.; Ohmayer, U.; Gamalinda, M.; Talkish, J.; Alexander, L.; Linnemann, J.; Milkereit, P.; Woolford, J.L. Ribosomal Proteins L7 and L8 Function in Concert with Six A₃ Assembly Factors to Propagate Assembly of Domains I and II of 25S rRNA in Yeast 60S Ribosomal Subunits. *RNA* **2012**, *18*, 1805–1822. [CrossRef]
41. Koš, M.; Tollervey, D. Yeast Pre-rRNA Processing and Modification Occur Cotranscriptionally. *Mol. Cell* **2010**, *37*, 809–820. [CrossRef] [PubMed]
42. Zisser, G.; Ohmayer, U.; Mauerhofer, C.; Mitterer, V.; Klein, I.; Rechberger, G.N.; Wolinski, H.; Prattes, M.; Pertschy, B.; Milkereit, P.; et al. Viewing Pre-60S Maturation at a Minute's Timescale. *Nucleic Acids Res.* **2018**, *46*, 3140–3151. [CrossRef] [PubMed]
43. Kofler, L.; Prattes, M.; Bergler, H. From Snapshots to Flipbook—Resolving the Dynamics of Ribosome Biogenesis with Chemical Probes. *Int. J. Mol. Sci.* **2020**, *21*, 2998. [CrossRef] [PubMed]
44. Rodgers, M.L.; Woodson, S.A. Transcription Increases the Cooperativity of Ribonucleoprotein Assembly. *Cell* **2019**, *179*, 1370–1381.e12. [CrossRef]
45. Tutuncuoglu, B.; Jakovljevic, J.; Wu, S.; Gao, N.; Woolford, J.L. The N-Terminal Extension of Yeast Ribosomal Protein L8 Is Involved in Two Major Remodeling Events during Late Nuclear Stages of 60S Ribosomal Subunit Assembly. *RNA* **2016**, *22*, 1386–1399. [CrossRef] [PubMed]
46. Gamalinda, M.; Ohmayer, U.; Jakovljevic, J.; Kumcuoglu, B.; Woolford, J.; Mbom, B.; Lin, L.; Woolford, J.L. A Hierarchical Model for Assembly of Eukaryotic 60S Ribosomal Subunit Domains. *Genes Dev.* **2014**, *28*, 198–210. [CrossRef]
47. Ben-Shem, A.; Garreau De Loubresse, N.; Melnikov, S.; Jenner, L.; Yusupova, G.; Yusupov, M. The Structure of the Eukaryotic Ribosome at 30 Å Resolution. *Science* **2011**, *334*, 1524–1529. [CrossRef]
48. Chen, W.; Xie, Z.; Yang, F.; Ye, K. Stepwise Assembly of the Earliest Precursors of Large Ribosomal Subunits in Yeast. *Nucleic Acids Res.* **2017**, *45*, 6837–6847. [CrossRef] [PubMed]
49. Fernández-Fernández, J.; Martín-Villanueva, S.; Perez-Fernandez, J.; De La Cruz, J. The Role of Ribosomal Proteins eL15 and eL36 in the Early Steps of Yeast 60S Ribosomal Subunit Assembly. *J. Mol. Biol.* **2023**, *435*, 168321. [CrossRef] [PubMed]
50. Sahasranaman, A.; Dembowski, J.; Strahler, J.; Andrews, P.; Maddock, J.; Woolford, J.L. Assembly of Saccharomyces Cerevisiae 60S Ribosomal Subunits: Role of Factors Required for 27S Pre-rRNA Processing: Analysing Ribosome Assembly One Step at a Time. *EMBO J.* **2011**, *30*, 4020–4032. [CrossRef]
51. Dembowski, J.A.; Ramesh, M.; McManus, C.J.; Woolford, J.L. Identification of the Binding Site of Rlp7 on Assembling 60S Ribosomal Subunits in Saccharomyces Cerevisiae. *RNA* **2013**, *19*, 1639–1647. [CrossRef]
52. Klinge, S.; Voigts-Hoffmann, F.; Leibundgut, M.; Ban, N. Atomic Structures of the Eukaryotic Ribosome. *Trends Biochem. Sci.* **2012**, *37*, 189–198. [CrossRef] [PubMed]
53. Zhou, D.; Zhu, X.; Zheng, S.; Tan, D.; Dong, M.-Q.; Ye, K. Cryo-EM Structure of an Early Precursor of Large Ribosomal Subunit Reveals a Half-Assembled Intermediate. *Protein Cell* **2019**, *10*, 120–130. [CrossRef] [PubMed]
54. Espinar-Marchena, F.J.; Fernández-Fernández, J.; Rodríguez-Galán, O.; Fernández-Pevida, A.; Babiano, R.; De La Cruz, J. Role of the Yeast Ribosomal Protein L16 in Ribosome Biogenesis. *FEBS J.* **2016**, *283*, 2968–2985. [CrossRef]
55. Espinar-Marchena, F.; Rodríguez-Galán, O.; Fernández-Fernández, J.; Linnemann, J.; de la Cruz, J. Ribosomal Protein L14 Contributes to the Early Assembly of 60S Ribosomal Subunits in Saccharomyces Cerevisiae. *Nucleic Acids Res.* **2018**, *46*, 4715–4732. [CrossRef]
56. Babiano, R.; Gamalinda, M.; Woolford, J.L.; De La Cruz, J. Saccharomyces Cerevisiae Ribosomal Protein L26 Is Not Essential for Ribosome Assembly and Function. *Mol. Cell. Biol.* **2012**, *32*, 3228–3241. [CrossRef] [PubMed]
57. Babiano, R.; De La Cruz, J. Ribosomal Protein L35 Is Required for 27SB Pre-rRNA Processing in Saccharomyces Cerevisiae. *Nucleic Acids Res.* **2010**, *38*, 5177–5192. [CrossRef]
58. Gamalinda, M.; Jakovljevic, J.; Babiano, R.; Talkish, J.; De La Cruz, J.; Woolford, J.L. Yeast Polypeptide Exit Tunnel Ribosomal Proteins L17, L35 and L37 Are Necessary to Recruit Late-Assembling Factors Required for 27SB Pre-rRNA Processing. *Nucleic Acids Res.* **2013**, *41*, 1965–1983. [CrossRef] [PubMed]
59. Rosado, I.V.; Kressler, D.; De La Cruz, J. Functional Analysis of Saccharomyces Cerevisiae Ribosomal Protein Rpl3p in Ribosome Synthesis. *Nucleic Acids Res.* **2007**, *35*, 4203–4213. [CrossRef]
60. Joret, C.; Capeyrou, R.; Belhabich-Baumas, K.; Plisson-Chastang, C.; Ghandour, R.; Humbert, O.; Fribourg, S.; Leulliot, N.; Lebaron, S.; Henras, A.K.; et al. The Npa1p Complex Chaperones the Assembly of the Earliest Eukaryotic Large Ribosomal Subunit Precursor. *PLoS Genet.* **2018**, *14*, e1007597. [CrossRef]
61. Kisly, I.; Gulay, S.P.; Mäeorg, U.; Dinman, J.D.; Remme, J.; Tamm, T. The Functional Role of eL19 and eB12 Intersubunit Bridge in the Eukaryotic Ribosome. *J. Mol. Biol.* **2016**, *428*, 2203–2216. [CrossRef] [PubMed]
62. Rutgers, C.A.; Rientjes, J.M.J.; Van 'T Riet, J.; Raué, H.A. rRNA Binding Domain of Yeast Ribosomal Protein L25. *J. Mol. Biol.* **1991**, *218*, 375–385. [CrossRef] [PubMed]
63. Biedka, S.; Micic, J.; Wilson, D.; Brown, H.; Diorio-Toth, L.; Woolford, J.L. Hierarchical Recruitment of Ribosomal Proteins and Assembly Factors Remodels Nucleolar Pre-60S Ribosomes. *J. Cell Biol.* **2018**, *217*, 2503–2518. [CrossRef]

64. Baßler, J.; Kallas, M.; Pertschy, B.; Ulbrich, C.; Thoms, M.; Hurt, E. The AAA-ATPase Rea1 Drives Removal of Biogenesis Factors during Multiple Stages of 60S Ribosome Assembly. *Mol. Cell* **2010**, *38*, 712–721. [CrossRef] [PubMed]
65. LaPeruta, A.J.; Hedayati, S.; Micic, J.; Fitzgerald, F.; Kim, D.; Oualline, G.; Woolford, J.L. Yeast Ribosome Biogenesis Factors Puf6 and Nog2 and Ribosomal Proteins uL2 and eL43 Act in Concert to Facilitate the Release of Nascent Large Ribosomal Subunits from the Nucleolus. *Nucleic Acids Res.* **2023**, *51*, 11277–11290. [CrossRef] [PubMed]
66. LaPeruta, A.J.; Micic, J.; Woolford Jr., J. L. Additional Principles That Govern the Release of Pre-Ribosomes from the Nucleolus into the Nucleoplasm in Yeast. *Nucleic Acids Res.* **2023**, *51*, 10867–10883. [CrossRef] [PubMed]
67. Melnikov, S.; Mailliot, J.; Shin, B.-S.; Rigger, L.; Yusupova, G.; Micura, R.; Dever, T.E.; Yusupov, M. Crystal Structure of Hypusine-Containing Translation Factor eIF5A Bound to a Rotated Eukaryotic Ribosome. *J. Mol. Biol.* **2016**, *428*, 3570–3576. [CrossRef] [PubMed]
68. Malyutin, A.G.; Musalgaonkar, S.; Patchett, S.; Frank, J.; Johnson, A.W. Nmd3 Is a Structural Mimic of EIF 5A, and Activates the Cp GTP Ase Lsg1 during 60S Ribosome Biogenesis. *EMBO J.* **2017**, *36*, 854–868. [CrossRef]
69. Musalgaonkar, S.; Black, J.J.; Johnson, A.W. The L1 Stalk Is Required for Efficient Export of Nascent Large Ribosomal Subunits in Yeast. *RNA* **2019**, *25*, 1549–1560. [CrossRef] [PubMed]
70. Wu, S.; Tutuncuoglu, B.; Yan, K.; Brown, H.; Zhang, Y.; Tan, D.; Gamalinda, M.; Yuan, Y.; Li, Z.; Jakovljevic, J.; et al. Diverse Roles of Assembly Factors Revealed by Structures of Late Nuclear Pre-60S Ribosomes. *Nature* **2016**, *534*, 133–137. [CrossRef]
71. Saveanu, C.; Bienvenu, D.; Namane, A.; Gleizes, P.; Gas, N.; Jacquier, A.; Fromont-Racine, M. Nog2p, a Putative GTPase Associated with Pre-60S Subunits and Required for Late 60S Maturation Steps. *EMBO J.* **2001**, *20*, 6475–6484. [CrossRef]
72. De La Cruz, J.; Sanz-Martinez, E.; Remacha, M. The Essential WD-Repeat Protein Rsa4p Is Required for rRNA Processing and Intra-Nuclear Transport of 60S Ribosomal Subunits. *Nucleic Acids Res.* **2005**, *33*, 5728–5739. [CrossRef]
73. Wilson, D.M.; Li, Y.; LaPeruta, A.; Gamalinda, M.; Gao, N.; Woolford, J.L. Structural Insights into Assembly of the Ribosomal Nascent Polypeptide Exit Tunnel. *Nat. Commun.* **2020**, *11*, 5111. [CrossRef]
74. Kressler, D.; Bange, G.; Ogawa, Y.; Stjepanovic, G.; Bradatsch, B.; Pratte, D.; Amlacher, S.; Strauß, D.; Yoneda, Y.; Katahira, J.; et al. Synchronizing Nuclear Import of Ribosomal Proteins with Ribosome Assembly. *Sci.* **2012**, *338*, 666–671. [CrossRef] [PubMed]
75. Matsuo, Y.; Granneman, S.; Thoms, M.; Manikas, R.-G.; Tollervey, D.; Hurt, E. Coupled GTPase and Remodelling ATPase Activities Form a Checkpoint for Ribosome Export. *Nature* **2014**, *505*, 112–116. [CrossRef] [PubMed]
76. Leidig, C.; Thoms, M.; Holdermann, I.; Bradatsch, B.; Berninghausen, O.; Bange, G.; Sinning, I.; Hurt, E.; Beckmann, R. 60S Ribosome Biogenesis Requires Rotation of the 5S Ribonucleoprotein Particle. *Nat. Commun.* **2014**, *5*, 3491. [CrossRef] [PubMed]
77. Dong, X.; Doerfel, L.K.; Sheng, K.; Rabuck-Gibbons, J.N.; Popova, A.M.; Lyumkis, D.; Williamson, J.R. Near-Physiological in Vitro Assembly of 50S Ribosomes Involves Parallel Pathways. *Nucleic Acids Res.* **2023**, *51*, 2862–2876. [CrossRef] [PubMed]
78. Duss, O.; Stepanyuk, G.A.; Puglisi, J.D.; Williamson, J.R. Transient Protein-RNA Interactions Guide Nascent Ribosomal RNA Folding. *Cell* **2019**, *179*, 1357–1369. [CrossRef]

Disclaimer/Publisher's Note: The statements, opinions and data contained in all publications are solely those of the individual author(s) and contributor(s) and not of MDPI and/or the editor(s). MDPI and/or the editor(s) disclaim responsibility for any injury to people or property resulting from any ideas, methods, instructions or products referred to in the content.

Review

The Beak of Eukaryotic Ribosomes: Life, Work and Miracles

Sara Martín-Villanueva [1,2,†], Carla V. Galmozzi [1,2,†], Carmen Ruger-Herreros [1,2], Dieter Kressler [3] and Jesús de la Cruz [1,2,*]

1. Instituto de Biomedicina de Sevilla, Hospital Universitario Virgen del Rocío/CSIC/Universidad de Sevilla, E-41013 Seville, Spain; smartin9@us.es (S.M.-V.); cgalmozzi@us.es (C.V.G.); carmenruger@us.es (C.R.-H.)
2. Departamento de Genética, Facultad de Biología, Universidad de Sevilla, E-41012 Seville, Spain
3. Department of Biology, University of Fribourg, CH-1700 Fribourg, Switzerland; dieter.kressler@unifr.ch
* Correspondence: jdlcd@us.es; Tel.: +34-955-923-126
† These authors contributed equally to this work.

Abstract: Ribosomes are not totally globular machines. Instead, they comprise prominent structural protrusions and a myriad of tentacle-like projections, which are frequently made up of ribosomal RNA expansion segments and N- or C-terminal extensions of ribosomal proteins. This is more evident in higher eukaryotic ribosomes. One of the most characteristic protrusions, present in small ribosomal subunits in all three domains of life, is the so-called beak, which is relevant for the function and regulation of the ribosome's activities. During evolution, the beak has transitioned from an all ribosomal RNA structure (helix h33 in 16S rRNA) in bacteria, to an arrangement formed by three ribosomal proteins, eS10, eS12 and eS31, and a smaller h33 ribosomal RNA in eukaryotes. In this review, we describe the different structural and functional properties of the eukaryotic beak. We discuss the state-of-the-art concerning its composition and functional significance, including other processes apparently not related to translation, and the dynamics of its assembly in yeast and human cells. Moreover, we outline the current view about the relevance of the beak's components in human diseases, especially in ribosomopathies and cancer.

Keywords: ribosomal protuberances; eS12; eS31; Ubi3; eS10; translation accuracy; ribosome biogenesis

Citation: Martín-Villanueva, S.; Galmozzi, C.V.; Ruger-Herreros, C.; Kressler, D.; de la Cruz, J. The Beak of Eukaryotic Ribosomes: Life, Work and Miracles. *Biomolecules* **2024**, *14*, 882. https://doi.org/10.3390/biom14070882

Academic Editor: Leonard B. Maggi Jr.

Received: 19 June 2024
Revised: 19 July 2024
Accepted: 21 July 2024
Published: 22 July 2024

Copyright: © 2024 by the authors. Licensee MDPI, Basel, Switzerland. This article is an open access article distributed under the terms and conditions of the Creative Commons Attribution (CC BY) license (https://creativecommons.org/licenses/by/4.0/).

1. Introduction: Protuberances of the Ribosomal Subunits

Ribosomes are intricate molecular machinery found in the cytoplasm of all organisms. They play crucial roles during the translation of cellular mRNAs by efficiently and accurately decoding their genetic information. In addition, ribosomes can be found inside two types of organelles, chloroplasts and mitochondria [1–4].

All ribosomes are universally composed of two subunits, the small and large ribosomal subunits (r-subunits), which comprise ribosomal RNAs (rRNAs) and ribosomal proteins (r-proteins). The large r-subunit is about twice the size of the small r-subunit [2,5]. In a model eukaryote, such as the yeast *Saccharomyces cerevisiae*, the small r-subunit or 40S is composed of a single 18S rRNA and 33 different r-proteins; in turn, the large r-subunit or 60S contains three rRNAs (5S, 5.8S and 25S rRNAs) and 46 (47 in humans) r-proteins [6,7]. The general shape of the ribosomes has been known since the early 1970s; however, structures at high resolution have only been obtained since the beginning of the twenty-first century, following significant advances in structural technologies, such as X-ray crystallography and cryo-electron microscopy (cryo-EM) (e.g., [7–10]). In agreement with the functional conservation of ribosomes in all organisms [2,5,11,12], ribosomes display considerable structural similarity in prokaryotes, organelles and eukaryotes, although organellar ribosomes have substantially diverged from bacterial ones, with which they share a common ancestor (e.g., [3,13]). Moreover, structural studies have clearly shown that eukaryotic ribosomes

(25–30 nm in diameter) are larger and more complex than their prokaryotic (20 nm in diameter) and organellar counterparts [2,3,5–7]. When observed at low resolution, ribosomes appear flattened and spherical in shape; however, this observation is far from being correct. In fact, both r-subunits are extremely irregular complexes containing different domains and exhibiting different discernible protuberances. The small r-subunit is formed by four different structural domains, the head, the platform, the body and the beak; the beak domain is the main protrusion found in small r-subunits (Figure 1A). The large r-subunit, which is a more compact unit, displays three characteristic protuberances, the L1- and the P-stalk and the central protuberance (Figure 1B). Importantly, the structure of several r-subunit protuberances and protrusions has only been determined at high resolution in a few organisms, a fact that is most likely due to the high mobility of these regions, which is clearly connected to their functions (e.g., [6]). Thus, the movement of the beak as part of the head of 40S r-subunits is important to allow the loading of mRNA on the ribosome and the interaction with translation factors, as we will discuss in detail later on. In turn, the primary role of the P-stalk is to provide a flexible hook outside of the core of the ribosome to recruit and then stimulate the activity of different translational GTPases during various stages of translation (e.g., [14–16]). The P-stalk also enables the activation of the Gcn2 kinase and mediates the interaction with distinct ribosome-inactivating proteins (e.g., [17–20]). The L1-stalk acts as a flexible protuberance at the antipodes of the P-stalk, facilitating the binding, movement and release of the deacylated tRNAs [21,22].

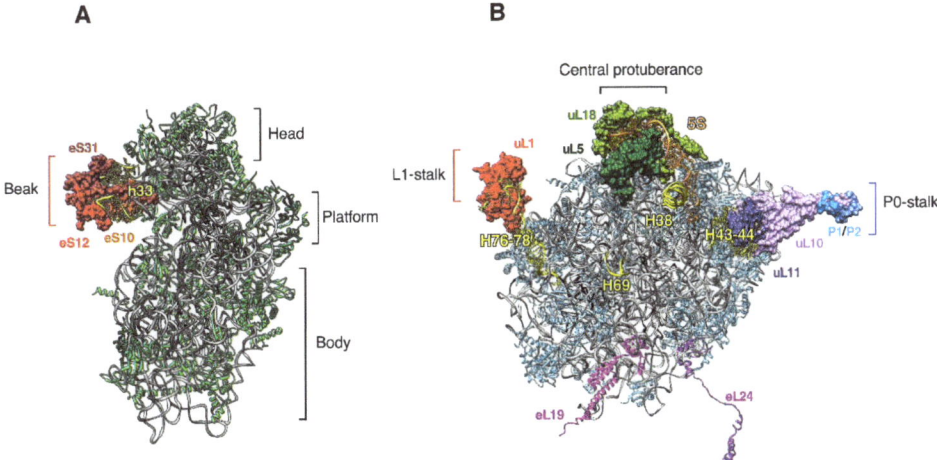

Figure 1. Protuberances of the r-subunits. (**A**) Structure of the mature 40S r-subunit of *Saccharomyces cerevisiae* showing its different structural domains (head, body, platform and beak) and highlighting the beak elements that include helix h33 of the 18S rRNA and the r-proteins eS10, eS12 and eS31. The 40S r-subunit is shown from the interface view; 18S rRNA is colored in gray (except h33 in yellow) and r-proteins not belonging to the beak structure are colored in green. Protein Data Bank (PDB) code: 4V88. (**B**) Structure of the mature 60S r-subunit of *S. cerevisiae* showing its different substructures (L1 and P0 stalks, and central protuberance) and highlighting the different protuberances: L1-stalk formed by helices H76-78 of 25S rRNA and the r-protein uL1; central protuberance formed by 5S rRNA and r-proteins uL5 and uL18; P0-stalk formed by helices H43-44 of 25S rRNA and the r-proteins uL11, uL10 (P0), P1 and P2 (two copies each). As other examples of protrusions found in the 60S r-subunits, we highlight helices H38 and H69 (yellow) and the long extensions of the r-proteins eL19 and eL24 (purple). The 60S r-subunit is shown from the interface view; 25S and 5.8S rRNAs are colored in gray and the rest of the r-proteins are colored in light blue. PDB code: 6OIG. Cartoons were generated using UCSF Chimera (https://www.rbvi.ucsf.edu/chimera; accessed on 1 June 2024).

In addition to being part of the prominent, above-described protuberances, many r-proteins have long non-globular extensions, typically serpentine ones that become deeply embedded inside the rRNA core of the r-subunits and stabilize their overall structure. There are also r-protein extensions (see Figure 1B for examples) that project out of the ribosome and may serve as additional regions for interactions of the ribosome with factors or cellular structures [23–25]. Frequently, these extensions, which are located at the N- and/or C-terminal ends of r-proteins, are not well ordered and are highly mobile due to the lack of stable interactions with the core of the ribosome; consequently, it has not been possible to model all of them from the current cryo-EM density maps.

Protuberances are not only flexible components of ribosomes, but also dynamic structures. The eukaryotic P-stalk is a pentameric complex, composed of the essential r-protein uL10 (previously named P0) and two non-essential heterodimers made up of another two r-proteins (named P1 and P2), that is connected to the body of the 60S r-subunit by the uL11 r-protein [15,26]. P1 and P2, when phosphorylated, can exchange during translation with their equivalent cytoplasmic pool of non-phosphorylated P1 and P2 variants [27–29]. Interestingly, in yeast, P1 and P2 r-proteins are not essential for cell viability, but their absence affects differently the translation of specific mRNAs [15,30]. Furthermore, it has been described that the amount of P1/P2 r-proteins bound to the ribosome changes under different physiological conditions. For instance, when yeast cells enter the stationary phase, the P1/P2 r-proteins are practically absent from ribosomes [31]. This and other observations made in different eukaryotes (e.g., [32,33]) have allowed different authors to propose that the eukaryotic P-stalk acts as a regulator of translation and that changes in its composition could selectively control the translation of specific groups of mRNAs [34]. In agreement with a regulatory role of the P-stalk during translation, mutation of the phosphorylation sites of the yeast acidic P1/P2 r-proteins does not alter their interaction with the ribosome but influences the translation of specific mRNAs, among them, those related to osmotic stress [35]. Further experiments, using high-resolution techniques (e.g., ribosome profiling), are required to provide details on such a regulatory process and on the specific proteome translated upon changes in P-stalk abundance and phosphorylation status.

To the best of our knowledge, aside from the acidic P1/P2 r-proteins, there is no evidence of exchangeability of r-proteins from the other ribosomal protuberances, including the beak; however, there are interesting reports about several other r-proteins whose nascent forms can replace an older copy of themselves on a mature ribosome. (i) Thus, uL16 has been described as an r-protein potentially able to cycle on and off large r-subunits [36–38]. uL16 is strategically positioned on the surface of the evolutionarily conserved core of the 60S r-subunit, near the corridor through which aminoacyl-tRNAs move during accommodation and also near other functional centers, such as the GTPase-associated center (GAC) and the peptidyl transferase center (PTC). uL16 is required for the joining of r-subunits during translation initiation and the rotation status of the ribosome [39–41]; importantly, the availability to assemble uL16 onto large r-subunits could also be used as a translational regulatory mechanism to limit global translation under unfavorable circumstances [37,41]. (ii) Another interesting r-protein is RACK1 (Asc1 in yeast). RACK1 is a WD40-domain protein located at the head region of the 40S r-subunit near the mRNA exit site, where it interacts with other r-proteins, among them uS3, several kinases, and translation initiation factors [42,43]. RACK1 has an important role in different aspects of the translation process: it is required for efficient translation of mRNAs with short open reading frames (e.g., those of r-protein genes), it is critical for translation during heat stress, and it facilitates ribosome-associated quality control (RQC) mechanisms such as those that are involved in the rescue of stalled collided ribosomes at consecutive rare codons, such as CGA (Arg) in yeast [44–49]. Although there is no experimental demonstration for RACK1 to cycle on and off ribosomes [50], it is clear that RACK1 is a non-essential r-protein whose loss does not disrupt ribosome integrity and translation [46]. Moreover, as RACK1 protein levels can be modulated by a variety of environmental insults, such as hypoxic stress, glucose deprivation or amino acid starvation or by the physiological cellular status (exponential versus stationary growth phase) (e.g., [51]), it is possible that the RACK1's ribosome association could be regulated, thereby, promoting differential translation. (iii) Another important inductor of exchangeability is chemical damage. Originally reported in prokaryotes [52], yeast ribosomes containing chemically damaged r-proteins can also be repaired by exchanging these with undamaged r-proteins, as convincingly demonstrated by the Karbstein laboratory [38,53,54]. The repair of damaged ribosomes might represent an important mechanism for maintaining the translational activity of cells following different insults as, for example, those produced by oxidative stress [53,55]. An analogous repair process via protein replacement has also been suggested to occur in neurons; thus, implying that this mechanism has been evolutionarily conserved (e.g., [56]). In yeast, the molecular details for a ribosome repair mechanism have been provided upon oxidation of eS26 and uL16, which are released from damaged ribosomes by their respective dedicated chaperones, Tsr2 and Sqt1, generating transiently eS26- and uL16-deficient ribosomes that are subsequently repaired with newly made r-proteins [38]. Ribosomes lacking eS26 can also be generated in a Tsr2-dependent manner upon the exposure of yeast cells to high salt or high pH conditions. Ribosome repair is extremely relevant from the physiological point of view as eS26-lacking ribosomes preferentially translate specific transcripts bearing Kozak sequence variations, including mRNAs enabling the biological response to high salt and high pH insults [55]. The recovery from stress is concomitant to the reincorporation of eS26 into ribosomes, again in a Tsr2-dependent manner [38,53]. This sophisticated system of autoregulation resembles that previously reported for the translation circuit of leaderless mRNAs (lmRNAs) in bacteria. These lmRNAs can be generated in response to adverse environmental conditions, some of them (e.g., the presence of antibiotics such as kasugamycin in the culture media) also being able to reprogram ribosomes to translate preferentially lmRNAs. Interestingly, this reprogramming involves the formation of stable r-particles (referred to as 61S particles) deficient in almost a dozen r-proteins from the small r-subunit, among them bS1 and other r-proteins associated along the path of the mRNA through this r-subunit [57,58].

Of special attention is the central protuberance (CP), where the 5S ribonucleoprotein particle (RNP), which is composed of 5S rRNA and r-proteins uL5 and uL18, plays an important regulatory role (Figure 1B). The whole 5S RNP, rather than its individual components, is incorporated as a prefabricated complex into early pre-60S r-particles during the nucleolar ribosome biogenesis phase, and it temporally adopts a conformation that is different from the one in the mature 60S r-subunit (e.g., [59–61]). From yeast to humans, ribosome biogenesis is tightly coupled to cell growth and proliferation, with the assembly of the 5S RNP playing a central regulatory role. Thus, in yeast, an imbalanced production of rRNAs and r-proteins generates defects in ribosome biogenesis leading to the accumulation of ribosome-unbound uL18, likely as part of the 5S RNP, which induces a delay in the G1/S phase of the cell cycle [62]. This behavior has been interpreted as part of a protective mechanism that prevents cell cycle progression when ribosome biogenesis is impaired, i.e., when not all necessary components are sufficiently available to ensure a complete and satisfactory assembly of ribosomes. In metazoans, the 5S RNP clearly accumulates when ribosome biogenesis is impaired [63–66]. The free 5S RNP binds to MDM2 (HDM2 in humans), which is an E3 ubiquitin ligase that ubiquitinates p53 and thereby channels it to degradation via the proteasome. Upon binding of the 5S RNP to MDM2, which occurs mutually exclusively to its binding to pre-60S r-particles, p53 escapes from MDM2-mediated degradation and accumulates [67,68]. Concomitantly, p53 is activated and, thus, exerts its different anti-proliferative functions, ranging from temporary cell cycle arrest to apoptosis [69]. The implications of the involvement of ribosome biogenesis in the regulation of p53 in human health and disease have been extensively discussed in other reviews (e.g., [68,70–72]) and will also be examined later in this work.

This review is aimed at giving insights into the composition, structure, role, biogenesis and dysfunction of the components of the beak, which is the most prominent protuberance found in all small r-subunits and so-called because of its resemblance to a bird's beak. We discuss all these features of the eukaryotic beak by specially focusing on the current knowledge about the beak of 40S r-subunits from the yeast *S. cerevisiae* and by highlighting similarities and differences compared to the beak of human ribosomes.

2. Composition of the Beak of the Eukaryotic Ribosome

Despite the fact that the beak is overall an easily recognizable structure in the small r-subunits of all ribosomes, the composition of the beak is quite different within the three domains of life. In bacteria, the beak is composed exclusively of rRNA, specifically, the helix h33 of 16S rRNA [9,73,74]. In contrast, the beak of eukaryotic ribosomes has been transformed into a mixture of rRNA and specific r-proteins not found in bacteria, with the biological reasons for this transformation remaining unsolved. The r-proteins bound to helix h33 of eukaryotic 18S rRNA, which itself is shorter than the bacterial h33, are eS10, eS12 and eS31 [1,2,6]. It is accepted that the structural core components of the archaeal ribosomes are of prokaryotic origin, to which specific elements, some shared with eukaryotes, have been added; therefore, archaeal ribosomes represent intermediate steps towards the evolution of eukaryotic ribosomes [75]. In this sense, it is interesting to mention that the beak of archaeal small r-subunits has a transitional complexity from an all-rRNA to an rRNA/r-protein protrusion; accordingly, many archaeal genomes encode clear homologs of the eukaryotic eS31 r-protein, which in all cases, however, lack the eukaryote-specific N-terminal extension [76,77]. In addition, cryo-EM has shown that ribosomes of distinct archaea contain at least two copies of eL8, one at the canonical location on the large r-subunit and another one bound at a position on h33 that is equivalent to the one occupied by eS12 on the eukaryotic beak [78]; further predictions suggest that this feature, i.e., the presence of eL8 in the beak, occurs in all archaeal ribosomes [78]. Moreover, this observation suggests that eS12 evolved from eL8, as both proteins share conserved regions and belong to the same family (InterPro entry IPR004038). Finally, it is clear that eukaryotic eS10 has no counterpart in archaeal ribosomes, as evidenced by different database searches, including BLAST [77,79]. A comparison of the beak composition and structure of ribosomes from prototypical bacteria, archaea and eukaryotes is shown in Figure 2.

Ribosomes present in organelles also contain all the structural landmarks that are characteristic of cytoplasmic ribosomes of prokaryotes and eukaryotes, including the beak. However, mito- and chlororibosomes have been found to be extremely diverse in terms of their composition, including the acquisition of organelle-specific r-proteins that has an impact on their overall structures. In general, chlororibosomes resemble bacterial ribosomes [80–82]; hence, the beak of these ribosomes is formed by the h33 rRNA protrusion and is devoid of r-proteins (Figure 3). In marked contrast to chlororibosomes, the characterization of mitoribosomes from diverse species representing the different major groups of eukaryotes has revealed that these have diverged considerably from each other and from prokaryotic and eukaryotic ribosomes [3,83,84]. In mitoribosomes, the beak can vary from the all-RNA prototype found in yeast to the massive protein-based beak found in the kinetoplastid *Trypanosoma brucei* (Figure 3).

Cytoplasmic ribosomes of the yeast *S. cerevisiae* contain a standard beak, composed of a helix h33 of 52 nucleotides and a single copy of three r-proteins, eS10, eS12 and eS31 (Figure 4) [6]. Yeast eS10 is encoded by two paralogous genes, *RPS10A* (YOR293W) and *RPS10B* (YMR230W). The two genes code for the virtually identical eS10A and eS10B r-proteins of 105 amino acids and ca. 12.7 kDa that only differ in three solvent-exposed amino acids (E6, D7 and T98 in eS10A versus Q6, E7 and S98 in eS10B). Mutants harboring individual deletions of the *RPS10A* and *RPS10B* genes are viable in different yeast backgrounds; moreover, while the *rps10B*Δ null mutant grows practically identical to the wild-type strain, the *rps10A*Δ null mutant exhibits only a mild increase in the doubling time [85]. In all genetic backgrounds, eS10 is an essential protein as the *rps10A*Δ *rps10B*Δ double mutant is inviable [85,86]. Yeast eS10 is a mostly globular protein; however, it contains an unstructured C-terminal extension of about 20 amino acids, which interacts with uS3 at the base of the beak in the mRNA entry channel [87]. In contrast to eS10 and most yeast r-proteins, eS12 and eS31 are non-essential r-proteins that are encoded by single-copy genes (*RPS12* or YOR369C, and *RPS31* or *UBI3* or YLR167W, respectively). However, in most genetic backgrounds, both the *rps12Δ* and the *ubi3Δ* mutant display a severe growth impairment [85,88,89]. Yeast eS12 is a small globular protein of 143 amino acids and ca. 15.5 kDa, containing an unstructured N-terminal extension of around 25 amino acids whose deletion causes a slow growth phenotype of still uncertain significance (our unpublished results). On the other hand, yeast eS31 is a small r-protein of 76 amino acids consisting of a globular domain, which is well conserved from archaea to eukaryotes, and a eukaryote-specific N-terminal extension of about 25 amino acids, which extends toward the ribosomal A-site and has relevant functions in translation and small r-subunit assembly ([1,76,90]; see later). More interestingly, it has been well reported that in most eukaryotes eS31 as well as eL40 are produced as C-terminal parts of ubiquitin-fused precursor proteins, which are rapidly processed to individual ubiquitin and r-protein moieties before assembly of the corresponding r-protein into the small and large r-subunit, respectively [91]. The biological relevance of maintaining these fusions during evolution for the correct production and assembly of these r-proteins and for the possible co-regulation of two related cellular functions, protein synthesis and protein degradation, has been previously covered and will not be further discussed in this review [91,92].

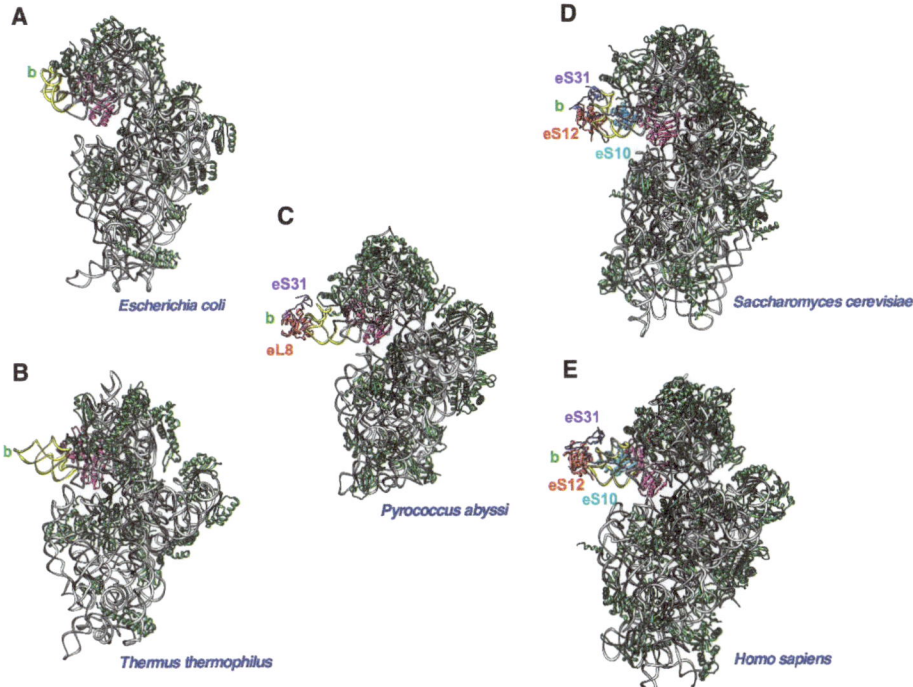

Figure 2. Comparison of small r-subunits of bacteria, archaea and eukaryotes. (**A**) Small r-subunit of *Escherichia coli*; PDB code: 7OE1; (**B**) small r-subunit of *Thermus thermophilus*; PDB code: 1J5E; (**C**) small r-subunit of *Pyrococcus abyssi*; PDB code: 7ZHG; (**D**) small r-subunit of *S. cerevisiae*; PDB code: 4V88; (**E**) small r-subunit of *Homo sapiens*; PDB code: 7R4X. In all cases, the interface view of the individual r-subunits is shown. The rRNA is colored in gray and the r-proteins in green. The beak (b) of all r-subunits is highlighted; helix h33 of rRNA is colored in yellow, and the r-proteins eS31, eL8, eS10 and eS12 in the indicated colors. The r-protein uS3, which is located at the base of the beak, is colored in pink. Cartoons were generated using UCSF Chimera (https://www.rbvi.ucsf.edu/chimera; accessed on 1 June 2024).

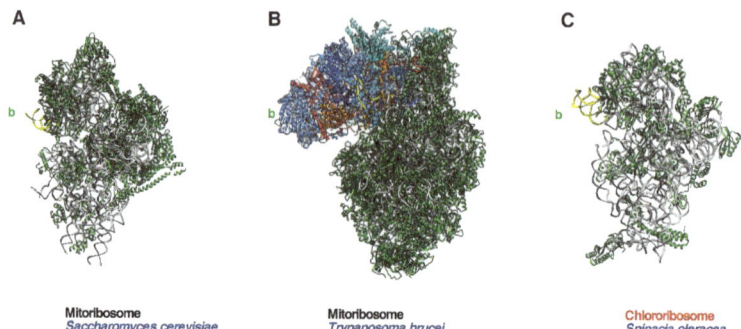

Figure 3. Comparison of the structural features of the small subunit of the mitochondrial ribosome from *S. cerevisiae* (**A**) and *Trypanosoma brucei* (**B**), and of the chloroplast ribosome from *Spinacia oleracea* (**C**). The PDB codes are 5MRC, 6HIW, and 5MMJ, respectively. In all cases, the interface view of the individual r-subunits is shown. The rRNA is colored in gray and the r-proteins in green. The beak (b) of all r-subunits is highlighted; helix h33 of rRNA is colored in yellow, and the set of specific r-proteins present in the *T. brucei* small r-subunit are shown in different colors. Cartoons were generated using UCSF Chimera (https://www.rbvi.ucsf.edu/chimera; accessed on 1 June 2024).

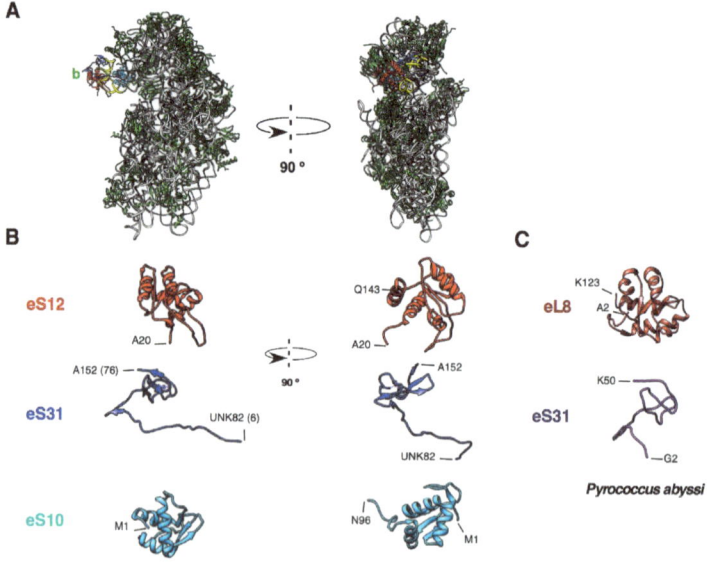

Figure 4. The r-proteins from the beak (b) of the yeast 40S r-subunit and the equivalent ones from the beak of the *Pyrococcus abyssi* 30S r-subunit. (**A**) Structure of the yeast 40S r-subunit (PDB code 4V88). (**B**) Structure of yeast eS12, eS31 and eS10 r-proteins are shown as found in the 40S r-subunit before or after 90° rotation on the Y-axis. (**C**) Structure of the equivalent r-proteins eL8 and eS31 from *P. abyssi* are also shown as found in the 30S r-subunit (PDB code 7ZHG). Note that the first 19 N-terminal residues of yeast eS12, the six first residues of yeast eS31, the last nine C-terminal residues of yeast eS10, the first residue of *P. abyssi* eL8 and the first and the last residue of *P. abyssi* eS31 are not present in the structures. Cartoons were generated using UCSF Chimera (https://www.rbvi.ucsf.edu/chimera; accessed on 1 June 2024).

In consonance with the critical importance of yeast eS10, eS12 and eS31 r-proteins for cell growth, it has also been reported that loss-of-function mutations in the genes coding for these proteins in other model eukaryotes (*Caenorhabditis elegans*, *Drosophila melanogaster*, *Danio rerio*, *Mus musculus*, *Homo sapiens*) lead to a myriad of adverse phenotypes, including lethality, increased cell death, cell cycle arrest, reduced fertility, organ development defects and tumorigenesis [93].

3. Roles of the Beak during Translation

The head of the small r-subunit is a flexible and dynamic structure involved in the engagement of the mRNAs and tRNAs during translation. Taking into consideration the strategic position of the beak at the entrance of the mRNA channel in the small r-subunit, it is not surprising that the beak has been linked to diverse functions during the translation process:

(i) During translation initiation, the beak is an important site for the interaction of *trans*-acting factors both in prokaryotic and eukaryotic ribosomes. For example, cryo-EM has revealed that the bacterial aldehyde-alcohol dehydrogenase E (AdhE) enzyme interacts with ribosomes in the beak region [94]. This enzyme provides a further RNA helicase activity, in addition to the intrinsic one of the ribosome [95], in order to ensure the linear configuration of structured mRNAs at the mRNA entrance to facilitate their translation [94]. In eukaryotes, several RNA helicases play roles during translation initiation. The canonical initiation factor eIF4A and the Ded1 (DDX3 in mammals) RNA helicase assist in the unwinding of the 5′-UTR secondary structure of most mRNAs [96–98]. Interestingly, in mammals, the translation initiation of endogenous and viral mRNAs with highly structured 5′-UTRs requires an additional RNA helicase, named DHX29 [99,100]. As revealed by cryo-EM, DHX29 contacts the beak and adjacent regions by interacting with at least uS3, eS10 and eS12 [101]. In other examples, the beak has been described as being important for the recognition of specific mRNAs. Accordingly, mammalian eS10 has been found to specifically interact with a class of cellular mRNAs containing the so-called TISU-element in their short 5′-UTRs [102].

(ii) The loading of the mRNA itself into the mRNA channel of the small r-subunit during translation initiation is regulated by the opening and closing of an mRNA latch situated below the beak that connects the body and the head of the small r-subunit [74]. The open conformation of this structure is promoted by the binding of distinct initiation factors (eIF1 and eIF1A in eukaryotes, IF1 in prokaryotes) to the small r-subunit during the formation of the eukaryotic 43S pre-initiation complex [103]. eIF1A is a globular protein harboring unstructured N- and C- terminal extensions of ca. 25 amino acids. During translation initiation, the globular domain of eIF1A is positioned at the A-site, while its extensions seem to project out of this site of the ribosome; thus, preventing tRNA binding to this site [103]. From X-ray crystallography, it can be inferred that the N-terminus of eIF1A directly contacts the eukaryote-specific N-terminal extension of eS31 and approaches extensions of other r-proteins, such as eS10, uS3 and uS19, that are adjacent to each other in the A-site [104]. These interactions seem to be crucial for translation as mutations in the N-terminal tail of eIF1A, which are frequently observed in several types of cancers [105], result in reduced binding of eIF1A to its r-protein partners and a hyperaccurate recognition of AUG codons that are embedded in an optimal sequence context [106,107]. Importantly, the phenotypic analysis of yeast *ubi3* mutations (e.g., *ubi3G75,76A*) that interfere with the cleavage of the ubiquitin-eS31 fusion protein indicates that the non-cleaved protein can still assemble into mature 40S r-subunits, which are active in translation but mildly defective at the translation initiation stage. This defect is likely due to the interference of the ubiquitin moiety with the binding and proper activity of the initiator tRNA and the eIF1A factor [89,104]. Moreover, interactions between several subunits of the eIF3 complex and beak components, including eS10 (e.g., [108]), have been described to occur within the yeast 43S pre-initiation complex; these are expected to be of functional relevance during translation initiation (e.g., [109]).

(iii) The beak also participates in the formation of the binding surface for the internal ribosome entry site (IRES) of some viral mRNAs on human 40S r-subunits. For instance, among other 40S r-proteins, eS10 contributes to the binding of the hepatitis C virus (HCV) IRES [110]. Viral proteins also interact with the beak to hijack the host's translation machinery. One interesting example concerns the SARS-CoV-2 coronavirus, whose non-structural protein NSP1 contains a globular N-terminal domain that binds the base of the beak, while its C-terminal extension blocks the mRNA channel entry site and thereby prevents any mRNA accommodation; thus, inhibiting translation of host mRNAs [111–113]. However, NSP1 does not impede the translation of viral mRNAs, which is promoted by the presence of a *cis*-acting RNA hairpin in the 5′-UTR of these mRNAs [114,115].

(iv) Translation is reversibly shutdown upon nutrient starvation in a variety of ways, including the accumulation of inactive or hibernating vacant 80S ribosomes. These dormant ribosomes contain eEF2·GTP in the A-site and the hibernation factor Stm1 (in yeast) or SERBP1 (in mammals) in the mRNA channel, thereby impeding mRNA binding [87,116]. This mechanism of blocking the mRNA entry tunnel resembles that mentioned above for the coronavirus NSP1 protein. It has been described that the C-terminal region of Stm1/SERBP1 also stably associates with the head of 40S r-subunits, likely via binding to eS10, eS12 and eS31 [87,116].

(v) In all ribosomes, the interaction of the different translation elongation factors with the ribosome leads to specific movements of the head domain of the small r-subunit, including that of its associated beak towards the shoulder of the body of the same r-subunit (e.g., [117,118]). In eukaryotes, the beak components themselves interact with distinct domains of the translation elongation factors, as exemplified by the interaction of eS12 and eS31 with domains II and IV of eEF2 [7]. In agreement with the important role of beak r-proteins in the fidelity of translation elongation, the depletion of the essential yeast eS10 as well as the mutation or deletion of the genes encoding yeast eS12 and eS31 result in translation defects, including misreading [76,88,119]. Moreover, due to the specific position of the N-terminal extension of eS31 in the A-site of the ribosome, the assembly of non-cleaved yeast Ubi3 is expected to sterically interfere with the binding of the translation elongation factors to the ribosomal GTPase-associated center [89,91].

(vi) The beak is also expected to be functionally relevant during translation termination. Cryo-EM structures have revealed how eukaryotic translation termination factor 1 (eRF1), whose overall shape resembles a tRNA molecule, interacts with a stop codon in the A-site of the ribosome via its N-terminal lobe (e.g., [120] and references therein). Notably, in the structures of pre-termination complexes, a short segment of the N-terminal lobe of eRF1 is in close proximity to the initial residues of the N-terminal extension of eS31 [120,121]. Moreover, the mini-domain of eRF1, which is an insertion within the C-terminal domain, also interacts with the N-terminal extension of eS31 and protrudes toward the beak where it contacts helix h33 [120–122].

(vii) Another example of the role of the beak components in translation comes from studies of the cellular responses to elongation stalls induced by different stresses (including oxidative stress, heat shock or starvation) as well as by particular sequences, strong secondary structures and chemical damage within mRNAs. Normally, when exposed to stressful conditions, cells adapt by halting or decreasing the global synthesis of new proteins, while, concomitantly, inducing the selective translation of mRNAs encoding proteins that are necessary for cell survival and stress recovery [123]. This translational reprogramming can be mediated by multiple parallel and independent signaling pathways that converge on the modulation of the function of a few key translation factors [124]. Relatively recent studies from the Silva laboratory showed that following oxidative stress, induced by an exposure of yeast cells to hydrogen peroxide, a set of r-proteins were K63-specifically polyubiquitinated at different residues by the ubiquitin-conjugating enzyme Rad6 and the ubiquitin-protein ligase Bre1, with the extent of this modification declining very rapidly during stress recovery [125,126]. Most of these r-proteins are located within the head of the small r-subunit of the ribosome and include uS3, the beak components eS10, eS12 and eS31 and the P-stalk proteins uL10, uL11 and P2 [127]. Although oxidative stress induces a rapid inhibition of translation initiation via activation of the Gcn2 kinase (see below), K63-linked ubiquitination of r-proteins leads to an additional response, which results in the stalling of translation at the elongation stage [127]. Using cryo-EM and cryo-electron tomography, the Silva laboratory was also able to demonstrate that K63-linked ubiquitination of ribosomes alters the conformation of distinct r-proteins, including eS31 and eS12, that are located at the interface of the two r-subunits where eEF2 binds, thereby interfering with its efficient binding and/or GTPase activity and promoting the translational halt at the elongation stage, specifically at the rotated pre-translocation stage 2 [128].

If a ribosome persistently stalls on an mRNA, collisions with the trailing ribosomes will eventually occur; this phenomenon triggers different ribosome-associated quality control (RQC) mechanisms. It is thought that these mechanisms have evolved in order to relieve stalled ribosomes, thus avoiding the depletion of active ribosome and tRNA pools, which would prevent their participation in new rounds of protein synthesis and could reduce cellular fitness or survival [129,130]. RQC mechanisms additionally target damaged mRNAs and incomplete polypeptide chains for degradation [129]. In prokaryotes, the rescue of ribosomes stalled at the 3′ end of mRNAs lacking a stop codon, thus containing an empty A-site, often involves the action of the long transfer-messenger RNA (tmRNA), whose interaction with the ribosome occurs through the formation of a ring of its large loop around the beak of the 30S r-subunit (for further information, see [131]). In eukaryotes, different rescue pathways center around the recognition of the empty A-site in the ribosome (e.g., [130,132]). The rescue of eukaryotic ribosomes stalled on truncated and aberrant mRNAs lacking stop codons (NSD, non-stop decay) relies on several factors, such as Dom34 (Pelota in mammals), Hbs1 (HBS1L in mammals), Rli1 (ABCE1 in mammals) and Ski7 (HBS1L3 in mammals). These factors interact or have the potential to interact with ribosomes in a similar manner as translation elongation and termination factors [133]; therefore, it is expected that the binding and function of these factors, and, thus, the fate of NSD, could be altered by mutations affecting beak components. Dom34/Pelota is structurally related to tRNAs and eRF1 and binds the ribosome in a similar way to these two; in turn, Hbs1, which interacts with Dom34/Pelota, is a member of the family of translational GTPases that includes eEF1, which delivers aminoacyl-tRNAs to the A-site, eRF3, which interacts with eRF1 in a similar manner to Hbs1 with Dom34, and Ski7, which is a paralog of Hbs1 [133]. The N-terminal domain of Ski7 mediates the recruitment of the exosome and the Ski2-Ski3-Ski8 complex (SKIV2L-TTC37-WDR61 in humans) [134–136], while its C-terminal part contains the GTPase-like domain that it is assumed to interact, similar to other translation GTPases, with the GAC site of the ribosome, but whose exact role is still unknown [137,138]. Cryo-EM structures of different ribosomes, Ski2/3/8 complex and exosome intermediates suggest a scenario where a stalled ribosome bound to the Ski2/3/8 complex recruits a pre-assembled exosome-Ski7 complex [136,139,140]. Through this triple (ribosome–Ski2/3/8 complex–exosome) interaction, aberrant mRNA substrates are unwound and guided into the exosome. The ribosome-bound Ski2/3/8 complex specifically recognizes the 40S r-subunit by binding near the entry of the mRNA channel and connecting the head and beak regions [140]. Concerning the beak, the Ski2 helicase interacts with several r-proteins, among them eS10 and uS3, while the N-terminal part of Ski3 contacts eS12 [139,140].

Prolonged ribosome stalling leads to a ribosome collision of the trailing ribosomes with the stalled ribosome [141,142]. Under these circumstances, the E3 ligase Hel2 (ZNF598 in mammals) recognizes the collided ribosomes and adds ubiquitin to a number of 40S r-proteins at precise lysine residues, among them eS10 and uS3 ([143] and references therein). This ubiquitination is assumed to serve as the starting signal for the progression of the RQC response in order to dissociate the stalled ribosomes into r-subunits and degrade their associated mRNAs and nascent peptides [144]. In addition, beside many other responses [142], ribosome collisions also activate the kinase Gcn2, and evidence suggests that this activation can occur independently of the presence of deacetylated tRNAs [145,146], which constitutes its classical activation pathway. This activation is dependent on Gcn1, and cryo-EM has nicely revealed how this long, tube-like HEAT repeat protein spans across a collided disome by forming an extensive network of interactions both with the leading and trailing ribosome [147]. Interestingly, along its interaction path, the region preceding the central eEF3-like HEAT repeats engages in contacts with eS10, eS12 and eS31 within the beak of the 40S r-subunit of the colliding ribosome [147,148]. Perturbations of the beak, elicited by absent or mutated beak r-proteins, could influence Gcn1 such that its ribosome association or Gcn2-binding capacity, both of which are required for Gcn2 activation, is affected. In turn, activated Gcn2 can then phosphorylate eIF2α to downregulate general translation initiation and to enable the translation of specific mRNAs, such as those encoding Gcn4 in yeast or ATF4 in mammals, in order to adequately respond to the stress that causes ribosomes to collide. In line with the structural integrity of the beak being necessary for an efficient Gcn1-mediated Gcn2 activation, lower levels of eS10 (individual deletion of *RPS10A* or *RPS10B*) or the absence of eS31 were shown to reduce the extent of eIF2α phosphorylation or to impair derepression of *GCN4* mRNA translation in response to amino acid starvation, respectively (e.g., [148,149]). Recently, the ubiquitination of eS31 has also been reported to occur in circumstances of translation elongation inhibition where the A-site is occluded by a trapped eEF1A factor bound to an aminoacyl-tRNA [150]. In this case, the reaction is dependent on an E3 ligase called RNF25. Ubiquitination of eS31 is required for the degradation of the trapped eEF1A, which itself is ubiquitinated both by RNF25 and an additional E3 ligase RNF14, with the latter directly interacting with GCN1, which is also essential for eEF1A degradation [150].

4. Other Cellular Functions of the Beak Components

The beak r-proteins, in addition to their clear role in translation, participate in other cellular processes, including ribosome biogenesis (see Section 5), activation of the p53-dependent pathway in response to nucleolar stress as well as oncogenesis (see Section 6), and cell competition (see below). Whether or not the effects that mutations in the beak r-proteins have on these processes are translation-dependent or independent is still unclear in some cases.

As mentioned above, a systematic study has analyzed the contribution of loss-of-function mutations for most r-proteins, including eS10, eS12 and eS31, to multiple phenotypic features in six relevant eukaryotic model organisms (*S. cerevisiae, C. elegans, D. melanogaster, D. rerio, M. musculus,* and *H. sapiens*) [93]. Several reports on the characterization of specific features of eS10 have highlighted the important role of this r-protein. These include the description of a hypo-proliferative phenotype, known as the Minute phenotype, associated with loss-of-function mutations in one of the two copies of the *RPS10* gene during development in *Drosophila*. The Minute phenotype is characterized by a prolongation of the developmental time, the presence of short and thin bristles, and reduced fertility [151]. In *Drosophila*, eS10 is encoded by duplicated genes, and, interestingly, the expression of one of the two genes is enriched in the germline cells of embryonic gonads, suggesting a germline-specific role [151]. In *Arabidopsis*, loss-of-function mutations in one of the genes encoding eS10 lead to a reduction in stamen number, shoot and floral meristem defects, and a leaf polarity deficiency [152].

The r-protein eS12 also plays interesting roles not directly related to ribosome biogenesis or translation. In specific neurons, *RPS12* mRNA levels, among other r-protein transcripts, seem to be reduced by an acute period of sleep deprivation (hippocampus) or injury of the sciatic nerve (dorsal root ganglion), suggesting dynamic changes in ribosome composition following these insults (reviewed in [153]). In *S. cerevisiae*, a specific mutation in the *RPS12* gene leads to the suppression of phenotypes elicited by rDNA instability upon Fob1 overexpression [154]. Whether this phenomenon is the result of an extra-ribosomal function of yeast eS12 remains to be explored. Undoubtedly, the most interesting function of eS12 besides its orthodox roles in ribosome biogenesis and translation is that related to cell competition in *D. melanogaster*. In the classical form of cell competition, wild-type cells (homozygotic cells for r-protein genes; hereafter $Rp^{+/+}$ cells) in genetic mosaic flies are able to actively eliminate their adjacent heterozygotic cells (heterozygotic cells for r-protein genes; hereafter $Rp^{+/-}$ cells) from imaginal discs via apoptosis [155–157]. In this process, eS12 plays a specific role as a sensor of an imbalance of r-proteins to allow the elimination of $Rp^{+/-}$ cells [158]. Thus, the viable missense *rps12*[G97D] mutant allele of *RPS12* in homozygosis prevents cell competition of $Rp^{+/-}$ cells by wild-type $Rp^{+/+}$ cells [158,159]. In other words, *rps12*[G97D]$^{+/+}$ $Rp^{+/-}$ cells are not eliminated by wild-type $Rp^{+/+}$ cells. Moreover, the relative copy number of the wild-type *RPS12* allele in $Rp^{+/-}$ cells is apparently what determines the competitiveness [158]. It has been shown that the eS12[G97D] variant efficiently assembles into 40S r-subunits [158]. Moreover, the yeast *rps12*[G102D] allele (equivalent to *Drosophila rps12*[G97D] allele), when it is the sole cellular source of eS12 r-protein, neither confers a growth defect nor a global impairment of translation (S. M.-V., unpublished results). Interestingly, it has been shown that *Drosophila* eS12 is required to increase the transcription of the gene encoding the transcription factor Xrp1 [159,160], which itself also directly regulates cell competition [156,159]. Consistently, loss-of-function mutations in Xrp1 also prevent competition of $Rp^{+/-}$ by wild-type $Rp^{+/+}$ cells [159]. Whether the function of eS12 in promoting Xrp1 expression is extra-ribosomal still needs confirmation.

As for eS10 and eS12, there are reports indicating that eS31 could also have ribosome-independent functions (for a recent review, see [161]). First, eS31 has been identified as a regulator of the LMP1 protein encoded by the Epstein-Barr virus (EBV); thus, eS31 binds directly to LMP1 and increases its stability by reducing its proteasome-mediated degradation [162]. Moreover, overexpression of eS31 leads to increased cell growth and survival as the result of LMP1-mediated oncogenic events (e.g., epithelial to mesenchymal transition, motility, migration and invasion). In addition, as *RPS31* (also known as *RPS27A*) mRNA levels are reduced in sperm with low motility, eS31 might be necessary for optimal sperm functionality in humans [163]. In plants, eS31 seems to be highly expressed in meristematic tissues, pollen and ovules [164], and flowers of *RPS31*-silenced plants exhibit abnormal development [165]. Finally, it has been observed that double-strand break DNA damage results in the MDM2-independent proteasomal degradation of eS31 in HEK293 human cells, and as a consequence, these cells contain ribosomes that specifically lack eS31 and exhibit lower global translation activity [166]. Whether this phenomenon is part of an adaptive response to deal with DNA damage remains to be determined [166].

5. Assembly and Maturation of the Beak Structure

The assembly of the beak has been analyzed in the context of the general maturation of the 40S r-subunit, which begins in the nucleolus and ends in the cytoplasm, both in yeast and in human cell lines (for a review, see [167]). Moreover, given the fact that the beak is a pronounced protrusion in the structure of the 40S r-subunit, the assembly of the beak represents a challenge for the nucleocytoplasmic transport of this r-subunit. Using genetics in yeast and siRNA technology in human cell lines, the role of the beak r-proteins in pre-rRNA processing and r-subunit assembly has also been well examined. In both cases, it has been described that the three r-proteins that form the eukaryotic beak are required for the production and the stability of mature 40S r-subunits.

In yeast, eS10 is an essential r-protein whose contribution to ribosome biogenesis has been assessed by the use of a yeast strain conditionally expressing this r-protein [86,168]. These studies showed that eS10 is required for the efficient maturation of the 20S pre-rRNA, which accumulates to high levels upon eS10 depletion both within nucleoplasmic pre-40S r-particles, as a consequence of a delay in the export of pre-40S r-particles and cytoplasmic pre-40S r-particles, and as a consequence of inefficient 20S pre-rRNA processing at site D [86,168]. Interestingly, knocking down the expression of human *RPS10* leads to cytoplasmic accumulation of the 18S-E pre-rRNA, which is the equivalent human form of the yeast 20S pre-rRNA [169,170]. In yeast, we and others have demonstrated that the quasi-essential r-proteins eS12 and eS31 are also crucial for the efficient cytoplasmic processing of the 20S pre-rRNA into mature 18S rRNA [88,89,92]. Similarly, the cytoplasmic accumulation of 18S-E pre-rRNA has also been reported upon siRNA-mediated knockdown of human *RPS12* or *RPS27A* expression [170–172].

The precise timing of the assembly of the beak rRNA and r-proteins has also been analyzed in both yeast and humans at a reasonable resolution. Assembly of this structure involves compositional and structural changes of both the beak rRNA and r-proteins. In general terms, while the timing of eS31 assembly (nucle(ol)ar or cytoplasmic) is still controversial, it is likely that eS12 is incorporated early during the formation of 90S pre-ribosomal particles and that eS10 assembly occurs within late cytoplasmic pre-40S r-particles (see below). In yeast and the fungus *Chaetomium thermophilum*, the structural analysis of 90S pre-ribosomal particles suggests that the four subdomains of the 18S rRNA (5′, central, 3′ major, and 3′ minor) fold independently and associate co-transcriptionally with a set of r-proteins and ribosome assembly factors (RAFs), before being compacted into a defined pre-ribosomal particle. Analysis of the first reported structures of 90S pre-ribosomal particles indicated that the folding of the 3′ major domain of the 18S rRNA, the helix h33 included, requires the prior co-transcriptional structuring of the 5′ domain [173–175]. However, a more recent structural determination of a series of 90S assembly intermediates from *C. thermophilum* provides evidence that the formation of the 90S does not follow a strict 5′ to 3′ co-transcriptional direction; instead, the 3′ major and 3′ minor domains seem to assemble first with the 5′-ETS domain of 35S pre-rRNA, preceding the incorporation of the 5′ and central domains of pre-18S rRNA into 90S pre-ribosomal particles [176]. In any case, from the diverse collection of structurally stable 90S r-particles available in the literature, it is clear that these particles contain a clearly identifiable, immature beak structure. In most of these particles, the nascent beak structure comprises eS12, while only a few of them also contain eS31. As an example of this, in the 90S structure reported by Sun et al. [173], the beak forms a protrusion and it is composed of helices h32-34 and the r-proteins eS12 and eS31, connected to the body of the particle by the RAF Emg1 (see Figure 5). At this level, it is also clear that the presence of Enp1, which stabilizes the beak by binding to helices h32-34, impedes the incorporation of eS10, which can only occur after the release of Enp1 from late pre-40S r-particles in the cytoplasm [177,178]. Moreover, the fact that eS31 (and to a lesser extent eS12) is not present in many of the structural maps of 90S r-particles available in the literature, as well as in a variety of further pre-40S r-particles (see below), clearly indicates that the association of these r-proteins with early precursors of 40S r-subunits might be highly labile and should only become stable during late and cytoplasmic steps of 40S r-subunit maturation, concomitant with the formation of a more rigid beak structure. In this regard, another RAF, Tsr1, apparently blocks the correct binding of eS31 until its repositioning at a late maturation step occurring on cytoplasmic pre-40S r-particles [179]. Alternatively, as favored by other authors, eS12 and especially eS31 are only incorporated into cytoplasmic pre-40S r-particles [179,180]. However, at least in the case of eS31, we have identified a functional nuclear localization signal (NLS) within the first 25 amino acids of the N-terminal extension of yeast eS31, which is conserved in other eukaryotes [76]. This sequence is sufficient to target a triple GFP reporter to the nucleus, and most importantly, a functional GFP-tagged eS31 protein notably accumulates in the nucleus upon depletion of different 90S RAFs, among them Emg1, which leads to the nuclear retention of pre-40S r-particles ([76] and S. M.-V., unpublished results). Whatever the case may be, the N-terminal ubiquitin moiety present in the linear precursor of eS31 in many eukaryotes, including yeast and humans, is very rapidly and efficiently processed. Accordingly, under wild-type conditions, the Ubi3 precursor has so far never been detected; hence, it must be processed prior to the incorporation of eS31 into pre-40S r-subunits [89,92,181]. Consequently, it is unlikely that the ubiquitin moiety fused to eS31 directly participates in the ribosomal assembly of eS31. Moreover, when a wild-type and a cleavage-deficient Ubi3 variant are co-expressed in the same cells, eS31 derived from wild-type Ubi3 is preferentially incorporated into pre-40S r-particles compared to the non-cleaved ubiquitin-eS31 fusion protein, which in turn is rapidly degraded [89]. Forcing the assembly of non-cleaved Ubi3 into nascent 40S r-subunits only mildly impairs their biogenesis, but, as mentioned above, may lead to translation initiation defects [89].

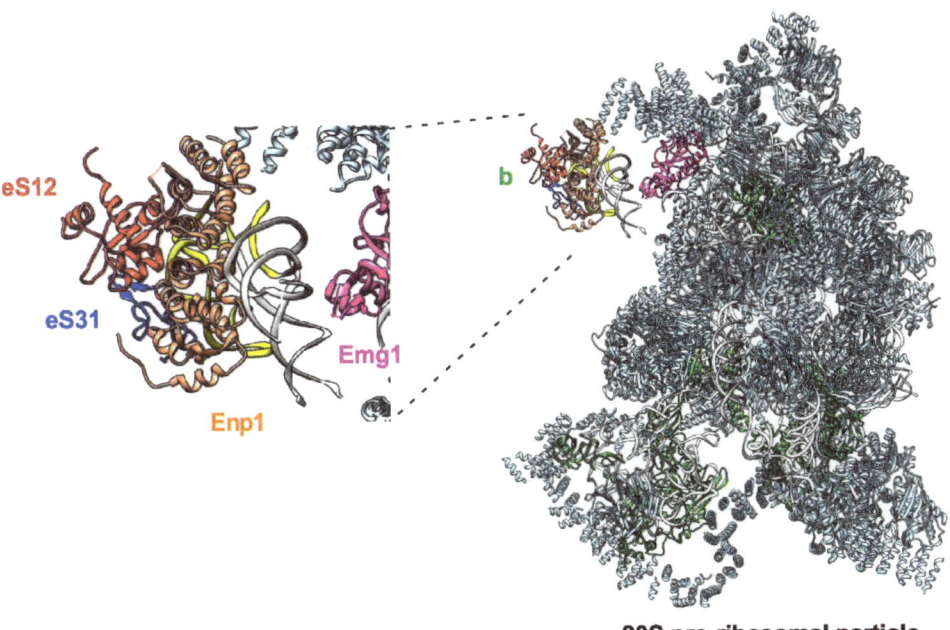

Figure 5. Formation of the early beak structure. Structural model of the yeast 90S pre-ribosomal particle (PDB code 5WYJ). The pre-rRNA is colored in gray, all r-proteins except eS12 and eS31 are colored in green and all ribosome assembly factors in light blue. Helix h33 is highlighted in yellow, eS12 in red and eS31 in blue. The interaction of Enp1 with the beak is shown. The beak is connected to the rest of the 90S particle by its interaction with Emg1/Nep1 (colored in pink). Left, close-up view of the beak (b) region. Cartoons were generated using UCSF Chimera (https://www.rbvi.ucsf.edu/chimera; accessed on 1 June 2024).

Following the sequential cleavages at sites A_0–A_2 within the 35S pre-rRNA, the yeast 90S r-particle is dismantled and converted into an early nuclear pre-40S r-particle, which is rapidly exported to the cytoplasm. At this step, and before export, most 90S RAFs have disassembled and only a few others have been recruited, among them Rio2, Tsr1, Ltv1 and Rrp12 [182]. Perhaps the characteristic that defines best the nucleoplasmic pre-40S intermediates is the high flexibility of their head domain, which becomes more structured as the particles transition through their maturation [183–185]. The incorporation of Ltv1 is relevant for beak formation as it interacts with Enp1 and the r-protein uS3, which binds at the base of the beak structure [119,186,187]. The recruitment of uS3 is initiated just before or concomitant with that of Ltv1 [188,189]. The r-protein uS3 consists of two distinct N- and C-terminal domains and is delivered to nuclear pre-40S r-particles by its dedicated chaperone Yar1 [190]. Yar1 binds only the N-terminal domain of uS3; thus, initial interaction of uS3 with pre-40S r-particles likely occurs through its C-terminal domain [188,191,192]. The release of Yar1 is concomitant with the interaction of the N-terminal domain of delivered uS3 with Ltv1. This interaction also contributes to preventing uS3 from prematurely acquiring its final and stable position within cytoplasmic pre-40S r-particles, which is only achieved upon the global structural changes occurring in these r-particles after Ltv1 release [119,179,190,191]. Indeed, a subcomplex formed by Ltv1, Enp1 and uS3 can be untethered from purified yeast pre-40S r-particles at high salt concentrations, while uS3 cannot be extracted from mature 40S r-subunits by the same treatment [193], indicating that uS3 is less stably integrated into pre-40S than mature 40S r-subunits. It has been suggested that a certain degree of flexibility in the beak is required at this nucleoplasmic stage because a rigid beak structure close to the head of the pre-40S r-subunit might hinder export through the nuclear pore complex (NPC) [179,193–195].

Once exported, the early cytoplasmic pre-40S r-particles undergo a cascade of maturation events, the first ones being essential for beak formation. The precise chronology of these events remains to be elucidated at high resolution, but it seems that it first involves the recruitment of the casein kinase Hrr25, also favored by the previous binding of the uS3 r-protein [190,192]. Moreover, the direct interaction between Hrr25 and Ltv1 appears to weaken the association between Ltv1 and Enp1 [192]. Hrr25, which is an essential protein, then phosphorylates Ltv1 on specific conserved serine residues, leading to the release of Ltv1 from pre-40S r-particles [188,189,192]. Strikingly, Hrr25 is no longer essential in the absence of Ltv1 or upon phosphomimetic substitutions of the specific Ltv1 serine residues [188,189], indicating that the essential function of Hrr25 is linked to Ltv1 in ribosome biogenesis. Interestingly, the release of Ltv1 is coordinated with that of Rio2 on the intersubunit side of the head domain of the pre-40S r-particle; a process that is mediated by the correct assembly of the uS10 r-protein [192]. The release of Ltv1 now provokes the dissociation of Enp1, which is also phosphorylated by the Hrr25 orthologue (CK1δ/ϵ) in humans [196]; phosphorylation of yeast Enp1 by Hrr25 is still controversial [189,193]. The dissociation of Enp1 and Ltv1 is absolutely required for nascent 40S r-subunits to become translationally competent, as their interaction with the beak environment would hinder the opening of the mRNA channel [177]. Another consequence of the dissociation of these factors is that eS10 gains access to its binding position and is integrated into the beak structure [180,186,197]. Concomitantly, uS19 and the two domains of uS3 are fitted into their mature position [119,179,192]. All these events promote the structural organization of the beak, which then enables the progression of the maturation events in other regions of the pre-40S r-particles [167,180].

Formation of the beak in human 40S r-subunits seems to occur in a similar way to that described in yeast, albeit with certain peculiarities [167,197,198]. Orthologs for all key factors mentioned above have been described in humans, including Enp1, Ltv1 and Hrr25 (Bystin, LTV1 and CK1δ/ϵ, respectively) [199]. Most importantly, despite differences in pre-rRNA processing between yeast and humans [200], the positioning, timing of interaction and dissociation, and function of all these RAFs have been well conserved [199]. Moreover, the structures of several nuclear and cytoplasmic pre-40S r-particles have been described, providing detailed insights into the maturation steps [184,197,198,201]. As an example, the cryo-EM structures of apparently nucleoplasmic human pre-40S r-particles contain, in addition to the orthologs of Enp1 and Ltv1, the r-proteins eS12, eS31 and uS3, despite the fact that, as in yeast, the assembly time point of eS31 and eS12 is again controversial [184,197].

Finally, although beyond of the scope of this review, the assembly of the prokaryotic beak is also an important late event during the maturation of 30S r-subunits; in this regard, it is interesting to mention that some authors have proposed that the bacterial assembly factor RimM, which is involved in the maturation of the 3' domain of the head of 30S r-subunits (see [202] and references therein), works as a functional analog of Ltv1 in bacteria [119,189]. Strikingly, in the absence of RimM, the 30S beak (h33) is not correctly folded, and several r-proteins, including uS19 as well as the tertiary binders uS3 and uS10, do not efficiently assemble into pre-30S r-particles [203,204]. Moreover, the assembly of the small subunit of the trypanosomal mitoribosome represents a major challenge for those researchers who are studying this process (e.g., [205]).

6. Beak Components and Human Diseases

It is evident that mutations and dysregulation of the majority of r-protein genes are linked to a range of human genetic diseases, such as ribosomopathies and cancer [206,207]; in this regard, beak r-proteins are not an exception.

Ribosomopathies are a group of rare inherited or acquired genetic diseases linked to defects in r-proteins or ribosome biogenesis factors [207–209]. Despite the importance of ribosomes in all cell types, these diseases result mainly in tissue-specific manifestations, especially in the hematopoietic system [207]. Intriguingly, inherited ribosomopathies are congenital and normally exhibit a paradoxical transition from early symptoms related to cellular hypo-proliferation to a hyper-proliferative oncogenic state later in life [210]. The best studied ribosomopathy, which is also one of the most prevalent ones (10 individuals per million live births) is the so-called Diamond–Blackfan Anemia (DBA) [211]. DBA is mostly a dominant genetic disorder (autosomal or X-linked) that is characterized by the reduced formation of red blood cells and is also associated with a series of other congenital anomalies, such as skeletal abnormalities, heart and genitourinary malformations, and an increased cancer susceptibility [212]. Most patients diagnosed with DBA harbor heterozygous loss-of-function mutations in particular genes encoding r-proteins, either of the small or the large r-subunit [212]. About 3% of all DBA patients have been reported to carry mutations in the *RPS10* gene, most likely leading in all cases to the production of non-functional eS10 variant proteins [212,213]; these mutations mostly consist of (i) insertion mutations that cause a frameshift and the appearance of a premature termination codon, (ii) nonsense mutations, namely, changes of particular codons to a stop codon (often, the R113Stop mutation), (iii) missense mutations that transform the *RPS10* start codon into an isoleucine or a threonine codon (M1I or M1T), and (iv) different other missense mutations (e.g., L14F, P30L) of so far unknown biological significance ([213,214]; for more details, check the UniProt entry P46783 and the OMIM entry 603632). Mutations in the human *RPS27A* gene, which codes for human eS31, have also been identified in patients with DBA (e.g., S57P); however, whether these mutations are indeed pathogenic genetic variants remains to be determined [215]. To our knowledge, no DBA-linked mutations have so far been reported in the *RPS12* gene. However, as the underlying mutations in at least 20% of patients with DBA syndromes have not yet been identified [212], it is still possible that *RPS12* alleles could be responsible for DBA manifestations; in line with this possibility, *RPS12* haploinsufficiency in mice leads to an erythropoiesis defect that recapitulates the one found in DBA patients (discussed in [216]). From a very simplistic point of view, as the DBA disease is mostly caused by loss-of-function mutations in several r-protein and a few RAF genes, all associated DBA symptoms and manifestations must come from common dysfunctions of the same molecular process that, in this case, can be no other than ribosome biogenesis [217], ultimately leading to an impairment or a limitation of translation. Thus, in a non-exclusive manner, it has been proposed that (i) the hypo-proliferative, pro-apoptotic anemia associated with DBA could be the consequence of a global reduction in translation, limiting below a critical threshold the synthesis of critical proteins, such as the globins and the transcription factor GATA1, with the latter being essential for normal erythropoiesis [218]. In agreement with this possibility, *GATA1* translation is reduced in erythroid precursor cells of DBA patients with mutations in different r-protein genes (e.g., [219–221]), and loss-of-function mutations in *GATA1* result in a DBA-like phenotype (e.g., [222,223] and references therein). (ii) DBA cells display elevated levels of reactive oxygen species (ROS), which, by generating a high oxidative stress, inhibit cell proliferation [206,207]. In this regard, lowering cellular ROS levels by antioxidants can rescue the proliferation defects in cells subjected to r-protein haploinsufficiency or carrying selected r-protein mutations [207,224]. (iii) As a consequence of the ribosome biogenesis deficiency occurring in DBA cells, the so-called nucleolar stress response is triggered in these cells, which induces p53 stabilization and enables p53 to transactivate its target genes, leading to cell cycle arrest, apoptosis, autophagy, and senescence [225,226]. In normal growth conditions, cellular levels of p53 are maintained low due to its efficient recognition and ubiquitination by the E3 ubiquitin ligase MDM2 (HDM2 in humans) and the subsequent degradation of ubiquitinated p53 by the proteasome. However, when ribosome biogenesis is impaired, non-assembled r-proteins tend to accumulate and can be released from the nucleolus to the nucleoplasm. Several different free r-proteins, but primarily uL5 and uL18 as part of the 5S RNP, can bind and sequester MDM2, thereby preventing the degradation of p53; thus, the upregulation of p53 explains many of the hypo-proliferative phenotypes displayed by DBA patients, including bone marrow erythroid hypoplasia [206,227,228]. In consonance with the relevant role of nucleolar stress in DBA, genetic or pharmacological inactivation of p53 can rescue disease-associated phenotypes [206,209,225,229]. Interestingly, it has been shown that eS31 is able to regulate the MDM2-p53 loop in response to nucleolar stress [230,231]. Moreover, eS31 apparently interacts with the central acidic domain of MDM2 through its eukaryote-specific N-terminal extension [230]. Therefore, this interaction seems not to be mutually exclusive from the ones of MDM2 with p53, which used a short, N-terminal segment to bind to the N-terminal domain of MDM2 [232], and with the r-proteins uL5 and uL18, which mostly involve the Zn-finger and RING domains of MDM2 [59,233]. Importantly, the overexpression of eS31 reduces MDM2-mediated ubiquitination of p53, thereby leading to its stabilization and activation [230,234]. The induction of p53 by eS31 may likely be additionally fueled by the observation that the *RPS27A* gene is apparently also transcriptionally activated by p53 [234]. Moreover, it has also been shown that MDM2 mediates the ubiquitination and proteasomal degradation of eS31 in response to nucleolar stress, indicating that free eS31 could be a physiological substrate of MDM2 [230]. It has been proposed that this mutual inhibitory regulation between MDM2 and eS31 may contribute to cellular recovery after the experienced stress [230]. Another report has suggested that the RAF PICT1 seems to regulate the interaction between eS31 and MDM2, as low levels of PICT1 are apparently required for the efficient translocation of eS31 from the nucleolus to the nucleoplasm, so that it can bind MDM2 [235]. It has also been described in several human lung cancer cell lines that eS31 could interact with uL5 in a way that might weaken the strength of the interaction between uL5 and MDM2; thus, knockdown of *RPS27A* stabilizes p53 in a uL5-dependent manner, promoting the p53 tumor suppressor functions [231]. Altogether, the above-mentioned data highlight that eS31 is connected to the MDM2-p53 axis and suggest that eS31 may be relevant for fine-tuning the cellular response to and recovery from nucleolar stress. However, its importance is apparently cell-type dependent, as recently shown by the Schneider group, who demonstrated that knockdown of *RPS27A* robustly induced p53 in certain cell lines but not in others [181].

Cancer cells require a high production of ribosomes to sustain boundless growth and cell division (e.g., [236]). Moreover, many r-proteins, including the beak ones, have been implicated in cancer development (e.g., [206,237–239]). Mutations and altered expression of beak r-proteins have been described in many cancer types, in some cases likely displaying an extra-ribosomal function: (i) high expression of *RPS10* has been found in colorectal, renal and prostate cancer [240], whereas it has been reported that ribosomes purified from MDA-MB-231 breast cancer cells contain substoichiometric levels of eS10, among other r-proteins [119]. (ii) Overexpression of *RPS12* has been observed in colon adenomatous polyps and carcinomas as well as in gastric cancers [241]. Deletions of *RPS12* are frequently observed in diffuse large B cell lymphomas [242], and eS12 has been reported to play a role as a stimulator of WNT secretion in cancer cells, which is particularly important in the context of triple-negative breast cancer initiation and progression [243]. (iii) It has been reported that eS31 is overexpressed in renal, colon, cervical, and breast cancers, chronic myeloid leukemia and lung adenocarcinoma [161,231,244,245]. In most of these cases, the overexpression of eS31 correlates with poor prognosis for the patients. A recent review has highlighted the expression and role of eS31 in cancer cells and tumor tissues [161].

7. Concluding Remarks and Future Perspectives

Herein, we have discussed the relevance of the beak, a structurally conserved region of the small r-subunit of cytoplasmic ribosomes in all three domains of life. Notably, the beak has transitioned from a structure composed exclusively of rRNA in bacteria to a protuberance comprising three specific r-proteins, eS10, eS12 and eS31, in eukaryotes, while an intermediate situation prevails in archaea as the beaks of many species only contain two r-proteins, eS31 and eL8, with the latter being clearly the ancestor of the eukaryotic eS12 r-protein. As also discussed, the beak has diverged considerably in the mitoribosomes of some organisms, such as in trypanosomatids, perhaps owing to the particular translation requirements inside these organelles [246,247]. As outlined in this review, the beak has important roles in the three major phases of translation (initiation, elongation and termination) and is involved in many other events of the translation process, including ribosome stalling and collisions, as well as other seemingly translation-unrelated processes, such as cell competition in *Drosophila*. In many of these processes, the biological significance of post-translational modifications, such as ubiquitination, is still poorly understood. We have also highlighted the implication of the beak r-proteins in the stepwise assembly of nascent 40S r-subunits. Regarding the maturation of the eukaryotic beak, it is evident that further research is required to precisely define the assembly timing of the beak r-proteins; for instance, it is still controversial whether eS31 associates with 40S r-subunit maturation intermediates in the nucleus or the cytoplasm. More studies are also required to elucidate whether the beak r-proteins play active or passive roles during the assembly and nuclear export of pre-40S r-subunits. Finally, we discussed the relevance of the beak r-proteins in human diseases, especially ribosomopathies and cancer. A deeper understanding of the connection between these diseases and the ribosome biogenesis process is expected to offer new perspectives for therapeutic approaches.

Author Contributions: S.M.-V., C.V.G., C.R.-H., D.K. and J.d.l.C. wrote the manuscript. D.K. and J.d.l.C. edited the draft and the final version of the manuscript. All authors have read and agreed to the published version of the manuscript.

Funding: This publication is part of the project R+D+i PID2022-136564NB-I00 funded by MCIN/AEI/10.13039/501100011033/ERDF, EU to J.d.l.C. We also acknowledge the Andalusian Platform for Biomodels and Resources in Genomic Edition, the Fortalece Program (FORT 2023) from MCIN, and the Translacore (CA21154) and ProteoCure (CA20113) COST Actions from the EU for support. D. K. is supported by the Swiss National Science Foundation (project grant 310030_204801). S.M.-V. was an academic research staff member of the Andalusian Research, Development, and Innovation Plan by the Andalusian Regional Government (PAIDI 2020). C.V.G. was supported by a Marie Sklodowska-Curie Individual Fellowship (H2020-MSCA-IF-2020, Grant 101024158 from the EU). C.R.-H. is supported by a María Zambrano Grant from the Spanish Ministry of Universities.

Acknowledgments: We thank members of the de la Cruz laboratory for valuable discussions, critical reading, and suggestions.

Conflicts of Interest: The authors declare no conflicts of interest.

References

1. Wilson, D.N.; Doudna Cate, J.H. The structure and function of the eukaryotic ribosome. *Cold Spring Harb. Perspect. Biol.* **2012**, *4*, a011536. [CrossRef] [PubMed]
2. Melnikov, S.; Ben-Shem, A.; Garreau de Loubresse, N.; Jenner, L.; Yusupova, G.; Yusupov, M. One core, two shells: Bacterial and eukaryotic ribosomes. *Nat. Struct. Mol. Biol.* **2012**, *19*, 560–567. [CrossRef] [PubMed]
3. Greber, B.J.; Ban, N. Structure and function of the mitochondrial ribosome. *Annu. Rev. Biochem.* **2016**, *85*, 103–132. [CrossRef] [PubMed]
4. Robles, P.; Quesada, V. Unveiling the functions of plastid ribosomal proteins in plant development and abiotic stress tolerance. *Plant Physiol. Biochem.* **2022**, *189*, 35–45. [CrossRef] [PubMed]
5. Sulima, S.O.; Dinman, J.D. The expanding riboverse. *Cells* **2019**, *8*, 1205. [CrossRef] [PubMed]
6. Yusupova, G.; Yusupov, M. High-resolution structure of the eukaryotic 80S ribosome. *Annu. Rev. Biochem.* **2014**, *83*, 467–486. [CrossRef] [PubMed]
7. Anger, A.M.; Armache, J.P.; Berninghausen, O.; Habeck, M.; Subklewe, M.; Wilson, D.N.; Beckmann, R. Structures of the human and *Drosophila* 80S ribosome. *Nature* **2013**, *497*, 80–85. [CrossRef] [PubMed]
8. Klinge, S.; Voigts-Hoffmann, F.; Leibundgut, M.; Ban, N. Atomic structures of the eukaryotic ribosome. *Trends Biochem. Sci.* **2012**, *37*, 189–198. [CrossRef]
9. Noeske, J.; Wasserman, M.R.; Terry, D.S.; Altman, R.B.; Blanchard, S.C.; Cate, J.H. High-resolution structure of the *Escherichia coli* ribosome. *Nat. Struct. Mol. Biol.* **2015**, *22*, 336–341. [CrossRef]
10. Amunts, A.; Brown, A.; Toots, J.; Scheres, S.H.W.; Ramakrishnan, V. Ribosome. The structure of the human mitochondrial ribosome. *Science* **2015**, *348*, 95–98. [CrossRef]

11. Lafontaine, D.L.; Tollervey, D. The function and synthesis of ribosomes. *Nat. Rev. Mol. Cell Biol.* **2001**, *2*, 514–520. [CrossRef] [PubMed]
12. Bernier, C.R.; Petrov, A.S.; Kovacs, N.A.; Penev, P.I.; Williams, L.D. Translation: The universal structural core of life. *Mol. Biol. Evol.* **2018**, *35*, 2065–2076. [CrossRef] [PubMed]
13. Agrawal, R.K.; Majumdar, S. Evolution: Mitochondrial ribosomes across species. *Methods Mol. Biol.* **2023**, *2661*, 7–21. [PubMed]
14. Gonzalo, P.; Reboud, J.P. The puzzling lateral flexible stalk of the ribosome. *Biol. Cell* **2003**, *95*, 179–193. [CrossRef] [PubMed]
15. Ballesta, J.P.G.; Remacha, M. The large ribosomal subunit stalk as a regulatory element of the eukaryotic translational machinery. *Prog. Nucleic Acid Res. Mol. Biol.* **1996**, *55*, 157–193. [PubMed]
16. Diaconu, M.; Kothe, U.; Schlunzen, F.; Fischer, N.; Harms, J.M.; Tonevitsky, A.G.; Stark, H.; Rodnina, M.V.; Wahl, M.C. Structural basis for the function of the ribosomal L7/12 stalk in factor binding and GTPase activation. *Cell* **2005**, *121*, 991–1004. [CrossRef]
17. Kulczyk, A.W.; Sorzano, C.O.S.; Grela, P.; Tchorzewski, M.; Tumer, N.E.; Li, X.P. Cryo-EM structure of Shiga toxin 2 in complex with the native ribosomal P-stalk reveals residues involved in the binding interaction. *J. Biol. Chem.* **2023**, *299*, 102795. [CrossRef]
18. Fan, X.; Zhu, Y.; Wang, C.; Niu, L.; Teng, M.; Li, X. Structural insights into the interaction of the ribosomal P stalk protein P2 with a type II ribosome-inactivating protein ricin. *Sci. Rep.* **2016**, *6*, 37803. [CrossRef]
19. Gupta, R.; Hinnebusch, A.G. Differential requirements for P stalk components in activating yeast protein kinase Gcn2 by stalled ribosomes during stress. *Proc. Natl. Acad. Sci. USA* **2023**, *120*, e2300521120. [CrossRef]
20. Inglis, A.J.; Masson, G.R.; Shao, S.; Perisic, O.; McLaughlin, S.H.; Hegde, R.S.; Williams, R.L. Activation of GCN2 by the ribosomal P-stalk. *Proc. Natl. Acad. Sci. USA* **2019**, *116*, 4946–4954. [CrossRef]
21. Trabuco, L.G.; Schreiner, E.; Eargle, J.; Cornish, P.; Ha, T.; Luthey-Schulten, Z.; Schulten, K. The role of L1 stalk-tRNA interaction in the ribosome elongation cycle. *J. Mol. Biol.* **2010**, *402*, 741–760. [CrossRef] [PubMed]
22. Mohan, S.; Noller, H.F. Recurring RNA structural motifs underlie the mechanics of L1 stalk movement. *Nat. Commun.* **2017**, *8*, 14285. [CrossRef] [PubMed]
23. Ghosh, A.; Komar, A.A. Eukaryote-specific extensions in ribosomal proteins of the small subunit: Structure and function. *Translation* **2015**, *3*, e999576. [CrossRef] [PubMed]
24. Timsit, Y.; Sergeant-Perthuis, G.; Bennequin, D. Evolution of ribosomal protein network architectures. *Sci. Rep.* **2021**, *11*, 625. [CrossRef] [PubMed]
25. Kisly, I.; Tamm, T. Archaea/eukaryote-specific ribosomal proteins-guardians of a complex structure. *Comput. Struct. Biotechnol. J.* **2023**, *21*, 1249–1261. [CrossRef] [PubMed]
26. Liljas, A.; Sanyal, S. The enigmatic ribosomal stalk. *Q. Rev. Biophys.* **2018**, *51*, e12. [CrossRef] [PubMed]
27. Zinker, S.; Warner, J.R. The ribosomal proteins of *Saccharomyces cerevisiae*. Phosphorylated and exchangeable proteins. *J. Biol. Chem.* **1976**, *251*, 1799–1807. [CrossRef] [PubMed]
28. Tsurugi, K.; Ogata, K. Evidence for the exchangeability of acidic ribosomal proteins on cytoplasmic ribosomes in regenerating rat liver. *J. Biochem.* **1985**, *98*, 1427–1431. [CrossRef] [PubMed]
29. Bautista-Santos, A.; Zinker, S. The P1/P2 protein heterodimers assemble to the ribosomal stalk at the moment when the ribosome Is committed to translation but not to the native 60S ribosomal subunit in *Saccharomyces cerevisiae*. *Biochemistry* **2014**, *53*, 4105–4112. [CrossRef]
30. Remacha, M.; Jiménez-Díaz, A.; Bermejo, B.; Rodríguez-Gabriel, M.A.; Guarinos, E.; Ballesta, J.P.G. Ribosomal acidic phosphoproteins P1 and P2 are not required for cell viability but regulate the pattern of protein expression in *Saccharomyces cerevisiae*. *Mol. Cell. Biol.* **1995**, *15*, 4754–4762. [CrossRef]
31. Sáenz-Robles, M.T.; Remacha, M.; Vilella, M.D.; Zinker, S.; Ballesta, J.P. The acidic ribosomal proteins as regulators of the eukaryotic ribosomal activity. *Biochim. Biophys. Acta* **1990**, *1050*, 51–55. [CrossRef] [PubMed]
32. Derylo, K.; Michalec-Wawiorka, B.; Krokowski, D.; Wawiorka, L.; Hatzoglou, M.; Tchorzewski, M. The uL10 protein, a component of the ribosomal P-stalk, is released from the ribosome in nucleolar stress. *Biochim. Biophys. Acta Mol. Cell Res.* **2018**, *1865*, 34–47. [CrossRef] [PubMed]
33. Siodmak, A.; Martínez-Seidel, F.; Rayapuram, N.; Bazin, J.; Alhoraibi, H.; Gentry-Torfer, D.; Tabassum, N.; Sheikh, A.H.; Kise, J.K.G.; Blilou, I.; et al. Dynamics of ribosome composition and ribosomal protein phosphorylation in immune signaling in *Arabidopsis thaliana*. *Nucleic Acids Res.* **2023**, *51*, 11876–11892. [CrossRef]
34. Ballesta, J.P.G.; Rodriguez-Gabriel, M.A.; Bou, G.; Briones, E.; Zambrano, R.; Remacha, M. Phosphorylation of the yeast ribosomal stalk. Functional affects and enzymes involved in the process. *FEMS Microbiol. Rev.* **1999**, *23*, 537–550. [CrossRef] [PubMed]
35. Zambrano, R.; Briones, E.; Remacha, M.; Ballesta, J.P. Phosphorylation of the acidic ribosomal P proteins in *Saccharomyces cerevisiae*: A reappraisal. *Biochemistry* **1997**, *36*, 14439–14446. [CrossRef] [PubMed]
36. Subramanian, A.R.; van Duin, J. Exchange of individual ribosomal proteins between ribosomes as studied by heavy isotope-transfer experiments. *Mol. Gen. Genet.* **1977**, *158*, 1–9. [CrossRef] [PubMed]
37. Dick, F.A.; Eisinger, D.P.; Trumpower, B.L. Exchangeability of Qsr1p, a large ribosomal subunit protein required for subunit joining, suggests a novel translational regulatory mechanism. *FEBS Lett.* **1997**, *419*, 1–3. [CrossRef] [PubMed]
38. Yang, Y.M.; Jung, Y.; Abegg, D.; Adibekian, A.; Carroll, K.S.; Karbstein, K. Chaperone-directed ribosome repair after oxidative damage. *Mol. Cell* **2023**, *83*, 1527–1537.e1525. [CrossRef] [PubMed]
39. Eisinger, D.P.; Dick, F.A.; Trumpower, B.L. Qsr1p, a 60S ribosomal subunit protein, is required for joining of 40S and 60S subunits. *Mol. Cell. Biol.* **1997**, *17*, 5136–5145. [CrossRef]

40. Sulima, S.O.; Gulay, S.P.; Anjos, M.; Patchett, S.; Meskauskas, A.; Johnson, A.W.; Dinman, J.D. Eukaryotic rpL10 drives ribosomal rotation. *Nucleic Acids Res.* **2014**, *42*, 2049–2063. [CrossRef]
41. Pollutri, D.; Penzo, M. Ribosomal protein L10: From function to dysfunction. *Cells* **2020**, *9*, 2503. [CrossRef]
42. Adams, D.R.; Ron, D.; Kiely, P.A. RACK1, A multifaceted scaffolding protein: Structure and function. *Cell Commun. Signal.* **2011**, *9*, 22. [CrossRef] [PubMed]
43. Singh, N.; Jindal, S.; Ghosh, A.; Komar, A.A. Communication between RACK1/Asc1 and uS3 (Rps3) is essential for RACK1/Asc1 function in yeast *Saccharomyces cerevisiae*. *Gene* **2019**, *706*, 69–76. [CrossRef]
44. Nilsson, J.; Sengupta, J.; Frank, J.; Nissen, P. Regulation of eukaryotic translation by the RACK1 protein: A platform for signalling molecules on the ribosome. *EMBO Rep.* **2004**, *5*, 1137–1141. [CrossRef]
45. Thompson, M.K.; Rojas-Duran, M.F.; Gangaramani, P.; Gilbert, W.V. The ribosomal protein Asc1/RACK1 is required for efficient translation of short mRNAs. *eLife* **2016**, *5*, e11154. [CrossRef] [PubMed]
46. Gerbasi, V.R.; Weaver, C.M.; Hill, S.; Friedman, D.B.; Link, A.J. Yeast Asc1p and mammalian RACK1 are functionally orthologous core 40S ribosomal proteins that repress gene expression. *Mol. Cell. Biol.* **2004**, *24*, 8276–8287. [CrossRef]
47. Gerbasi, V.R.; Browne, C.M.; Samir, P.; Shen, B.; Sun, M.; Hazelbaker, D.Z.; Galassie, A.C.; Frank, J.; Link, A.J. Critical role for *Saccharomyces cerevisiae* Asc1p in translational initiation at elevated temperatures. *Proteomics* **2018**, *18*, e1800208. [CrossRef]
48. Ikeuchi, K.; Inada, T. Ribosome-associated Asc1/RACK1 is required for endonucleolytic cleavage induced by stalled ribosome at the 3' end of nonstop mRNA. *Sci. Rep.* **2016**, *6*, 28234. [CrossRef]
49. Wolf, A.S.; Grayhack, E.J. Asc1, homolog of human RACK1, prevents frameshifting in yeast by ribosomes stalled at CGA codon repeats. *RNA* **2015**, *21*, 935–945. [CrossRef] [PubMed]
50. Johnson, A.G.; Lapointe, C.P.; Wang, J.; Corsepius, N.C.; Choi, J.; Fuchs, G.; Puglisi, J.D. RACK1 on and off the ribosome. *RNA* **2019**, *25*, 881–895. [CrossRef]
51. Rachfall, N.; Schmitt, K.; Bandau, S.; Smolinski, N.; Ehrenreich, A.; Valerius, O.; Braus, G.H. RACK1/Asc1p, a ribosomal node in cellular signaling. *Mol. Cell. Proteom.* **2013**, *12*, 87–105. [CrossRef] [PubMed]
52. Pulk, A.; Liiv, A.; Peil, L.; Maivali, U.; Nierhaus, K.; Remme, J. Ribosome reactivation by replacement of damaged proteins. *Mol. Microbiol.* **2010**, *75*, 801–814. [CrossRef] [PubMed]
53. Yang, Y.M.; Karbstein, K. The chaperone Tsr2 regulates Rps26 release and reincorporation from mature ribosomes to enable a reversible, ribosome-mediated response to stress. *Sci. Adv.* **2022**, *8*, eabl4386. [CrossRef] [PubMed]
54. Yang, Y.M.; Karbstein, K. Ribosome assembly and repair. *Annu. Rev. Cell Dev. Biol.* **2024**. [CrossRef] [PubMed]
55. Ferretti, M.B.; Ghalei, H.; Ward, E.A.; Potts, E.L.; Karbstein, K. Rps26 directs mRNA-specific translation by recognition of Kozak sequence elements. *Nat. Struct. Mol. Biol.* **2017**, *24*, 700–707. [CrossRef]
56. Fusco, C.M.; Desch, K.; Dorrbaum, A.R.; Wang, M.; Staab, A.; Chan, I.C.W.; Vail, E.; Villeri, V.; Langer, J.D.; Schuman, E.M. Neuronal ribosomes exhibit dynamic and context-dependent exchange of ribosomal proteins. *Nat. Commun.* **2021**, *12*, 6127. [CrossRef] [PubMed]
57. Leiva, L.E.; Katz, A. Regulation of leaderless mRNA translation in bacteria. *Microorganisms* **2022**, *10*, 723. [CrossRef] [PubMed]
58. Kaberdina, A.C.; Szaflarski, W.; Nierhaus, K.H.; Moll, I. An unexpected type of ribosomes induced by kasugamycin: A look into ancestral times of protein synthesis? *Mol. Cell* **2009**, *33*, 227–236. [CrossRef] [PubMed]
59. Castillo Duque de Estrada, N.M.; Thoms, M.; Flemming, D.; Hammaren, H.M.; Buschauer, R.; Ameismeier, M.; Bassler, J.; Beck, M.; Beckmann, R.; Hurt, E. Structure of nascent 5S RNPs at the crossroad between ribosome assembly and MDM2-p53 pathways. *Nat. Struct. Mol. Biol.* **2023**, *30*, 1119–1131. [CrossRef] [PubMed]
60. Lau, B.; Huang, Z.; Kellner, N.; Niu, S.; Berninghausen, O.; Beckmann, R.; Hurt, E.; Cheng, J. Mechanism of 5S RNP recruitment and helicase-surveilled rRNA maturation during pre-60S biogenesis. *EMBO Rep.* **2023**, *24*, e56910. [CrossRef]
61. Leidig, C.; Thoms, M.; Holdermann, I.; Bradatsch, B.; Berninghausen, O.; Bange, G.; Sinning, I.; Hurt, E.; Beckmann, R. 60S ribosome biogenesis requires rotation of the 5S ribonucleoprotein particle. *Nat. Commun.* **2014**, *5*, 3491. [CrossRef]
62. Gómez-Herreros, F.; Rodríguez-Galán, O.; Morillo-Huesca, M.; Maya, D.; Arista-Romero, M.; de la Cruz, J.; Chávez, S.; Muñoz-Centeno, M.C. Balanced production of ribosome components is required for proper G1/S transition in *Saccharomyces cerevisiae*. *J. Biol. Chem.* **2013**, *288*, 31689–31700. [CrossRef]
63. Sloan, K.E.; Bohnsack, M.T.; Watkins, N.J. The 5S RNP couples p53 homeostasis to ribosome biogenesis and nucleolar stress. *Cell Rep.* **2013**, *5*, 237–247. [CrossRef] [PubMed]
64. Donati, G.; Peddigari, S.; Mercer, C.A.; Thomas, G. 5S ribosomal RNA is an essential component of a nascent ribosomal precursor complex that regulates the Hdm2-p53 checkpoint. *Cell Rep.* **2013**, *4*, 87–98. [CrossRef] [PubMed]
65. Gentilella, A.; Morón-Durán, F.D.; Fuentes, P.; Zweig-Rocha, G.; Riano-Canalias, F.; Pelletier, J.; Ruiz, M.; Turón, G.; Castano, J.; Tauler, A.; et al. Autogenous control of 5'TOP mRNA stability by 40S ribosomes. *Mol. Cell* **2017**, *67*, 55–70.e54. [CrossRef] [PubMed]
66. Eastham, M.J.; Pelava, A.; Wells, G.R.; Lee, J.K.; Lawrence, I.R.; Stewart, J.; Deichner, M.; Hertle, R.; Watkins, N.J.; Schneider, C. The induction of p53 correlates with defects in the production, but not the levels, of the small ribosomal subunit and stalled large ribosomal subunit biogenesis. *Nucleic Acids Res.* **2023**, *51*, 9397–9414. [CrossRef] [PubMed]
67. Zheng, J.; Lang, Y.; Zhang, Q.; Cui, D.; Sun, H.; Jiang, L.; Chen, Z.; Zhang, R.; Gao, Y.; Tian, W.; et al. Structure of human MDM2 complexed with RPL11 reveals the molecular basis of p53 activation. *Genes Dev.* **2015**, *29*, 1524–1534. [CrossRef] [PubMed]

68. Pelava, A.; Schneider, C.; Watkins, N.J. The importance of ribosome production, and the 5S RNP-MDM2 pathway, in health and disease. *Biochem. Soc. Trans.* **2016**, *44*, 1086–1090. [CrossRef] [PubMed]
69. Chen, J. The cell-cycle arrest and apoptotic functions of p53 in tumor initiation and progression. *Cold Spring Harb. Perspect. Med.* **2016**, *6*, a026104. [CrossRef] [PubMed]
70. Bustelo, X.R.; Dosil, M. Ribosome biogenesis and cancer: Basic and translational challenges. *Curr. Opin. Genet. Dev.* **2018**, *48*, 22–29. [CrossRef]
71. Penzo, M.; Montanaro, L.; Trere, D.; Derenzini, M. The ribosome biogenesis-cancer connection. *Cells* **2019**, *8*, 55. [CrossRef] [PubMed]
72. Lindström, M.S.; Bartek, J.; Maya-Mendoza, A. p53 at the crossroad of DNA replication and ribosome biogenesis stress pathways. *Cell Death Differ.* **2022**, *29*, 972–982. [CrossRef] [PubMed]
73. Wimberly, B.T.; Brodersen, D.E.; Clemons, W.M.; Morgan-Warren, R.J.; Carter, A.P.; Vonrhein, C.; Hartsch, T.; Ramakrishnan, V. Structure of the 30S ribosomal subunit. *Nature* **2000**, *407*, 327–339. [CrossRef] [PubMed]
74. Schluenzen, F.; Tocilj, A.; Zarivach, R.; Harms, J.; Bashan, A.; Bartels, H.; Agmon, I.; Franceschi, F.; Yonath, A. Structure of functional activated small ribosomal subunit at 3.3 Å resolution. *Cell* **2000**, *102*, 615–623. [CrossRef]
75. Londei, P.; Ferreira-Cerca, S. Ribosome biogenesis in archaea. *Front Microbiol.* **2021**, *12*, 686977. [CrossRef] [PubMed]
76. Fernández-Pevida, A.; Martín-Villanueva, S.; Murat, G.; Lacombe, T.; Kressler, D.; de la Cruz, J. The eukaryote-specific N-terminal extension of ribosomal protein S31 contributes to the assembly and function of 40S ribosomal subunits. *Nucleic Acids Res.* **2016**, *44*, 7777–7791. [CrossRef] [PubMed]
77. Lecompte, O.; Ripp, R.; Thierry, J.C.; Moras, D.; Poch, O. Comparative analysis of ribosomal proteins in complete genomes: An example of reductive evolution at the domain scale. *Nucleic Acids Res.* **2002**, *30*, 5382–5390. [CrossRef] [PubMed]
78. Armache, J.P.; Anger, A.M.; Marquez, V.; Franckenberg, S.; Frohlich, T.; Villa, E.; Berninghausen, O.; Thomm, M.; Arnold, G.J.; Beckmann, R.; et al. Promiscuous behaviour of archaeal ribosomal proteins: Implications for eukaryotic ribosome evolution. *Nucleic Acids Res.* **2013**, *41*, 1284–1293. [CrossRef] [PubMed]
79. Mendler, K.; Chen, H.; Parks, D.H.; Lobb, B.; Hug, L.A.; Doxey, A.C. AnnoTree: Visualization and exploration of a functionally annotated microbial tree of life. *Nucleic Acids Res.* **2019**, *47*, 4442–4448. [CrossRef]
80. Manuell, A.L.; Quispe, J.; Mayfield, S.P. Structure of the chloroplast ribosome: Novel domains for translation regulation. *PLoS Biol.* **2007**, *5*, e209. [CrossRef]
81. Bieri, P.; Leibundgut, M.; Saurer, M.; Boehringer, D.; Ban, N. The complete structure of the chloroplast 70S ribosome in complex with translation factor pY. *EMBO J.* **2017**, *36*, 475–486. [CrossRef] [PubMed]
82. Perez-Boerema, A.; Aibara, S.; Paul, B.; Tobiasson, V.; Kimanius, D.; Forsberg, B.O.; Wallden, K.; Lindahl, E.; Amunts, A. Structure of the chloroplast ribosome with chl-RRF and hibernation-promoting factor. *Nat. Plants* **2018**, *4*, 212–217. [CrossRef]
83. Desai, N.; Brown, A.; Amunts, A.; Ramakrishnan, V. The structure of the yeast mitochondrial ribosome. *Science* **2017**, *355*, 528–531. [CrossRef]
84. Scaltsoyiannes, V.; Corre, N.; Waltz, F.; Giegé, P. Types and functions of mitoribosome-specific ribosomal proteins across eukaryotes. *Int. J. Mol. Sci.* **2022**, *23*, 3474. [CrossRef] [PubMed]
85. Steffen, K.K.; McCormick, M.A.; Pham, K.M.; Mackay, V.L.; Delaney, J.R.; Murakami, C.J.; Kaeberlein, M.; Kennedy, B.K. Ribosome deficiency protects against ER stress in *Saccharomyces cerevisiae*. *Genetics* **2012**, *191*, 107–118. [CrossRef] [PubMed]
86. Ferreira-Cerca, S.; Pöll, G.; Gleizes, P.E.; Tschochner, H.; Milkereit, P. Roles of eukaryotic ribosomal proteins in maturation and transport of pre-18S rRNA and ribosome function. *Mol. Cell* **2005**, *20*, 263–275. [CrossRef] [PubMed]
87. Ben-Shem, A.; Garreau de Loubresse, N.; Melnikov, S.; Jenner, L.; Yusupova, G.; Yusupov, M. The structure of the eukaryotic ribosome at 3.0 Å resolution. *Science* **2011**, *334*, 1524–1529. [CrossRef]
88. Martín-Villanueva, S.; Fernández-Fernández, J.; Rodríguez-Galán, O.; Fernández-Boraita, J.; Villalobo, E.; de la Cruz, J. Role of the 40S beak ribosomal protein eS12 in ribosome biogenesis and function in *Saccharomyces cerevisiae*. *RNA Biol.* **2020**, *17*, 1261–1276. [CrossRef]
89. Lacombe, T.; García-Gómez, J.J.; de la Cruz, J.; Roser, D.; Hurt, E.; Linder, P.; Kressler, D. Linear ubiquitin fusion to Rps31 and its subsequent cleavage are required for the efficient production and functional integrity of 40S ribosomal subunits. *Mol. Microbiol.* **2009**, *72*, 69–84. [CrossRef]
90. Rössler, I.; Weigl, S.; Fernández-Fernández, J.; Martín-Villanueva, S.; Strauss, D.; Hurt, E.; de la Cruz, J.; Pertschy, B. The C-terminal tail of ribosomal protein Rps15 is engaged in cytoplasmic pre-40S maturation. *RNA Biol.* **2022**, *19*, 560–574. [CrossRef]
91. Martín-Villanueva, S.; Gutiérrez, G.; Kressler, D.; de la Cruz, J. Ubiquitin and Ubiquitin-Like proteins and domains in ribosome production and function: Chance or necessity? *Int. J. Mol. Sci.* **2021**, *22*, 4359. [CrossRef] [PubMed]
92. Finley, D.; Bartel, B.; Varshavsky, A. The tails of ubiquitin precursors are ribosomal proteins whose fusion to ubiquitin facilitates ribosome biogenesis. *Nature* **1989**, *338*, 394–401. [CrossRef] [PubMed]
93. Polymenis, M. Ribosomal proteins: Mutant phenotypes by the numbers and associated gene expression changes. *Open Biol.* **2020**, *10*, 200114. [CrossRef] [PubMed]
94. Shasmal, M.; Dey, S.; Shaikh, T.R.; Bhakta, S.; Sengupta, J. *E. coli* metabolic protein aldehyde-alcohol dehydrogenase-E binds to the ribosome: A unique moonlighting action revealed. *Sci. Rep.* **2016**, *6*, 19936. [CrossRef] [PubMed]
95. Qu, X.; Wen, J.D.; Lancaster, L.; Noller, H.F.; Bustamante, C.; Tinoco, I., Jr. The ribosome uses two active mechanisms to unwind messenger RNA during translation. *Nature* **2011**, *475*, 118–121. [CrossRef] [PubMed]

96. Valentini, M.; Linder, P. Happy birthday: 30 years of RNA helicases. *Methods Mol. Biol.* **2021**, *2209*, 17–34. [PubMed]
97. Yourik, P.; Aitken, C.E.; Zhou, F.; Gupta, N.; Hinnebusch, A.G.; Lorsch, J.R. Yeast eIF4A enhances recruitment of mRNAs regardless of their structural complexity. *eLife* **2017**, *6*, e31476. [CrossRef] [PubMed]
98. Parsyan, A.; Svitkin, Y.; Shahbazian, D.; Gkogkas, C.; Lasko, P.; Merrick, W.C.; Sonenberg, N. mRNA helicases: The tacticians of translational control. *Nat. Rev. Mol. Cell Biol.* **2011**, *12*, 235–245. [CrossRef] [PubMed]
99. Pisareva, V.P.; Pisarev, A.V. DHX29 and eIF3 cooperate in ribosomal scanning on structured mRNAs during translation initiation. *RNA* **2016**, *22*, 1859–1870. [CrossRef]
100. Pisareva, V.P.; Pisarev, A.V.; Komar, A.A.; Hellen, C.U.; Pestova, T.V. Translation initiation on mammalian mRNAs with structured 5′UTRs requires DExH-box protein DHX29. *Cell* **2008**, *135*, 1237–1250. [CrossRef]
101. Hashem, Y.; des Georges, A.; Dhote, V.; Langlois, R.; Liao, H.Y.; Grassucci, R.A.; Hellen, C.U.; Pestova, T.V.; Frank, J. Structure of the mammalian ribosomal 43S preinitiation complex bound to the scanning factor DHX29. *Cell* **2013**, *153*, 1108–1119. [CrossRef]
102. Haimov, O.; Sinvani, H.; Martin, F.; Ulitsky, I.; Emmanuel, R.; Tamarkin-Ben-Harush, A.; Vardy, A.; Dikstein, R. Efficient and accurate translation initiation directed by TISU involves RPS3 and RPS10e binding and differential eukaryotic initiation factor 1A regulation. *Mol. Cell. Biol.* **2017**, *37*, e00150-00117. [CrossRef]
103. Passmore, L.A.; Schmeing, T.M.; Maag, D.; Applefield, D.J.; Acker, M.G.; Algire, M.A.; Lorsch, J.R.; Ramakrishnan, V. The eukaryotic translation initiation factors eIF1 and eIF1A induce an open conformation of the 40S ribosome. *Mol. Cell* **2007**, *26*, 41–50. [CrossRef]
104. Weisser, M.; Voigts-Hoffmann, F.; Rabl, J.; Leibundgut, M.; Ban, N. The crystal structure of the eukaryotic 40S ribosomal subunit in complex with eIF1 and eIF1A. *Nat. Struct. Mol. Biol.* **2013**, *20*, 1015–1017. [CrossRef] [PubMed]
105. Sehrawat, U.; Koning, F.; Ashkenazi, S.; Stelzer, G.; Leshkowitz, D.; Dikstein, R. Cancer-associated eukaryotic translation initiation factor 1A mutants impair Rps3 and Rps10 binding and enhance scanning of cell cycle genes. *Mol. Cell. Biol.* **2019**, *39*, e00441-00418. [CrossRef] [PubMed]
106. Martin-Marcos, P.; Zhou, F.; Karunasiri, C.; Zhang, F.; Dong, J.; Nanda, J.; Kulkarni, S.D.; Sen, N.D.; Tamame, M.; Zeschnigk, M.; et al. eIF1A residues implicated in cancer stabilize translation preinitiation complexes and favor suboptimal initiation sites in yeast. *eLife* **2017**, *6*, e31250. [CrossRef]
107. Fekete, C.A.; Mitchell, S.F.; Cherkasova, V.A.; Applefield, D.; Algire, M.A.; Maag, D.; Saini, A.K.; Lorsch, J.R.; Hinnebusch, A.G. N- and C-terminal residues of eIF1A have opposing effects on the fidelity of start codon selection. *EMBO J.* **2007**, *26*, 1602–1614. [CrossRef]
108. Zeman, J.; Itoh, Y.; Kukačka, Z.; Rosůlek, M.; Kavan, D.; Kouba, T.; Jansen, M.E.; Mohammad, M.P.; Novák, P.; Valášek, L.S. Binding of eIF3 in complex with eIF5 and eIF1 to the 40S ribosomal subunit is accompanied by dramatic structural changes. *Nucleic Acids Res.* **2019**, *47*, 8282–8300. [CrossRef] [PubMed]
109. Poncová, K.; Wagner, S.; Jansen, M.E.; Beznosková, P.; Gunisova, S.; Herrmannova, A.; Zeman, J.; Dong, J.; Valášek, L.S. uS3/Rps3 controls fidelity of translation termination and programmed stop codon readthrough in co-operation with eIF3. *Nucleic Acids Res.* **2019**, *47*, 11326–11343. [CrossRef] [PubMed]
110. Malygin, A.A.; Shatsky, I.N.; Karpova, G.G. Proteins of the human 40S ribosomal subunit involved in hepatitis C IRES binding as revealed from fluorescent labeling. *Biochemistry* **2013**, *78*, 53–59. [CrossRef]
111. Tidu, A.; Janvier, A.; Schaeffer, L.; Sosnowski, P.; Kuhn, L.; Hammann, P.; Westhof, E.; Eriani, G.; Martin, F. The viral protein NSP1 acts as a ribosome gatekeeper for shutting down host translation and fostering SARS-CoV-2 translation. *RNA* **2020**, *27*, 253–264. [CrossRef] [PubMed]
112. Thoms, M.; Buschauer, R.; Ameismeier, M.; Koepke, L.; Denk, T.; Hirschenberger, M.; Kratzat, H.; Hayn, M.; Mackens-Kiani, T.; Cheng, J.; et al. Structural basis for translational shutdown and immune evasion by the Nsp1 protein of SARS-CoV-2. *Science* **2020**, *369*, 1249–1255. [CrossRef] [PubMed]
113. Schubert, K.; Karousis, E.D.; Jomaa, A.; Scaiola, A.; Echeverria, B.; Gurzeler, L.A.; Leibundgut, M.; Thiel, V.; Muhlemann, O.; Ban, N. SARS-CoV-2 Nsp1 binds the ribosomal mRNA channel to inhibit translation. *Nat. Struct. Mol. Biol.* **2020**, *27*, 959–966. [CrossRef] [PubMed]
114. Schubert, K.; Karousis, E.D.; Ban, I.; Lapointe, C.P.; Leibundgut, M.; Baumlin, E.; Kummerant, E.; Scaiola, A.; Schonhut, T.; Ziegelmuller, J.; et al. Universal features of Nsp1-mediated translational shutdown by coronaviruses. *Mol. Cell* **2023**, *83*, 3546–3557.e3548. [CrossRef] [PubMed]
115. Bujanic, L.; Shevchuk, O.; von Kugelgen, N.; Kalinina, A.; Ludwik, K.; Koppstein, D.; Zerna, N.; Sickmann, A.; Chekulaeva, M. The key features of SARS-CoV-2 leader and NSP1 required for viral escape of NSP1-mediated repression. *RNA* **2022**, *28*, 766–779. [CrossRef] [PubMed]
116. Brown, A.; Baird, M.R.; Yip, M.C.; Murray, J.; Shao, S. Structures of translationally inactive mammalian ribosomes. *eLife* **2018**, *7*, e40486. [CrossRef] [PubMed]
117. Spahn, C.M.; Gomez-Lorenzo, M.G.; Grassucci, R.A.; Jorgensen, R.; Andersen, G.R.; Beckmann, R.; Penczek, P.A.; Ballesta, J.P.; Frank, J. Domain movements of elongation factor eEF2 and the eukaryotic 80S ribosome facilitate tRNA translocation. *EMBO J.* **2004**, *23*, 1008–1119. [CrossRef] [PubMed]
118. Hassan, A.; Byju, S.; Freitas, F.C.; Roc, C.; Pender, N.; Nguyen, K.; Kimbrough, E.M.; Mattingly, J.M.; Gonzalez, R.L., Jr.; de Oliveira, R.J.; et al. Ratchet, swivel, tilt and roll: A complete description of subunit rotation in the ribosome. *Nucleic Acids Res.* **2023**, *51*, 919–934. [CrossRef] [PubMed]

119. Collins, J.C.; Ghalei, H.; Doherty, J.R.; Huang, H.; Culver, R.N.; Karbstein, K. Ribosome biogenesis factor Ltv1 chaperones the assembly of the small subunit head. *J. Cell Biol.* **2018**, *217*, 4141–4154. [CrossRef]
120. Shao, S.; Murray, J.; Brown, A.; Taunton, J.; Ramakrishnan, V.; Hegde, R.S. Decoding mammalian ribosome-mRNA states by translational GTPase complexes. *Cell* **2016**, *167*, 1229–1240.e1215. [CrossRef]
121. Taylor, D.; Unbehaun, A.; Li, W.; Das, S.; Lei, J.; Liao, H.Y.; Grassucci, R.A.; Pestova, T.V.; Frank, J. Cryo-EM structure of the mammalian eukaryotic release factor eRF1-eRF3-associated termination complex. *Proc. Natl. Acad. Sci. USA* **2012**, *109*, 18413–18418. [CrossRef]
122. Preis, A.; Heuer, A.; Barrio-Garcia, C.; Hauser, A.; Eyler, D.E.; Berninghausen, O.; Green, R.; Becker, T.; Beckmann, R. Cryoelectron microscopic structures of eukaryotic translation termination complexes containing eRF1-eRF3 or eRF1-ABCE1. *Cell Rep.* **2014**, *8*, 59–65. [CrossRef]
123. Liu, B.; Qian, S.B. Translational reprogramming in cellular stress response. *Wiley Interdiscip. Rev. RNA* **2014**, *5*, 301–315. [CrossRef]
124. Roux, P.P.; Topisirovic, I. Signaling pathways involved in the regulation of mRNA translation. *Mol. Cell. Biol.* **2018**, *38*, e00070-00018. [CrossRef]
125. Silva, G.M.; Finley, D.; Vogel, C. K63 polyubiquitination is a new modulator of the oxidative stress response. *Nat. Struct. Mol. Biol.* **2015**, *22*, 116–123. [CrossRef]
126. Simoes, V.; Cizubu, B.K.; Harley, L.; Zhou, Y.; Pajak, J.; Snyder, N.A.; Bouvette, J.; Borgnia, M.J.; Arya, G.; Bartesaghi, A.; et al. Redox-sensitive E2 Rad6 controls cellular response to oxidative stress via K63-linked ubiquitination of ribosomes. *Cell Rep.* **2022**, *39*, 110860. [CrossRef]
127. Back, S.; Gorman, A.W.; Vogel, C.; Silva, G.M. Site-specific K63 ubiquitinomics provides insights into translation regulation under stress. *J. Proteome Res.* **2019**, *18*, 309–318. [CrossRef] [PubMed]
128. Zhou, Y.; Kastritis, P.L.; Dougherty, S.E.; Bouvette, J.; Hsu, A.L.; Burbaum, L.; Mosalaganti, S.; Pfeffer, S.; Hagen, W.J.H.; Forster, F.; et al. Structural impact of K63 ubiquitin on yeast translocating ribosomes under oxidative stress. *Proc. Natl. Acad. Sci. USA* **2020**, *117*, 22157–22166. [CrossRef] [PubMed]
129. Inada, T.; Beckmann, R. Mechanisms of translation-coupled quality control. *J. Mol. Biol.* **2024**, *436*, 168496. [CrossRef] [PubMed]
130. Filbeck, S.; Cerullo, F.; Pfeffer, S.; Joazeiro, C.A.P. Ribosome-associated quality-control mechanisms from bacteria to humans. *Mol. Cell* **2022**, *82*, 1451–1466. [CrossRef]
131. Guyomar, C.; D'Urso, G.; Chat, S.; Giudice, E.; Gillet, R. Structures of tmRNA and SmpB as they transit through the ribosome. *Nat. Commun.* **2021**, *12*, 4909. [CrossRef]
132. Joazeiro, C.A.P. Ribosomal stalling during translation: Providing substrates for ribosome-associated protein quality control. *Annu. Rev. Cell Dev. Biol.* **2017**, *33*, 343–368. [CrossRef] [PubMed]
133. Powers, K.T.; Szeto, J.A.; Schaffitzel, C. New insights into no-go, non-stop and nonsense-mediated mRNA decay complexes. *Curr. Opin. Struct. Biol.* **2020**, *65*, 110–118. [CrossRef]
134. Araki, Y.; Takahashi, S.; Kobayashi, T.; Kajiho, H.; Hoshino, S.; Katada, T. Ski7p G protein interacts with the exosome and the Ski complex for 3′-to-5′ mRNA decay in yeast. *EMBO J.* **2001**, *20*, 4684–4693. [CrossRef] [PubMed]
135. Kalisiak, K.; Kulinski, T.M.; Tomecki, R.; Cysewski, D.; Pietras, Z.; Chlebowski, A.; Kowalska, K.; Dziembowski, A. A short splicing isoform of HBS1L links the cytoplasmic exosome and SKI complexes in humans. *Nucleic Acids Res.* **2017**, *45*, 2068–2080. [CrossRef]
136. Keidel, A.; Kögel, A.; Reichelt, P.; Kowalinski, E.; Schafer, I.B.; Conti, E. Concerted structural rearrangements enable RNA channeling into the cytoplasmic Ski238-Ski7-exosome assembly. *Mol. Cell* **2023**, *83*, 4093–4105.e4097. [CrossRef]
137. Zinder, J.C.; Lima, C.D. Targeting RNA for processing or destruction by the eukaryotic RNA exosome and its cofactors. *Genes Dev.* **2017**, *31*, 88–100. [CrossRef] [PubMed]
138. Kowalinski, E.; Schuller, A.; Green, R.; Conti, E. *Saccharomyces cerevisiae* Ski7 Is a GTP-Binding protein adopting the characteristic conformation of active translational GTPases. *Structure* **2015**, *23*, 1336–1343. [CrossRef]
139. Kögel, A.; Keidel, A.; Bonneau, F.; Schafer, I.B.; Conti, E. The human SKI complex regulates channeling of ribosome-bound RNA to the exosome via an intrinsic gatekeeping mechanism. *Mol. Cell* **2022**, *82*, 756–769.e758. [CrossRef]
140. Schmidt, C.; Kowalinski, E.; Shanmuganathan, V.; Defenouillère, Q.; Braunger, K.; Heuer, A.; Pech, M.; Namane, A.; Berninghausen, O.; Fromont-Racine, M.; et al. The cryo-EM structure of a ribosome-Ski2-Ski3-Ski8 helicase complex. *Science* **2016**, *354*, 1431–1433. [CrossRef]
141. De, S.; Mühlemann, O. A comprehensive coverage insurance for cells: Revealing links between ribosome collisions, stress responses and mRNA surveillance. *RNA Biol.* **2022**, *19*, 609–621. [CrossRef]
142. Iyer, K.V.; Müller, M.; Tittel, L.S.; Winz, M.L. Molecular Highway Patrol for Ribosome Collisions. *Chembiochem* **2023**, *24*, e202300264. [CrossRef]
143. Matsuo, Y.; Ikeuchi, K.; Saeki, Y.; Iwasaki, S.; Schmidt, C.; Udagawa, T.; Sato, F.; Tsuchiya, H.; Becker, T.; Tanaka, K.; et al. Ubiquitination of stalled ribosome triggers ribosome-associated quality control. *Nat. Commun.* **2017**, *8*, 159. [CrossRef] [PubMed]
144. Dougherty, S.E.; Maduka, A.O.; Inada, T.; Silva, G.M. Expanding role of ubiquitin in translational control. *Int. J. Mol. Sci.* **2020**, *21*, 1151. [CrossRef] [PubMed]
145. Wu, C.C.; Peterson, A.; Zinshteyn, B.; Regot, S.; Green, R. Ribosome collisions trigger general stress responses to regulate cell fate. *Cell* **2020**, *182*, 404–416.e414. [CrossRef] [PubMed]

146. Yan, L.L.; Zaher, H.S. Ribosome quality control antagonizes the activation of the integrated stress response on colliding ribosomes. *Mol. Cell* **2021**, *81*, 614–628.e614. [CrossRef]
147. Pochopien, A.A.; Beckert, B.; Kasvandik, S.; Berninghausen, O.; Beckmann, R.; Tenson, T.; Wilson, D.N. Structure of Gcn1 bound to stalled and colliding 80S ribosomes. *Proc. Natl. Acad. Sci. USA* **2021**, *118*, e2022756118. [CrossRef]
148. Lee, S.J.; Swanson, M.J.; Sattlegger, E. Gcn1 contacts the small ribosomal protein Rps10, which is required for full activation of the protein kinase Gcn2. *Biochem. J.* **2015**, *466*, 547–559. [CrossRef]
149. Mueller, P.P.; Grueter, P.; Hinnebusch, A.G.; Trachsel, H. A ribosomal protein is required for translational regulation of GCN4 mRNA. Evidence for involvement of the ribosome in eIF2 recycling. *J. Biol. Chem.* **1998**, *273*, 32870–32877. [CrossRef]
150. Oltion, K.; Carelli, J.D.; Yang, T.; See, S.K.; Wang, H.Y.; Kampmann, M.; Taunton, J. An E3 ligase network engages GCN1 to promote the degradation of translation factors on stalled ribosomes. *Cell* **2023**, *186*, 346–362.e317. [CrossRef]
151. Marygold, S.J.; Roote, J.; Reuter, G.; Lambertsson, A.; Ashburner, M.; Millburn, G.H.; Harrison, P.M.; Yu, Z.; Kenmochi, N.; Kaufman, T.C.; et al. The ribosomal protein genes and *Minute* loci of *Drosophila melanogaster*. *Genome Biol.* **2007**, *8*, R216. [CrossRef] [PubMed]
152. Stirnberg, P.; Liu, J.P.; Ward, S.; Kendall, S.L.; Leyser, O. Mutation of the cytosolic ribosomal protein-encoding *RPS10B* gene affects shoot meristematic function in *Arabidopsis*. *BMC Plant Biol.* **2012**, *12*, 160. [CrossRef] [PubMed]
153. Islam, R.A.; Rallis, C. Ribosomal biogenesis and heterogeneity in development, disease, and aging. *Epigenomes* **2023**, *7*, 26. [CrossRef] [PubMed]
154. Yanagi, S.; Iida, T.; Kobayashi, T. *RPS12* and *UBC4* are related to senescence signal production in the ribosomal RNA gene cluster. *Mol. Cell. Biol.* **2022**, *42*, e0002822. [CrossRef] [PubMed]
155. Morata, G. Cell competition: A historical perspective. *Dev. Biol.* **2021**, *476*, 33–40. [CrossRef] [PubMed]
156. Kiparaki, M.; Baker, N.E. Ribosomal protein mutations and cell competition: Autonomous and nonautonomous effects on a stress response. *Genetics* **2023**, *224*, iyad080. [CrossRef] [PubMed]
157. Baker, N.E. Emerging mechanisms of cell competition. *Nat. Rev. Genet.* **2020**, *21*, 683–697. [CrossRef] [PubMed]
158. Kale, A.; Ji, Z.; Kiparaki, M.; Blanco, J.; Rimesso, G.; Flibotte, S.; Baker, N.E. Ribosomal protein S12e has a distinct function in cell competition. *Dev. Cell* **2018**, *44*, 42–55.e44. [CrossRef] [PubMed]
159. Lee, C.H.; Kiparaki, M.; Blanco, J.; Folgado, V.; Ji, Z.; Kumar, A.; Rimesso, G.; Baker, N.E. A regulatory response to ribosomal protein mutations controls translation, growth, and cell competition. *Dev. Cell* **2018**, *46*, 456–469.e454. [CrossRef]
160. Ji, Z.; Kiparaki, M.; Folgado, V.; Kumar, A.; Blanco, J.; Rimesso, G.; Chuen, J.; Liu, Y.; Zheng, D.; Baker, N.E. *Drosophila* RpS12 controls translation, growth, and cell competition through Xrp1. *PLoS Genet.* **2019**, *15*, e1008513. [CrossRef]
161. Luo, J.; Zhao, H.; Chen, L.; Liu, M. Multifaceted functions of RPS27a: An unconventional ribosomal protein. *J. Cell. Physiol.* **2023**, *238*, 485–497. [CrossRef] [PubMed]
162. Hong, S.W.; Kim, S.M.; Jin, D.H.; Kim, Y.S.; Hur, D.Y. RPS27a enhances EBV-encoded LMP1-mediated proliferation and invasion by stabilizing of LMP1. *Biochem. Biophys. Res. Commun.* **2017**, *491*, 303–309. [CrossRef] [PubMed]
163. Bansal, S.K.; Gupta, N.; Sankhwar, S.N.; Rajender, S. Differential genes expression between fertile and infertile spermatozoa revealed by transcriptome analysis. *PLoS ONE* **2015**, *10*, e0127007. [CrossRef] [PubMed]
164. Bäurle, I.; Laux, T. Apical meristems: The plant's fountain of youth. *Bioessays* **2003**, *25*, 961–970. [CrossRef] [PubMed]
165. Hanania, U.; Velcheva, M.; Sahar, N.; Flaishman, M.; Or, E.; Degani, O.; Perl, A. The ubiquitin extension protein S27a is differentially expressed in developing flower organs of Thompson seedless versus Thompson seeded grape isogenic clones. *Plant Cell Rep.* **2009**, *28*, 1033–1042. [CrossRef] [PubMed]
166. Riepe, C.; Zelin, E.; Frankino, P.A.; Meacham, Z.A.; Fernandez, S.G.; Ingolia, N.T.; Corn, J.E. Double stranded DNA breaks and genome editing trigger loss of ribosomal protein RPS27A. *FEBS J.* **2022**, *289*, 3101–3114. [CrossRef] [PubMed]
167. Cerezo, E.; Plisson-Chastang, C.; Henras, A.K.; Lebaron, S.; Gleizes, P.E.; O'Donohue, M.F.; Romeo, Y.; Henry, Y. Maturation of pre-40S particles in yeast and humans. *Wiley Interdiscip. Rev. RNA* **2019**, *10*, e1516. [CrossRef] [PubMed]
168. Ferreira-Cerca, S.; Pöll, G.; Kuhn, H.; Neueder, A.; Jakob, S.; Tschochner, H.; Milkereit, P. Analysis of the in vivo assembly pathway of eukaryotic 40S ribosomal proteins. *Mol. Cell* **2007**, *28*, 446–457. [CrossRef] [PubMed]
169. Rouquette, J.; Choesmel, V.; Gleizes, P.E. Nuclear and cytoplasmic processing of precursors to the 40S ribosomal subunits in mammalian cells. *EMBO J.* **2005**, *24*, 2862–2872. [CrossRef] [PubMed]
170. O'Donohue, M.F.; Choesmel, V.; Faubladier, M.; Fichant, G.; Gleizes, P.E. Functional dichotomy of ribosomal proteins during the synthesis of mammalian 40S ribosomal subunits. *J. Cell Biol.* **2010**, *190*, 853–866. [CrossRef]
171. Nicolas, E.; Parisot, P.; Pinto-Monteiro, C.; de Walque, R.; De Vleeschouwer, C.; Lafontaine, D.L. Involvement of human ribosomal proteins in nucleolar structure and p53-dependent nucleolar stress. *Nat. Commun.* **2016**, *7*, 11390. [CrossRef] [PubMed]
172. Wild, T.; Horvath, P.; Wyler, E.; Widmann, B.; Badertscher, L.; Zemp, I.; Kozak, K.; Csucs, G.; Lund, E.; Kutay, U. A protein inventory of human ribosome biogenesis reveals an essential function of exportin 5 in 60S subunit export. *PLoS Biol.* **2010**, *8*, e1000522. [CrossRef]
173. Sun, Q.; Zhu, X.; Qi, J.; An, W.; Lan, P.; Tan, D.; Chen, R.; Wang, B.; Zheng, S.; Zhang, C.; et al. Molecular architecture of the 90S small subunit pre-ribosome. *eLife* **2017**, *6*, e22086. [CrossRef] [PubMed]
174. Kornprobst, M.; Turk, M.; Kellner, N.; Cheng, J.; Flemming, D.; Kos-Braun, I.; Kos, M.; Thoms, M.; Berninghausen, O.; Beckmann, R.; et al. Architecture of the 90S pre-ribosome: A structural view on the birth of the eukaryotic ribosome. *Cell* **2016**, *166*, 380–393. [CrossRef] [PubMed]

175. Cheng, J.; Kellner, N.; Berninghausen, O.; Hurt, E.; Beckmann, R. 3.2-Å-resolution structure of the 90S preribosome before A$_1$ pre-rRNA cleavage. *Nat. Struct. Mol. Biol.* **2017**, *24*, 954–964. [CrossRef] [PubMed]
176. Cheng, J.; Bassler, J.; Fischer, P.; Lau, B.; Kellner, N.; Kunze, R.; Griesel, S.; Kallas, M.; Berninghausen, O.; Strauss, D.; et al. Thermophile 90S pre-ribosome structures reveal the reverse order of co-transcriptional 18S rRNA subdomain integration. *Mol. Cell* **2019**, *75*, 1256–1269.e1257. [CrossRef] [PubMed]
177. Strunk, B.S.; Loucks, C.R.; Su, M.; Vashisth, H.; Cheng, S.; Schilling, J.; Brooks, C.L., III; Karbstein, K.; Skiniotis, G. Ribosome assembly factors prevent premature translation initiation by 40S assembly intermediates. *Science* **2011**, *333*, 1449–1453. [CrossRef] [PubMed]
178. Heuer, A.; Thomson, E.; Schmidt, C.; Berninghausen, O.; Becker, T.; Hurt, E.; Beckmann, R. Cryo-EM structure of a late pre-40S ribosomal subunit from *Saccharomyces cerevisiae*. *eLife* **2017**, *6*, e30189. [CrossRef] [PubMed]
179. Blomqvist, E.K.; Huang, H.; Karbstein, K. A disease associated mutant reveals how Ltv1 orchestrates RP assembly and rRNA folding of the small ribosomal subunit head. *PLoS Genet.* **2023**, *19*, e1010862. [CrossRef] [PubMed]
180. Plassart, L.; Shayan, R.; Montellese, C.; Rinaldi, D.; Larburu, N.; Pichereaux, C.; Froment, C.; Lebaron, S.; O'Donohue, M.F.; Kutay, U.; et al. The final step of 40S ribosomal subunit maturation is controlled by a dual key lock. *eLife* **2021**, *10*, e61254. [CrossRef]
181. Eastham, M.J.; Pelava, A.; Wells, G.R.; Watkins, N.J.; Schneider, C. RPS27a and RPL40, which are produced as ubiquitin fusion proteins, are not essential for p53 signalling. *Biomolecules* **2023**, *13*, 898. [CrossRef] [PubMed]
182. Schäfer, T.; Strauss, D.; Petfalski, E.; Tollervey, D.; Hurt, E. The path from nucleolar 90S to cytoplasmic 40S pre-ribosomes. *EMBO J.* **2003**, *22*, 1370–1380. [CrossRef] [PubMed]
183. Cheng, J.; Lau, B.; La Venuta, G.; Ameismeier, M.; Berninghausen, O.; Hurt, E.; Beckmann, R. 90S pre-ribosome transformation into the primordial 40S subunit. *Science* **2020**, *369*, 1470–1476. [CrossRef] [PubMed]
184. Cheng, J.; Lau, B.; Thoms, M.; Ameismeier, M.; Berninghausen, O.; Hurt, E.; Beckmann, R. The nucleoplasmic phase of pre-40S formation prior to nuclear export. *Nucleic Acids Res.* **2022**, *50*, 11924–11937. [CrossRef] [PubMed]
185. Du, Y.; An, W.; Zhu, X.; Sun, Q.; Qi, J.; Ye, K. Cryo-EM structure of 90S small ribosomal subunit precursors in transition states. *Science* **2020**, *369*, 1477–1481. [CrossRef] [PubMed]
186. Scaiola, A.; Pena, C.; Weisser, M.; Bohringer, D.; Leibundgut, M.; Klingauf-Nerurkar, P.; Gerhardy, S.; Panse, V.G.; Ban, N. Structure of a eukaryotic cytoplasmic pre-40S ribosomal subunit. *EMBO J.* **2018**, *37*, e98499. [CrossRef] [PubMed]
187. Johnson, M.C.; Ghalei, H.; Doxtader, K.A.; Karbstein, K.; Stroupe, M.E. Structural heterogeneity in pre-40S ribosomes. *Structure* **2017**, *25*, 329–340. [CrossRef] [PubMed]
188. Mitterer, V.; Gantenbein, N.; Birner-Gruenberger, R.; Murat, G.; Bergler, H.; Kressler, D.; Pertschy, B. Nuclear import of dimerized ribosomal protein Rps3 in complex with its chaperone Yar1. *Sci. Rep.* **2016**, *6*, 36714. [CrossRef] [PubMed]
189. Ghalei, H.; Schaub, F.X.; Doherty, J.R.; Noguchi, Y.; Roush, W.R.; Cleveland, J.L.; Stroupe, M.E.; Karbstein, K. Hrr25/CK1δ-directed release of Ltv1 from pre-40S ribosomes is necessary for ribosome assembly and cell growth. *J. Cell Biol.* **2015**, *208*, 745–759. [CrossRef] [PubMed]
190. Mitterer, V.; Murat, G.; Rety, S.; Blaud, M.; Delbos, L.; Stanborough, T.; Bergler, H.; Leulliot, N.; Kressler, D.; Pertschy, B. Sequential domain assembly of ribosomal protein S3 drives 40S subunit maturation. *Nat. Commun.* **2016**, *7*, 10336. [CrossRef]
191. Holzer, S.; Ban, N.; Klinge, S. Crystal structure of the yeast ribosomal protein rpS3 in complex with its chaperone Yar1. *J. Mol. Biol.* **2013**, *425*, 4154–4160. [CrossRef] [PubMed]
192. Mitterer, V.; Shayan, R.; Ferreira-Cerca, S.; Murat, G.; Enne, T.; Rinaldi, D.; Weigl, S.; Omanic, H.; Gleizes, P.E.; Kressler, D.; et al. Conformational proofreading of distant 40S ribosomal subunit maturation events by a long-range communication mechanism. *Nat. Commun.* **2019**, *10*, 2754. [CrossRef] [PubMed]
193. Schäfer, T.; Maco, B.; Petfalski, E.; Tollervey, D.; Bottcher, B.; Aebi, U.; Hurt, E. Hrr25-dependent phosphorylation state regulates organization of the pre-40S subunit. *Nature* **2006**, *441*, 651–655. [CrossRef] [PubMed]
194. Hector, R.D.; Burlacu, E.; Aitken, S.; Bihan, T.L.; Tuijtel, M.; Zaplatina, A.; Cook, A.G.; Granneman, S. Snapshots of pre-rRNA structural flexibility reveal eukaryotic 40S assembly dynamics at nucleotide resolution. *Nucleic Acids Res.* **2014**, *42*, 12138–12154. [CrossRef] [PubMed]
195. Seiser, R.M.; Sundberg, A.E.; Wollam, B.J.; Zobel-Thropp, P.; Baldwin, K.; Spector, M.D.; Lycan, D.E. Ltv1 is required for efficient nuclear export of the ribosomal small subunit in *Saccharomyces cerevisiae*. *Genetics* **2006**, *174*, 679–691. [CrossRef] [PubMed]
196. Zemp, I.; Wandrey, F.; Rao, S.; Ashiono, C.; Wyler, E.; Montellese, C.; Kutay, U. CK1δ and CK1ε are components of human 40S subunit precursors required for cytoplasmic 40S maturation. *J. Cell Sci.* **2014**, *127*, 1242–1253. [PubMed]
197. Ameismeier, M.; Cheng, J.; Berninghausen, O.; Beckmann, R. Visualizing late states of human 40S ribosomal subunit maturation. *Nature* **2018**, *558*, 249–253. [CrossRef] [PubMed]
198. Larburu, N.; Montellese, C.; O'Donohue, M.F.; Kutay, U.; Gleizes, P.E.; Plisson-Chastang, C. Structure of a human pre-40S particle points to a role for RACK1 in the final steps of 18S rRNA processing. *Nucleic Acids Res.* **2016**, *44*, 8465–8478. [CrossRef] [PubMed]
199. Dörner, K.; Ruggeri, C.; Zemp, I.; Kutay, U. Ribosome biogenesis factors-from names to functions. *EMBO J.* **2023**, *42*, e112699. [CrossRef] [PubMed]
200. Henras, A.K.; Plisson-Chastang, C.; O'Donohue, M.F.; Chakraborty, A.; Gleizes, P.E. An overview of pre-ribosomal RNA processing in eukaryotes. *Wiley Interdiscip. Rev. RNA* **2015**, *6*, 225–242. [CrossRef]
201. Ameismeier, M.; Zemp, I.; van den Heuvel, J.; Thoms, M.; Berninghausen, O.; Kutay, U.; Beckmann, R. Structural basis for the final steps of human 40S ribosome maturation. *Nature* **2020**, *587*, 683–687. [CrossRef]

202. Clatterbuck Soper, S.F.; Dator, R.P.; Limbach, P.A.; Woodson, S.A. In vivo X-ray footprinting of pre-30S ribosomes reveals chaperone-dependent remodeling of late assembly intermediates. *Mol. Cell* **2013**, *52*, 506–516. [CrossRef] [PubMed]
203. Guo, Q.; Goto, S.; Chen, Y.; Feng, B.; Xu, Y.; Muto, A.; Himeno, H.; Deng, H.; Lei, J.; Gao, N. Dissecting the in vivo assembly of the 30S ribosomal subunit reveals the role of RimM and general features of the assembly process. *Nucleic Acids Res.* **2013**, *41*, 2609–2620. [CrossRef]
204. Leong, V.; Kent, M.; Jomaa, A.; Ortega, J. *Escherichia coli rimM* and *yjeQ* null strains accumulate immature 30S subunits of similar structure and protein complement. *RNA* **2013**, *19*, 789–802. [CrossRef]
205. Saurer, M.; Ramrath, D.J.F.; Niemann, M.; Calderaro, S.; Prange, C.; Mattei, S.; Scaiola, A.; Leitner, A.; Bieri, P.; Horn, E.K.; et al. Mitoribosomal small subunit biogenesis in trypanosomes involves an extensive assembly machinery. *Science* **2019**, *365*, 1144–1149. [CrossRef]
206. Kang, J.; Brajanovski, N.; Chan, K.T.; Xuan, J.; Pearson, R.B.; Sanij, E. Ribosomal proteins and human diseases: Molecular mechanisms and targeted therapy. *Signal Transduct. Target Ther.* **2021**, *6*, 323. [CrossRef]
207. Kampen, K.R.; Sulima, S.O.; Vereecke, S.; De Keersmaecker, K. Hallmarks of ribosomopathies. *Nucleic Acids Res.* **2020**, *48*, 1013–1028. [CrossRef] [PubMed]
208. Mills, E.W.; Green, R. Ribosomopathies: There's strength in numbers. *Science* **2017**, *358*, eaan2755. [CrossRef] [PubMed]
209. Farley-Barnes, K.I.; Ogawa, L.M.; Baserga, S.J. Ribosomopathies: Old concepts, new controversies. *Trends Genet.* **2019**, *35*, 754–767. [CrossRef]
210. De Keersmaecker, K.; Sulima, S.O.; Dinman, J.D. Ribosomopathies and the paradox of cellular hypo- to hyperproliferation. *Blood* **2015**, *125*, 1377–1382. [CrossRef]
211. Danilova, N.; Gazda, H.T. Ribosomopathies: How a common root can cause a tree of pathologies. *Dis. Models Mech.* **2015**, *8*, 1013–1026. [CrossRef] [PubMed]
212. Da Costa, L.; Mohandas, N.; David, N.L.; Platon, J.; Marie, I.; O'Donohue, M.F.; Leblanc, T.; Gleizes, P.E. Diamond-Blackfan anemia, the archetype of ribosomopathy: How distinct is it from the other constitutional ribosomopathies? *Blood Cells Mol. Dis.* **2024**, *106*, 102838. [CrossRef] [PubMed]
213. Doherty, L.; Sheen, M.R.; Vlachos, A.; Choesmel, V.; O'Donohue, M.F.; Clinton, C.; Schneider, H.E.; Sieff, C.A.; Newburger, P.E.; Ball, S.E.; et al. Ribosomal protein genes *RPS10* and *RPS26* are commonly mutated in Diamond-Blackfan anemia. *Am. J. Hum. Genet.* **2010**, *86*, 222–228. [CrossRef] [PubMed]
214. Boria, I.; Garelli, E.; Gazda, H.T.; Aspesi, A.; Quarello, P.; Pavesi, E.; Ferrante, D.; Meerpohl, J.J.; Kartal, M.; Da Costa, L.; et al. The ribosomal basis of Diamond-Blackfan Anemia: Mutation and database update. *Hum. Mutat.* **2010**, *31*, 1269–1279. [CrossRef] [PubMed]
215. Gazda, H.T.; Sheen, M.R.; Vlachos, A.; Choesmel, V.; O'Donohue, M.F.; Schneider, H.; Darras, N.; Hasman, C.; Sieff, C.A.; Newburger, P.E.; et al. Ribosomal protein L5 and L11 mutations are associated with cleft palate and abnormal thumbs in Diamond-Blackfan anemia patients. *Am. J. Hum. Genet.* **2008**, *83*, 769–780. [CrossRef] [PubMed]
216. Folgado-Marco, V.; Ames, K.; Chuen, J.; Gritsman, K.; Baker, N.E. Haploinsufficiency of the essential gene *Rps12* causes defects in erythropoiesis and hematopoietic stem cell maintenance. *eLife* **2023**, *12*, e69322. [CrossRef] [PubMed]
217. Choesmel, V.; Bacqueville, D.; Rouquette, J.; Noaillac-Depeyre, J.; Fribourg, S.; Cretien, A.; Leblanc, T.; Tchernia, G.; Da Costa, L.; Gleizes, P.E. Impaired ribosome biogenesis in Diamond-Blackfan anemia. *Blood* **2007**, *109*, 1275–1283. [CrossRef] [PubMed]
218. Ferreira, R.; Ohneda, K.; Yamamoto, M.; Philipsen, S. GATA1 function, a paradigm for transcription factors in hematopoiesis. *Mol. Cell. Biol.* **2005**, *25*, 1215–1227. [CrossRef] [PubMed]
219. Ludwig, L.S.; Gazda, H.T.; Eng, J.C.; Eichhorn, S.W.; Thiru, P.; Ghazvinian, R.; George, T.I.; Gotlib, J.R.; Beggs, A.H.; Sieff, C.A.; et al. Altered translation of GATA1 in Diamond-Blackfan anemia. *Nat. Med.* **2014**, *20*, 748–753. [CrossRef] [PubMed]
220. Khajuria, R.K.; Munschauer, M.; Ulirsch, J.C.; Fiorini, C.; Ludwig, L.S.; McFarland, S.K.; Abdulhay, N.J.; Specht, H.; Keshishian, H.; Mani, D.R.; et al. Ribosome levels selectively regulate translation and lineage commitment in human hematopoiesis. *Cell* **2018**, *173*, 90–103.e119. [CrossRef]
221. Rio, S.; Gastou, M.; Karboul, N.; Derman, R.; Suriyun, T.; Manceau, H.; Leblanc, T.; El Benna, J.; Schmitt, C.; Azouzi, S.; et al. Regulation of globin-heme balance in Diamond-Blackfan anemia by HSP70/GATA1. *Blood* **2019**, *133*, 1358–1370. [CrossRef]
222. van Dooijeweert, B.; Kia, S.K.; Dahl, N.; Fenneteau, O.; Leguit, R.; Nieuwenhuis, E.; van Solinge, W.; van Wijk, R.; Da Costa, L.; Bartels, M. GATA-1 defects in Diamond-Blackfan anemia: Phenotypic characterization points to a specific subset of disease. *Genes* **2022**, *13*, 447. [CrossRef]
223. Ling, T.; Crispino, J.D. GATA1 mutations in red cell disorders. *IUBMB Life* **2020**, *72*, 106–118. [CrossRef]
224. Sulima, S.O.; Kampen, K.R.; Vereecke, S.; Pepe, D.; Fancello, L.; Verbeeck, J.; Dinman, J.D.; De Keersmaecker, K. Ribosomal lesions promote oncogenic mutagenesis. *Cancer Res.* **2019**, *79*, 320–327. [CrossRef] [PubMed]
225. Golomb, L.; Volarevic, S.; Oren, M. p53 and ribosome biogenesis stress: The essentials. *FEBS Lett.* **2014**, *588*, 2571–2579. [CrossRef] [PubMed]
226. Bursac, S.; Brdovcak, M.C.; Donati, G.; Volarevic, S. Activation of the tumor suppressor p53 upon impairment of ribosome biogenesis. *Biochim. Biophys. Acta* **2014**, *1842*, 817–830. [CrossRef]
227. Liu, Y.; Deisenroth, C.; Zhang, Y. RP-MDM2-p53 pathway: Linking ribosomal biogenesis and tumor surveillance. *Trends Cancer* **2016**, *2*, 191–204. [CrossRef]

228. Dutt, S.; Narla, A.; Lin, K.; Mullally, A.; Abayasekara, N.; Megerdichian, C.; Wilson, F.H.; Currie, T.; Khanna-Gupta, A.; Berliner, N.; et al. Haploinsufficiency for ribosomal protein genes causes selective activation of p53 in human erythroid progenitor cells. *Blood* **2011**, *117*, 2567–2576. [CrossRef]
229. Barlow, J.L.; Drynan, L.F.; Trim, N.L.; Erber, W.N.; Warren, A.J.; McKenzie, A.N. New insights into 5q- syndrome as a ribosomopathy. *Cell Cycle* **2010**, *9*, 4286–4293. [CrossRef] [PubMed]
230. Sun, X.X.; DeVine, T.; Challagundla, K.B.; Dai, M.S. Interplay between ribosomal protein S27a and MDM2 protein in p53 activation in response to ribosomal stress. *J. Biol. Chem.* **2011**, *286*, 22730–22741. [CrossRef]
231. Li, H.; Zhang, H.; Huang, G.; Bing, Z.; Xu, D.; Liu, J.; Luo, H.; An, X. Loss of RPS27a expression regulates the cell cycle, apoptosis, and proliferation via the RPL11-MDM2-p53 pathway in lung adenocarcinoma cells. *J. Exp. Clin. Cancer Res.* **2022**, *41*, 33. [CrossRef] [PubMed]
232. Kussie, P.H.; Gorina, S.; Marechal, V.; Elenbaas, B.; Moreau, J.; Levine, A.J.; Pavletich, N.P. Structure of the MDM2 oncoprotein bound to the p53 tumor suppressor transactivation domain. *Science* **1996**, *274*, 948–953. [CrossRef] [PubMed]
233. Lindström, M.S.; Jin, A.; Deisenroth, C.; White Wolf, G.; Zhang, Y. Cancer-associated mutations in the MDM2 zinc finger domain disrupt ribosomal protein interaction and attenuate MDM2-induced p53 degradation. *Mol. Cell. Biol.* **2007**, *27*, 1056–1068. [CrossRef] [PubMed]
234. Nosrati, N.; Kapoor, N.R.; Kumar, V. DNA damage stress induces the expression of ribosomal protein S27a gene in a p53-dependent manner. *Gene* **2015**, *559*, 44–51. [CrossRef] [PubMed]
235. Wang, H.; Zhao, J.; Yang, J.; Wan, S.; Fu, Y.; Wang, X.; Zhou, T.; Zhang, Z.; Shen, J. PICT1 is critical for regulating the Rps27a-Mdm2-p53 pathway by microtubule polymerization inhibitor against cervical cancer. *Biochim. Biophys. Acta Mol. Cell Res.* **2021**, *1868*, 119084. [CrossRef] [PubMed]
236. Pelletier, J.; Thomas, G.; Volarevic, S. Ribosome biogenesis in cancer: New players and therapeutic avenues. *Nat. Rev. Cancer* **2018**, *18*, 51–63. [CrossRef] [PubMed]
237. de Las Heras-Rubio, A.; Perucho, L.; Paciucci, R.; Vilardell, J.; Lleonart, M.E. Ribosomal proteins as novel players in tumorigenesis. *Cancer Metastasis Rev.* **2014**, *33*, 115–141. [CrossRef]
238. Pecoraro, A.; Pagano, M.; Russo, G.; Russo, A. Ribosome Biogenesis and Cancer: Overview on Ribosomal Proteins. *Int. J. Mol. Sci.* **2021**, *22*, 5496. [CrossRef] [PubMed]
239. Ajore, R.; Raiser, D.; McConkey, M.; Joud, M.; Boidol, B.; Mar, B.; Saksena, G.; Weinstock, D.M.; Armstrong, S.; Ellis, S.R.; et al. Deletion of ribosomal protein genes is a common vulnerability in human cancer, especially in concert with TP53 mutations. *EMBO Mol. Med.* **2017**, *9*, 498–507. [CrossRef] [PubMed]
240. Ponten, F.; Jirström, K.; Uhlen, M. The Human Protein Atlas-a tool for pathology. *J. Pathol.* **2008**, *216*, 387–393. [CrossRef]
241. Chen, D.; Zhang, R.; Shen, W.; Fu, H.; Liu, S.; Sun, K.; Sun, X. RPS12-specific shRNA inhibits the proliferation, migration of BGC823 gastric cancer cells with S100A4 as a downstream effector. *Int. J. Oncol.* **2013**, *42*, 1763–1769. [CrossRef] [PubMed]
242. Derenzini, E.; Agostinelli, C.; Rossi, A.; Rossi, M.; Scellato, F.; Melle, F.; Motta, G.; Fabbri, M.; Diop, F.; Kodipad, A.A.; et al. Genomic alterations of ribosomal protein genes in diffuse large B cell lymphoma. *Br. J. Haematol.* **2019**, *185*, 330–334. [CrossRef] [PubMed]
243. Katanaev, V.L.; Kryuchkov, M.; Averkov, V.; Savitsky, M.; Nikolaeva, K.; Klimova, N.; Khaustov, S.; Solis, G.P. HumanaFly: High-throughput transgenesis and expression of breast cancer transcripts in Drosophila eye discovers the RPS12-Wingless signaling axis. *Sci. Rep.* **2020**, *10*, 21013. [CrossRef] [PubMed]
244. Wang, Q.; Cai, Y.; Fu, X.; Chen, L. High RPS27A expression predicts poor prognosis in patients with HPV type 16 cervical cancer. *Front. Oncol.* **2021**, *11*, 752974. [CrossRef] [PubMed]
245. Wang, H.; Yu, J.; Zhang, L.; Xiong, Y.; Chen, S.; Xing, H.; Tian, Z.; Tang, K.; Wei, H.; Rao, Q.; et al. RPS27a promotes proliferation, regulates cell cycle progression and inhibits apoptosis of leukemia cells. *Biochem. Biophys. Res. Commun.* **2014**, *446*, 1204–1210. [CrossRef] [PubMed]
246. Ramrath, D.J.F.; Niemann, M.; Leibundgut, M.; Bieri, P.; Prange, C.; Horn, E.K.; Leitner, A.; Boehringer, D.; Schneider, A.; Ban, N. Evolutionary shift toward protein-based architecture in trypanosomal mitochondrial ribosomes. *Science* **2018**, *362*, 422. [CrossRef]
247. Bochler, A.; Querido, J.B.; Prilepskaja, T.; Soufari, H.; Simonetti, A.; Del Cistia, M.L.; Kuhn, L.; Ribeiro, A.R.; Valášek, L.S.; Hashem, Y. Structural differences in translation initiation between pathogenic trypanosomatids and their mammalian hosts. *Cell Rep.* **2020**, *33*, 108534. [CrossRef]

Disclaimer/Publisher's Note: The statements, opinions and data contained in all publications are solely those of the individual author(s) and contributor(s) and not of MDPI and/or the editor(s). MDPI and/or the editor(s) disclaim responsibility for any injury to people or property resulting from any ideas, methods, instructions or products referred to in the content.

Article

Structural Insights into the Distortion of the Ribosomal Small Subunit at Different Magnesium Concentrations

Ting Yu [†], Junyi Jiang [†], Qianxi Yu, Xin Li and Fuxing Zeng *

Department of Systems Biology, School of Life Sciences, Southern University of Science and Technology, No. 1088 Xueyuan Avenue, Shenzhen 518055, China
* Correspondence: zengfx@sustech.edu.cn; Tel.: +86-0755-8801-8659
† These authors contributed equally to this work.

Abstract: Magnesium ions are abundant and play indispensable functions in the ribosome. A decrease in Mg^{2+} concentration causes 70S ribosome dissociation and subsequent unfolding. Structural distortion at low Mg^{2+} concentrations has been observed in an immature pre50S, while the structural changes in mature subunits have not yet been studied. Here, we purified the 30S subunits of *E. coli* cells under various Mg^{2+} concentrations and analyzed their structural distortion by cryo-electron microscopy. Upon systematically interrogating the structural heterogeneity within the 1 mM Mg^{2+} dataset, we observed 30S particles with different levels of structural distortion in the decoding center, h17, and the 30S head. Our model showed that, when the Mg^{2+} concentration decreases, the decoding center distorts, starting from h44 and followed by the shifting of h18 and h27, as well as the dissociation of ribosomal protein S12. Mg^{2+} deficiency also eliminates the interactions between h17, h10, h15, and S16, resulting in the movement of h17 towards the tip of h6. More flexible structures were observed in the 30S head and platform, showing high variability in these regions. In summary, the structures resolved here showed several prominent distortion events in the decoding center and h17. The requirement for Mg^{2+} in ribosomes suggests that the conformational changes reported here are likely shared due to a lack of cellular Mg^{2+} in all domains of life.

Keywords: structural distortion; magnesium concentration; ribosome; CryoEM

Citation: Yu, T.; Jiang, J.; Yu, Q.; Li, X.; Zeng, F. Structural Insights into the Distortion of the Ribosomal Small Subunit at Different Magnesium Concentrations. *Biomolecules* **2023**, *13*, 566. https://doi.org/10.3390/biom13030566

Academic Editors: Brigitte Pertschy and Ingrid Rössler

Received: 11 February 2023
Revised: 15 March 2023
Accepted: 17 March 2023
Published: 20 March 2023

Copyright: © 2023 by the authors. Licensee MDPI, Basel, Switzerland. This article is an open access article distributed under the terms and conditions of the Creative Commons Attribution (CC BY) license (https:// creativecommons.org/licenses/by/ 4.0/).

1. Introduction

Metal ions are the second most abundant component after water molecules in living cells and are involved in all fundamental biological processes, including protein synthesis, enzymatic reactions, and others [1]. Protein synthesis is mediated by ribosomes, in which the information carried by mRNA is translated into amino acid sequences. A ribosome requires metal ions, including Mg^{2+}, Zn^{2+}, and K^+, to maintain its structure and activity. Mg^{2+} is the most abundant multivalent cation in cells and plays an essential role in the assembly of ribosomes by neutralizing negative charges from phosphates present in the rRNA backbone and enabling the correct folding and compaction of rRNA [2,3].

The bacterial 70S ribosome is a complex macromolecule composed of small (30S) and large (50S) subunits. Recently, a high-resolution ribosome structure was determined, showing that a single ribosome in *Escherichia coli* (*E. coli*) contains at least 309 Mg^{2+} ions [4]. It has been known for decades that the structure and function of ribosomes are strongly influenced by the presence of Mg^{2+} [5]. For example, the in vitro association between small and large ribosomal subunits required to form intact ribosomes depends strongly on the Mg^{2+} concentration [6], and decreasing the Mg^{2+} concentration below 1 mM causes the dissociation and subsequent unfolding of 70S ribosomes [2,7,8]. Meanwhile, extremely low Mg^{2+} causes irreversible structural distortions and even disassembly into individual ribosomal constituents [3]. Research on ribosome unfolding showed that EDTA-dialysis could be used to progressively remove the Mg^{2+} from ribosomes and resulted in the conversion of 50S subunits into 21S particles via a 36S intermediate and the conversion of the 30S subunit into 16S particles via a 26S intermediate, in which the 36S and 26S particles were reversible through the readdition of Mg^{2+}, whereas the 21S and 16S particles containing only 23S and 16S rRNA were irreversible [2]. The growth of E. coli cells under conditions of Mg^{2+} starvation results in ribosome degradation [9]. Furthermore, Mg^{2+} stabilizes the codon–anticodon interaction at the A site and influences the binding of RRF to the ribosome [10,11]. In addition, Mg^{2+} can partly complement the functions of several ribosomal proteins, such as L1, L23, and L34 [6,12]. For instance, an increased Mg^{2+} concentration suppresses the defects in 70S ribosome formation caused by a lack of ribosomal protein L34. Previous studies have also suggested that lower Mg^{2+} concentrations greatly increase the susceptibility of ribosomes to attack by ribonuclease [13,14]. In E. coli cells, the concentration of intracellular free Mg^{2+} ranges from 1 mM to 5 mM [15,16]. Although Mg^{2+} is essential for ribosomes, excess Mg^{2+} reduces their translation activity and accuracy [17,18]. For instance, the error frequency measured in vitro at 10 mM Mg^{2+} is 10 times higher than that at 5 mM Mg^{2+} [19].

In order to examine the impact of low Mg^{2+} exposure on the 30S structure, we used cryo-electron microscopy (cryo-EM) to explore the structures of 30S particles and characterize a series of unnatural 30S structures at 1 mM Mg^{2+}. The 30S ribosomal subunit is composed of one rRNA molecule (16S rRNA) and approximately 21 r-proteins, which are organized into four distinct structural domains: the body (5′ domain), the platform (central domain), the head (3′ major domain), and helix 44 with h45 (3′ minor domain) [20]. The assembly of the 30S subunit is a robust process proceeding via multiple redundant parallel pathways, where the 5′ body domain forms first, followed by the central platform, head domains, and lastly, the 3′ minor domain with the functionally important decoding center [21–24]. Although ribosome unfolding has been studied with many different techniques, such as the measure of the sedimentation coefficient, viscosity, and diffusion constant, as well as ultraviolet absorption and laser Raman spectroscopy, and despite the fact that evidence has been presented to show that discrete intermediates exist in the unfolding reaction, little information about specific structural changes during unfolding has been provided [3,5]. A recent study of rRNA self-folding showed that in the absence of Mg^{2+} or with Mg^{2+} of up to ~1 mM, the tertiary interactions of the 16S central domain are disrupted, resulting in expanded conformations containing only secondary structures [8]. How does Mg^{2+} facilitate the folding of rRNA fragments, especially the formation of tertiary contacts? In this study, our cryoEM structures showed that the 30S particles collected at 1 mM showed several missing structural features and conformation changes, including missing h44 and S12 and the movement of h17 and h27. These structural distortions in unfolding intermediates provide insight into ribosome biogenesis and should be taken into account in vitro assembling studies.

2. Materials and Methods

2.1. E. coli Strains and Cell Culture

In this study, 30S particles were purified from an E. coli strain with the mutant DbpA protein overexpressed, obtained from another project conducted by our research group. Since the structural distortion observed in these $30S_{DbpA}$ at 1 mM Mg^{2+} was shown to be exactly the same as that of 30S purified from the wild-type BL21 strain (data not shown), the cryoEM data of $30S_{DbpA}$ were used for reconstruction and designated as $30S_{1mM}$ in this study. Wild-type E. coli BL21 and MRE600 were used for the sucrose gradients and purification of 30S at 2.5 and 10 mM Mg^{2+}.

For the sucrose gradient and 30S purification, E. coli cells of the strains BL21 and MRE600 were grown to an OD_{600} between 0.6 and 0.8 at 37 °C and 220 rpm in LB medium before harvesting. Yeast cells of the strain BY4742 were grown to an OD_{600} of 0.65 at 30 °C and 220 rpm in YPD medium. HEK293F cells were grown in suspension culture to approximately 2×10^6 cells per milliliter before harvesting.

2.2. Sucrose Gradient Centrifugation Analysis

For the sucrose gradient analysis, 50 mL of cells were collected by centrifugation at 3500 rpm at 4 °C for 10 min and were resuspended in 200 µL of lysis buffer (20 mM HEPES-KOH pH 7.5, 150 mM NH_4Cl, 4 mM β-mercaptoethanol) with different concentrations of Mg^{2+} and EDTA, according to the experimental requirements, and DNase I (RNase-free) was added at a final concentration of 20 U/mL. The cells were broken with a grinder, followed by centrifugation at 14,000 rpm at 4 °C for 20 min. Fifteen units of A_{260} of clarified lysates were loaded onto a twelve milliliters of sucrose gradient.

For the EDTA-treated assay, the clarified lysates were loaded onto a 10–50% sucrose gradient in lysis buffer with 10 mM EDTA and then centrifuged at 4 °C with an SW41 rotor for 4 h at 35,000 rpm. For the assay of different concentrations of Mg^{2+}, the clarified lysates were loaded onto a 10–50% sucrose gradient in lysis buffer with 0.5 mM, 1 mM, 2.5 mM, 5 mM, 10 mM, and 20 mM Mg^{2+} and then centrifuged at 4 °C with an SW40 rotor for 12 h at 32,000 rpm. The profiles were detected through the continuous monitoring of the absorbance at 260 nm using a Biocomp Gradient Master/AKTA pure.

2.3. Cryo-EM Sample Preparation and Data Collection

Samples of $30S_{1mM}$, $30S_{2.5mM}$, and $30S_{10mM}$ with the corresponding peaks were collected and dialyzed to remove the sucrose. The fresh samples were diluted to 300 nM in a corresponding buffer, and then 2.5 µL of isolated particles was applied to glow-discharged R1.2/1.3 holey carbon grids with 2–4 nm continuous carbon film on top. After 30 s of waiting, the grids were blotted for 3 s and plunged into liquid ethane using a Vitrobot device (FEI) operating at 4 °C and 100% humidity.

Micrographs were collected on a Titan Krios G3i operating at 300 kV with a Gatan K3 Summit. Data acquisition was performed using the software EPU, with a nominal magnification of 81,000× g, which yields a final pixel size of 1.095 Å on the object scale (defocus ranging from −1.5 µm to −2.5 µm). For each micrograph stack, 30 frames were collected, for a total dose of 30 electrons per pixel. For the EDTA-treated sample, micrographs were recorded on a Titan Krios G3i operating at 300 kV with a Gatan K2 Summit.

2.4. Image Processing

Motion correction on the micrograph level was performed with MotionCorr2 [25]. The program CTFFIND4 was used to estimate the contrast transfer function parameters [26]. Image processing, including micrograph screening, particle picking, 2D and 3D classification, refinement, and postprocessing, were performed with RELION 3.1.0 [27].

For the $30S_{1mM}$ sample, a total of 497,673 particles were subject to a cascade of 2D and 3D classification. After one round of 3D classification, 440,474 particles were subjected to 3D auto-refine and then subjected to 3D classification with a mask on the decoding centers, S12 and h17, respectively, in which the alignment of the particles was omitted.

For the $30S_{2.5mM}$ and $30S_{10mM}$ samples, a total of 360,621 and 349,073 particles were subjected to several rounds of 2D/3D classification to remove the non-ribosomes and bad particles. Finally, 139,475 and 146,689 particles, respectively, were used for the 3D reconstructions.

2.5. Model Building

A high-resolution cryo-EM structure of the *E. coli* ribosome (PDB:7k00) was used as the initial model, and rigid-body fitted to the density map using Chimera and Coot 0.8.9 [4,28,29]. The ribosomal protein and 16S rRNA were then fitted individually as rigid bodies and manually adjusted for the best fit between the map and the model. For the additional h17 density on the map, h17(437–497) was extracted from 7k00 23s rRNA and fitted as a rigid body. To illustrate the shifting of the rRNA helices, h6, h16, h18, h24, h27, and h45 were fitted to the maps using rigid-body fitting followed by real-space refinement in Coot [28].

2.6. CryoDRGN Analysis

To study the correlations of the structural distortion between different blocks, we exploited cryoDRGN's powerful generative model to analyze the structural heterogeneity [30]. The particles were processed in RELION 3.1.0 until 3D refinement was achieved with the mask on the body, which contained 440,474 particles. Then, the results were applied for cryoDRGN training, in which the particles were downsampled to a box size of 256 (1.638 Å per pixel). The networks for the datasets were trained with an eight-dimensional latent variable.

For the subunit occupancy analysis, 500 volumes were sampled from the latent space. Then, the body domain of 30S was split into 32 blocks using a PDB file that was rigid-body-fitted to the refined map, including coordinates for two h17 helices (h17$_{in}$ and h17$_{out}$). The generated atomic models were used to create masks corresponding to each of the rRNA helices and ribosomal proteins. Then, these 32 masks were applied to each of the 500 volumes in turn, and finally, the occupancy of the density was calculated for each block, and the correlation between them was calculated by hierarchical clustering analysis.

3. Results

3.1. Ribosomal Subunits Are Destroyed by EDTA Treatment

It is known that metal ions are essential for stabilizing the structure of ribosomes and maintaining their activity. EDTA-treated ribosomes of *E. coli* have been reported to be Y- and X-shaped for the unfolded 30S and 50S [31], respectively. To symmetrically study the structures of ribosome subunits under the conditions of low Mg^{2+} concentrations, crude ribosomes from the *E. coli* strains BL21 and MRE600, as well as yeast and human cells, were analyzed by sucrose gradient sedimentation containing 10 mM EDTA (Figure 1a–d). Gradients with 10 or 2.5 mM Mg^{2+} were used as controls (gray curves in Figure 1a–d). The profiles of the gradients with EDTA showed peaks corresponding to the small and large subunits shifted towards the top of the gradient, which indicated a dramatic decrease in the molecular weight or particle size of the subunits. Then, the two shifted peaks of *E. coli* BL21 and MRE600 were collected and pooled (Figure 1a,b). CryoEM imaging showed that the subunits obtained from the 10 mM EDTA gradient were largely destroyed, with most, if not all, of the ribosomal proteins being dissociated and the rRNA being exposed in extended states (Figure 1e,f, white arrows). The shifting of the subunit peaks in the sucrose gradient profiles and the extended shapes of the rRNA observed by cryoEM imaging indicate that the extraction of Mg^{2+} from both prokaryotic and eukaryotic ribosomes by EDTA destroys the structure of ribosomes. Following this, the destruction process was further studied using 30S subunits as a model with a series of cryoEM structures.

3.2. Structural Distortion of 30S at a Low Magnesium Ion Concentration

Magnesium ions are the most abundant ions present in ribosomes [1]. To further study the distortion process of the structure of ribosome subunits, relevant fractions in a sucrose gradient sedimentation of the *E. coli* strain were collected and subjected to cryoEM analysis (Figure 2a). In this study, we isolated the 30S particles from a reference-free 2D classification strategy in Relion 3.1.0 for intensive study (Figure 2d). Typical images of the $30S_{1mM}$ peak showed 30S profiles with a flexible h17 helix and smear tracks of the 30S head (Figure 2d, red arrows), showing that the low Mg^{2+} concentration affected the 30S structure. To further confirm this hypothesis, we collected the 30S peak from sucrose gradients with 2.5 and 10 mM Mg^{2+}, respectively, for cryoEM analysis (Figure 2b,c). The 2D averages for $30S_{2.5mM}$ showed a better head and h17 helix (Figure 2d, 2nd line), and the $30S_{10mM}$ showed a far more stable 30S head and a fold-in h17 helix (Figure 2d). These observations indicate the important role of Mg^{2+} ions in maintaining the structure of rRNA in the 30S subunit. Particles from these three datasets of $30S_{1mM}$, $30S_{2.5mM}$, and $30S_{10mM}$ were then extracted for further analysis (Supplementary Materials Figures S1–S3).

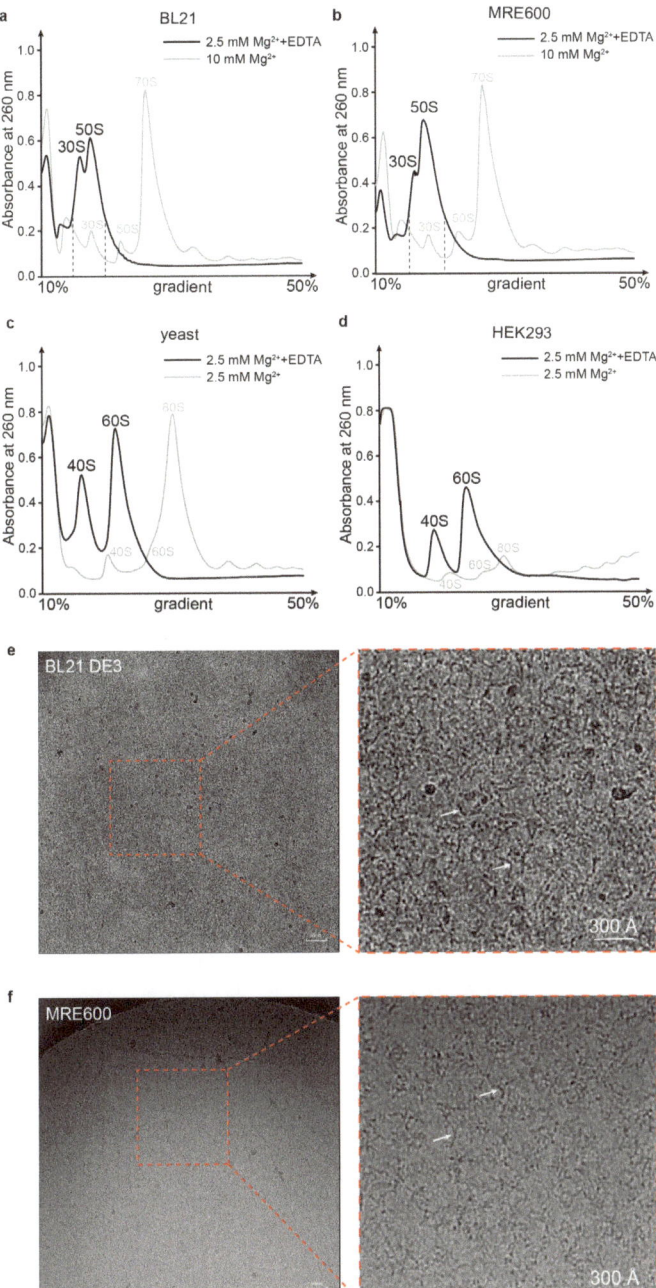

Figure 1. Subunits of the 70S ribosome were destroyed by EDTA. (**a**,**b**) *E. coli* cells of the BL21 DE3 (**a**) and MRE600 (**b**) strains were collected at OD$_{600}$ = 0.6 and further analyzed with a sucrose gradient in conditions with or without 10 mM EDTA. Shifting of the peaks was indicated, and the 30S and 50S peaks from the gradient with additional EDTA were collected (dashed line labeled) for

cryoEM analysis. (**c,d**) Yeast BY4742 cells (**c**) and HEK293F cells (**d**) grown to exponential phase were collected and disrupted by a French press and homogenizer, respectively. Cell extracts were then analyzed with a sucrose gradient under conditions of 2.5 mM Mg^{2+} or 10 mM EDTA. (**e,f**) CryoEM images representing the typical particle shapes (white arrows) for BL21 DE3 and MRE600 (**f**) strains. The scale bar is labeled in white.

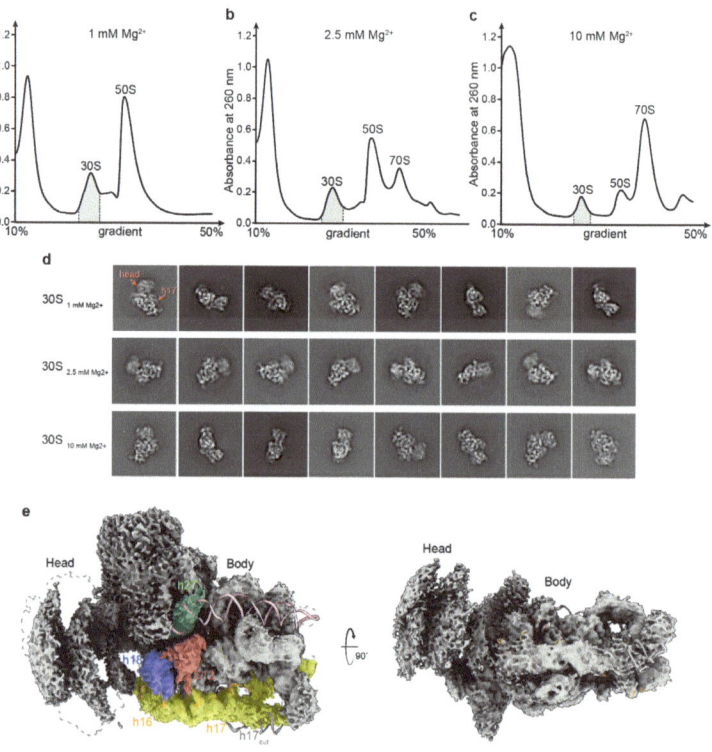

Figure 2. A 30S subunit at 1 mM Mg^{2+} has a flexible h44, h17, decoding center, and head. (**a–c**) Sucrose gradient sedimentation profile of *E. coli* ribosomes under 1 mM (**a**), 2.5 mM (**b**), and 10 mM (**c**) Mg^{2+} conditions. The 30S peak indicated in gray shadow was collected separately for cryoEM analysis. (**d**) Reference-free 2D classification averages for 30S particles under different conditions, as shown in (**a–c**). The flexible h17 and head are labeled with red arrows. (**e**) The overall structure and map of a 30S subunit at 1 mM Mg^{2+} concentration are represented in cartoon and surface, respectively. The rRNA helices h16/h17, h18, h27, and h44 and the S12 protein are colored in yellow, blue, green, magenta, and red. The dashed line represents the mature 30S under 10 mM Mg^{2+} conditions in this study.

3.3. Overall Structures of the 30S Subunits under Different Mg^{2+} Concentrations

As seen above, the 30S at 1 mM Mg^{2+} showed a flexible h17 and a blurry head. To explore the structural distortion in more detail, the particles were subjected to a 3D reconstruction using Relion 3.1.0 [27], with a mask on the 30S body to eliminate the interference of the blurry head-on alignment, resulting in a set of structures with a resolution of 3–5 Å (Figure 2e and Supplementary Materials Figures S1 and S2). Consistent with the 2D averages seen in Figure 2d, 30S in the 1 mM Mg^{2+} condition showed the same conformation, with a poorly aligned head and well-defined body (Figure 2e). Spahn et al. proposed that the rotation of the head of the small subunit directs the movement of the tRNAs to the P and E sites [32]. Hence, the head of 30S plays a vital role in translation, and a comparative structural analysis of 55 ribosome structures showed that the 30S head rotated by 0–21 degrees related to the body part, as determined by the E-R method [33,34]. Using multi-body refinement in Relion 3.1.0, a program used to classify heterogeneous cryo-EM structures, the 30S from 1 mM Mg^{2+} showed a much broader range of rotation in its head (Figure S4). Even with a mask on the head, the reconstructions still failed to show a clear map of the rRNA or proteins (Figure S4a), which means that low Mg^{2+} conditions can cause the 30S head to become even more flexible on its own accord. Meanwhile, compared to the 30S at 2.5 and 10 mM Mg^{2+} concentrations (Figure S3), the first thing that we noticed in the 1 mM Mg^{2+} reconstructions was the absence of h44 (Figure 2e), which has almost disappeared in 2.5 mM Mg^{2+} as well (Figure S3a). Flexible regions, including the decoding center, h16, h17, and platform, and a weak S12 density were also observed on the map (Figure 2e). These functional regions mature at different timepoints when 30S assemble, requiring plenty of Mg^{2+} to stabilize their positions [35].

3.4. Movement of Incompact Helices and Loss of S12 in the Decoding Center

In 30S, the decoding center, which is composed of h27, h28, h1, h2, the upper part of h44, and h45, offers a place for interaction between mRNA and tRNA, contributing to the fidelity of decoding through the monitoring of codon-anticodon base pairing [20,36]. To separate the different conformations contained in the 30S reconstructions of 1 mM Mg^{2+}, we first performed non-alignment 3D classification based on the auto-refined angles with a mask on the decoding center, resulting in different reconstructions with diverse structural distortions in the decoding center. Here, the maps are named as $h27_{in-1,2}$ and $h27_{out-1-4}$ according to their h27 positions, and three main classes are obtained (Figure 3a–c and Supplementary Materials Figures S1 and S5). Approximately 7% of particles ($h27_{out-3}$) showed long-distance (~26 Å) shifting towards h18, and, accordingly, h18, h24, and h45 became more flexible (Figure 3a). The $h27_{out-4}$, with 15% particles, showed short-distance (~19 Å) shifting compared to $h27_{out-3}$ (Figure 3b), whereas the $h17_{in-2}$, which contained approximately 56% of the total particles, showed a stable h27 that was fixed to the mature state (Figure 3c). The other three states showed a different level of movement, according to which $h27_{out-1}$ and $h27_{out-2}$ were similar to $h27_{out-4}$, containing a short-distance-shifted h27 helix (Supplementary Materials Figure S5a,b), and the $h27_{in-1}$ state had a near-mature h27 helix, as seen in $h17_{in-2}$ (Supplementary Materials Figure S5c).

In the states separated by the mask of the decoding center, in addition to the movement of h27 and the correlated swing of h18, h24, and h45, we also observed a weakened density in the S12 protein. S12 is located near the decoding center. It is composed of two distinct parts, including the N-terminal extension and the conserved C-terminal globular region [37,38]. S12 plays a pivotal role in decoding functions and is a key mediator in maintaining the fidelity of translation on the ribosome. Research has shown that S12 is important for the inspection of codon–anticodon pairings at the ribosomal A site [39]. The N-terminus of the protein binds the solvent surface of the SSU, with the extension in contact with the rRNA dense regions, ending with a C-terminal globular region localized at the inter-subunit face of the SSU, which means that S12 plays a vital role in maintaining the small subunit structure [40]. To further understand the binding of S12 in the decoding center, we then exploited 3D classification using a mask on the S12 protein alone, resulting in 10 different classes showing various occupancy levels of S12 (Figure 3d,e and Supplementary Materials Figures S1 and S6). In the states of $S12_4$ and $S12_9$, which contained approximately 14% of the total particles, the S12 protein was completely missing (Figure 3d and Supplementary Materials Figure S6d). Along with the disappearance of the S12 protein, helix h18 and h27 in these two states moved by approximately 7 and 18 Å, respectively (Figure 3d). State $S12_{10}$ had 53% of particles and showed a high level of occupancy at the S12 density, with slight shifting of h18 and h27 (Figure 3e). States $S12_{1-8}$, except $S12_4$ and $S12_6$, also had a partial density of S12, which means that low Mg^{2+} only partially destabilized the interactions between S12 and the rRNA (Supplementary Materials Figure S6). State $S12_6$ had no density in the position of S12 but showed extra density above S12, probably being in an intermediate state (Supplementary Materials Figure S6f). We should also note that, due to the steric hindrance, the helix h16 lying close to h18 moved further when S12 was missing compared to the states with S12 proteins (Figure 3d,e).

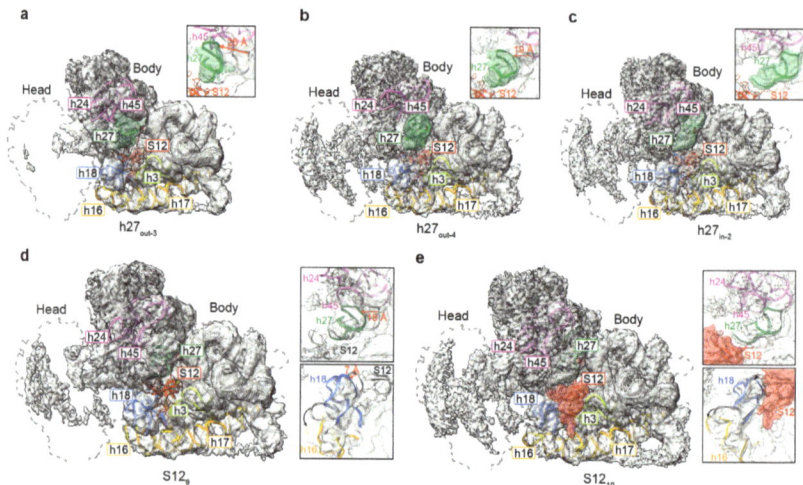

Figure 3. A low Mg^{2+} concentration destabilizes the decoding center and causes a loss of S12 protein. (**a–c**) Three representative reconstructions were classified from 30S particles at 1 mM Mg^{2+} by applying a mask to the decoding center. A completely out-shifted h27 ((**a**), h27$_{out-3}$), a partially out-shifted h27 ((**b**), h27$_{out-4}$), and h27 in its original position ((**c**), h27$_{in-2}$), are represented in cartoon and surface (green) for their rigid-body-fitted structure and density map, respectively. (**d**,**e**) Two representative reconstructions with missing (**d**) and fully occupied S12 protein ((**e**), red), were classified from 30S particles at 1 mM Mg^{2+} by applying a mask to S12, as shown in the cartoon and on the surface. The rRNA helices h3 (limon), h16/h17 (yellow), h18 (blue), and h24/h45 (magenta) are also labeled according to the positions of the decoding center. Details of the interactions and movement (black arrows) of h27 and S12 are represented as inserted subfigures. A dashed line represents the mature 30S under 10 mM Mg^{2+} conditions in this study.

3.5. Movement of h17 towards h6

In addition to the decoding center, we could also observe an extra density around h17 toward h16 (Figure 2d). Applying a mask to h17 alone, we identified four major classes from the particles of 30S at 1 mM Mg^{2+} (Figure S1). The map named h17$_{in}$ is identical to the mature 30S, in which h17 interacts with h10 through its tip from A465 to C470 and with h15 through the central part of h17. Interactions between the central part of h17 and the positively charged residues of the S16 protein were also observed (Figure 4a,b). The second one, containing 38% of the particles and named h17$_{out}$, had a clear helix-shaped density that moved outwards by approximately 54 degrees, in which the relocated h17 could be fitted, folding as a bridge that connected the tip of h6 and junction of h16/h17 (Figure 4c). In this conformation, the tip of h17 (G462–U464) is close to the helix h6 (U85–G86) and stabilized by the interactions between them (Figure 4d). Since h16 and h17 are connected and form a long helix in the 30S subunit, we also compared the h16 helix in these two states (Supplementary Materials Figure S7). A bent h16/h17, as one can observe in a mature 30S, was observed in the h17$_{in}$ state, and a long, straighter helix was found in the h17$_{out}$ state. Whether h17 is located within or shifted out, h16 swayed at the same angle in both states, which means that the junction of h16 and h17 is very flexible; thus, the movement of h16 is independent of h17 shifting. The third class, called h17$_{in-2}$, showed a blurry density between the h17$_{in}$ and h17$_{out}$ conformations, being closer to h17$_{in}$, meaning that it should be in the intermediate state (Supplementary Materials Figure S1). The remaining 23% of particles comprised a 30S subunit with an invisible h17 and most of its helices, probably due to its highly flexible rRNAs.

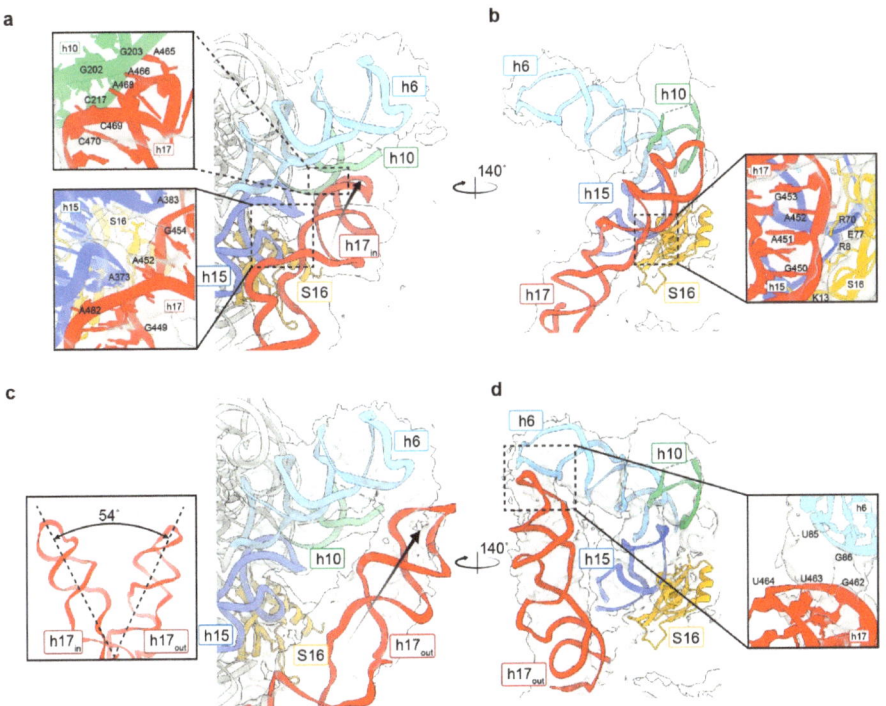

Figure 4. The h17 helix shifted outwards to the tip of h6. (**a**) Interactions between h17 (red), h10 (green), and h15 (blue) in the reconstruction of h17$_{in}$. The direction of h17 is labeled with a black arrow. (**b**) Interactions between h17 and S16 proteins in the reconstruction of h17$_{in}$. (**c**) Shifting of h17 from h17$_{in}$ to h17$_{out}$. The shifted angle was measured in Chimerax, and the direction of h17 in the h17$_{out}$ reconstruction is labeled with a black arrow. (**d**) The interactions between the out-shifted h17 and the tip of h6.

3.6. Correlation of the Structural Distortions between Different Blocks

Except for the decoding center and h17, as mentioned above, some other parts of the 30S also became less stable under a low Mg^{2+} concentration. To determine whether there are correlations between the destabilization of these rRNA helices and ribosomal proteins, we exploited cryoDRGN's powerful generative model by sampling 500 volumes from the latent space for a total of 440,474 particles with 35 learning epochs [30] (Figure 5a). In this strategy, the coordinates of the 30S head were first removed since no clear density could be assigned to it. Then, the body was split into 32 blocks using a PDB file that was rigid-body-fit to the refined map, including coordinates for two h17 helices (h17$_{in}$ and h17$_{out}$). Each structural block contained an individual rRNA helix, or ribosomal protein. The occupancy of the density was calculated for each block, and the correlation between them was calculated by hierarchical clustering analysis (Figure 5a). Here, we observed that h44 had totally disappeared, and the occupancy of h17, S11, and S21 was significantly decreased in most of the volumes. Additionally, approximately half of the 500 volumes had a poor overall density, revealing the global flexibility of 30S at 1 mM Mg^{2+}. For the other half of the volumes that had a higher occupancy for most of the blocks in the 30S body, we could identify correlations between the unstable helices and proteins (Figure 5a,b).

Hierarchical clustering showed that proteins S11 and S21 were omitted simultaneously in most cases, and h23, h24, and h45 showed a similar pattern. It is worth noting that the occupancies of h27 and protein S12 in the decoding center had different distributions in the 500 volumes compared to h17 (Figure 5a). We further analyzed the distributions of all the particles in the states identified using different masks (Figure 5c,d). The particle number in each state was calculated and analyzed by the chi-squared test. This showed that when h27 is in states $h27_{out-1,2,3}$ or $h27_{in-1}$ and S12 is in states $S12_{1-8}$, the h17 helix tends to be flexible, whereas in the $h17_{out}$ particles, the h27 tends to be fixed in the mature position ($h27_{in-2}$) and S12 is complete ($S12_{10}$). On the other hand, h27 tends to shift outward, and S12 shifts away in the two $h17_{in}$ states. These results showed a negative correlation between the movements of h17 and h27/S12, similar to the hierarchical clustering results (Figure 5a).

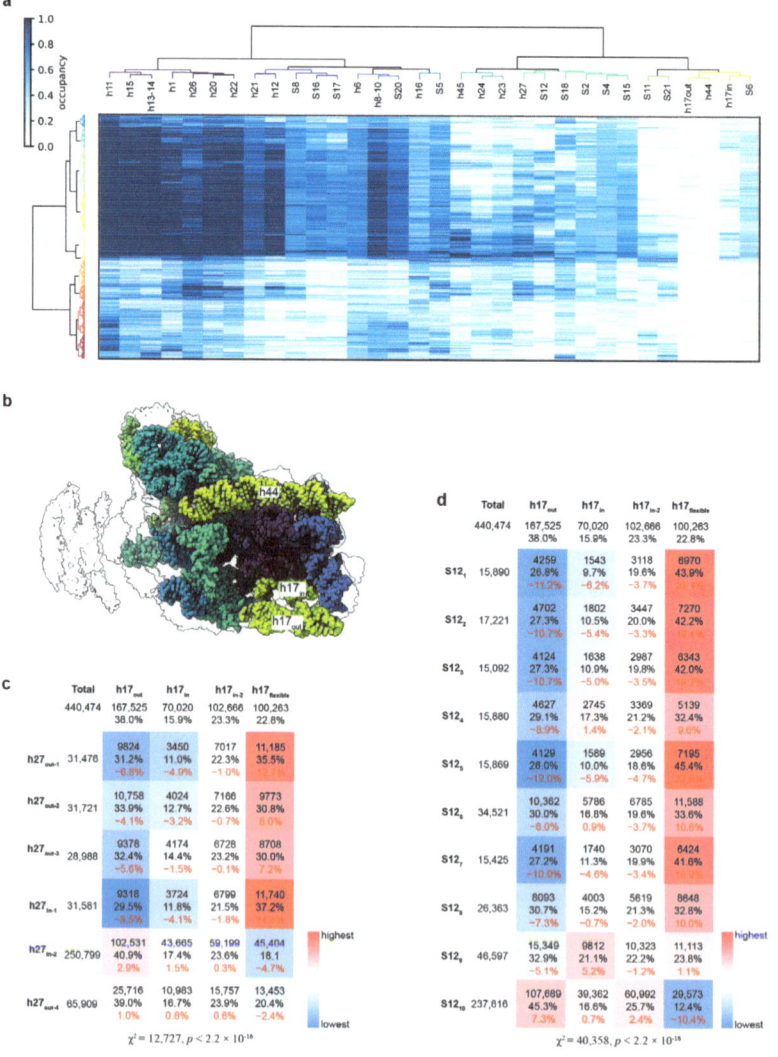

Figure 5. The movements of the different structural blocks are correlated. (**a**) Occupancy analysis of $30S_{1mM}$ particles displayed as a heatmap. Rows (500) correspond to sampled density maps, and

columns (32) correspond to structural elements defined by the atomic model. Only the 30S body was analyzed. (**b**) Atomic models of the 30S subunit used for subunit occupancy analysis are colored according to the structural blocks defined through hierarchical clustering in (**a**). Structural features of interest are annotated. (**c**) Correlation between h17 and h27 in the reconstructions classified with different masks. The mask on h17 defined four different states of h17, and the mask on the decoding center defined six states of h27. Particles in each state were selected, and the distribution of particles in different h17 conformations was calculated. For each h27 state, the difference in distribution compared to the total particles is colored in red. (**d**) Same as (**c**), but the correlation between h17 and S12 was calculated.

3.7. Mg^{2+} Is Essential for Ribosome Structure Stability

To further investigate the effect of the Mg^{2+} concentration on ribosome assembly, we performed sucrose gradient analysis with different levels of Mg^{2+} (Figure 6). As observed at 1 mM Mg^{2+} (Figure 2a), the 0.5 mM Mg^{2+} condition also completely separated the 50S and 30S subunits (Figure 6a), whereas at the 2.5 and 5 mM Mg^{2+} concentrations, the 70S ribosome formed gradually (Figures 2b and 6b). Our structural study of the 30S peak at 2.5 mM Mg^{2+} also showed a barely visible h44 and a flexible h17 (Figure S3a). With 10 mM Mg^{2+} or above, the 70S became very stable (Figures 2c and 6c). To determine whether the 30S and 50S purified from 1 mM Mg^{2+} were still functional, we collected and incubated these two peaks, followed by reloading the gradient with 10 mM Mg^{2+}. The results showed that most of the 30S and 50S could form an intact 70S (Figure 6d). The remaining 30S that could not interact with 50S might be the result of S12 protein dissociation, as observed in our structures (Figure 3). Analysis of the ribosomal proteins in the 30S peak by SDS-PAGE showed several weakened bands at 0.5 and 1 mM Mg^{2+} compared to 10 mM Mg^{2+} (Supplementary Materials Figure S8). These could be the results of weak binding to the rRNA at a low Mg^{2+} concentration, followed by a falloff from 30S when the concentration by centrifugation was applied.

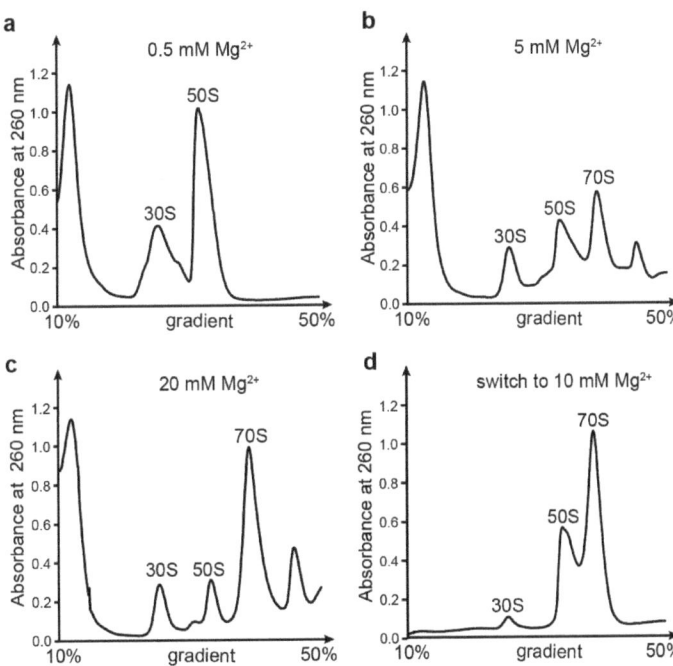

Figure 6. The effect of Mg^{2+} concentration on the assembly of 70S. (**a**–**d**) Sucrose gradient profiles for

crude 70S at 0.5 mM (**a**), 5 mM (**b**), and 20 mM (**c**) Mg^{2+} concentrations. (**d**) Profile of 30S and 50S purified from 1 mM Mg^{2+} in a gradient containing 10 mM Mg^{2+}. Fractions corresponding to 30S and 50S under 1 mM Mg^{2+} conditions (Figure 2a) were pooled, and sucrose was removed, followed by reloading onto a sucrose gradient containing 10 mM Mg^{2+}.

4. Discussion and Conclusions

The data presented here, ranging from 1 mM to 10 mM Mg^{2+}, represent a model of the process of *E. coli* 30S unfolding during Mg^{2+} decrease (Figure 7). Initially, the 30S structure is completely intact at 10 mM Mg^{2+}. When the concentration of magnesium ions decreases, the rotation angle of the head increases accordingly, along with the loosening of the proteins and RNA helices. Here, h44 is most sensitive to magnesium ions, first becoming unstable, and then h17 begins to swing away (Figure 7, dashed box at 2.5 mM Mg^{2+}). With the Mg^{2+} decreasing to 1 mM, the rRNA helices h16, h18, and h27 in the decoding center also become unstable, and protein S12 starts to dissociate from the 30S subunit (Figure 7, dashed boxes in 1 mM Mg^{2+}). When there are no magnesium ions at all, that is, after EDTA treatment, the 23S rRNA becomes completely unfolded, with a loss of ribosomal proteins. Most likely, the secondary structures are still stable, except for those helices with only a few canonical Watson–Crick base pairs [8]. Our cryoEM structures showed important tertiary interactions stabilizing the rRNA fragments formed/destroyed at different Mg^{2+} concentrations, reflecting differences in stability between different regions. This is consistent with the calculation showing that the midpoint, the Mg^{2+} concentration at which individual secondary or tertiary interactions occur, is not unique, and the coordination of Mg^{2+} with rRNA is nucleotide specific and does not occur in a random, diffusive manner [8].

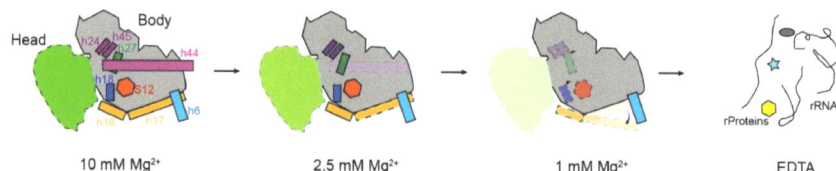

Figure 7. A model of 30S structural distortion at a low Mg^{2+} concentration. Here, 2.5 mM Mg^{2+} first causes the destabilization of h44 and partial loosening of h17, and the 30S head becomes more flexible. A further decrease in Mg^{2+} to 1 mM shifts h17 toward the tip of h6, and h16, h18, and h27 become flexible at the same time. Ribosomal S12 near the decoding center starts to leave, causing an irreversible 30S if no additional protein is supplied. EDTA incubation was used to extract all the Mg^{2+} from ribosomes and completely destroy the 30S subunit, release the ribosomal proteins, and linearize the 16S rRNA.

The magnesium ion, which is the most abundant of the divalent cations in living cells, has an irreplaceable function in stabilizing the secondary structure of ribosomal RNA, binding ribosomal proteins to the ribosome, and ribosomal interaction due to its high charge density and relatively small ionic radius (0.6 Å) [3,41,42]. It is known that the removal of Mg^{2+} from ribosomes results in a non-functional ribosome, especially in the case of EDTA-treated *E. coli* ribosomes, which show a different structural assembly manner [2]. In this study, we used cryoEM to study the disassembly process of the small subunit of the *E. coli* ribosome and showed a series of snapshots of the distortional 30S, which revealed a sequential movement of the rRNA helices and ribosomal proteins in the 30S subunits. Several cryoEM structures suggest that the pre30S/pre40S particles transit through a vibrating state [43–45]. We concluded that the movements observed in this study using a low Mg^{2+} concentration are correlated with the structural rearrangements observed during 30S/40S maturation.

Firstly, all the reconstructions obtained in this study lacked a clear h44, which means that this helix is totally moved out at a 1 mM Mg^{2+} concentration, although we could observe weak densities above the decoding center in some of the states responsible for the moved h44. The helix h44 is directly involved in mRNA decoding, as is the formation of two inter-subunit bridges (B2a and B3) that participate in association with the large ribosomal subunit [46,47]. In the process of the decoding center's maturation, h44 forms before the correct base pairing of h28 and the linker h28/h44. The assembly factors RimP and RsmA then lift h44 to access h28 for its refolding with the help of RbfA. In the last step, the decoding center, with an h44 in the final position, is checked by RsgA, which means that the placing of the h44 helix is the last step in 30S maturation. Thus, it makes sense that h44 is the first part to be dissociated when the 30S structure becomes unstable [45]. In the decoding center, we also observed a movement of h27 by approximately 26 degrees towards h18 due to the disappearance of h44. Helix h27, termed the switch helix, is packed groove-to-groove with the upper end of h44, which is the target of amino-glycoside antibiotics [20,48]. A recent study showed that h27 and h21 had larger fluctuations than other helices in the central domain of 16S rRNA due to their absence in extensive tertiary interactions [8]. In other words, h27, under conditions of low Mg^{2+} concentration, also becomes more flexible than other helices. In addition to h27 and h44, the surrounding helices h18, h24, and h45 also showed significant movement under the 1 mM Mg^{2+} conditions.

Secondly, in part of the reconstruction, the ribosomal protein S12 dissociated from the decoding center. S12 is the third binding r-protein according to the Nomura assembly map [49,50]. The C-terminal globular region of S12 is close to the decoding center, and it is unique among SSU r-proteins, because S12 is the only protein located on the RNA-rich surface that interacts with the large subunit [20]. It is clear that the ribosomal protein S12 plays a pivotal role in tRNA selection by the ribosome [39,51]. Crystal structures have revealed that the closed conformation of the 30S subunit is stabilized by interactions between the conserved amino acid residues of ribosomal protein S12 and 16S rRNA h44 at the decoding site [52]. Recent simulation studies showed that the central domain of the 16S rRNA without ribosomal proteins unfolds at low Mg^{2+} concentrations of roughly 2 mM.

Thirdly, an approximately 54 degree rotation was observed in h17, which generated novel interactions between the tip of h17 and h6. In the mature 30S, h16/h17 comprises one of the three long helices in the 30S subunit and interacts with h18/h15, which is situated beside it and forms the backbone of the entire body. The three major longitudinal elements, h44, h16/h17, and h7, act as structural pillars that extend over 110 Å, and not only stabilize the body but also transmit conformational changes and displacements over a very long distance [20]. Here, we observed that h17 started to shift towards the tip of h6, even at 2.5 mM Mg^{2+}, whereas this movement did not affect the connected h16. Furthermore, the 30S head and related helices in the neck region were too flexible to be defined in our reconstructions. In the translation process, large-scale rotation of the head domain is required for the movement of mRNA and tRNA translocation [33,34].

In *E. coli*, the Mg^{2+} concentration is reported to be approximately 1–5 mM, whereas the total Mg^{2+} concentration, including the Mg^{2+} chelated by biological molecules, is around 100 mM [16,53]. Recently, the ribosome itself was shown to be involved in Mg^{2+} homeostasis [5,54]. A high-resolution crystal structure of the 70S showed that it contains more than 170 Mg^{2+} ions bound tightly to the ribosome [40]. Additionally, a 2.0 Å cryoEM structure of the *E. coli* ribosome identified 309 Mg^{2+} in 70S with 93 Mg^{2+} in 30S [4]. Counting the Mg^{2+} associated loosely or through out-sphere interactions showed that more than 585 Mg^{2+} are predicted to bind the elongation complex of the ribosome [55]. Studies of *B. subtilis* showed that the total cellular Mg^{2+} concentration decreased in proportion to the amount of 70S ribosome when *B. subtilis* lacked an individual copy of the rRNA operons [12,15]. Combined with our observation of the sequential structure distortions of 30S rRNA in this study, a corollary is that the ribosome is a reservoir of Mg^{2+}, and when the cellular free Mg^{2+} decreases, the ribosome first releases the Mg^{2+} on the surface, which only involves a small portion of rRNA helices, such as h44 and h17 in the 30S, and this results in a reversible ribosome. Meanwhile, the mechanisms through which ribosomes actively participate in Mg^{2+} homeostasis should be elucidated in detail in future studies.

It has been reported that the most ancient parts of the ribosome are the PTC and the ribonucleotides, which depend heavily on metal ions for their structural stability [56,57]. Moreover, the twofold pseudo-symmetry in and around the PTC, which is composed of RNA and Mg^{2+}, has been suggested to be the structural origin of the ribosome [57–60]. Mg^{2+} is also essential for the small subunit functional regions. According to our structures, the *E. coli* 30S and 50S subunits separate and subsequently unfold when the Mg^{2+} concentration is below 1 mM. Ribosomes are usually purified at 10 mM Mg^{2+} to ensure that they are close to their natural state, but subunit separation occurs at 1 mM Mg^{2+} [61,62]. One should pay attention to the fact that protein S12 starts to dissociate from 30S under this condition, and long-time incubation at 1 mM Mg^{2+} should be avoided. These results can provide guidance for studies of ribosome structural and functional stability as well as the process of ribosome assembly.

Supplementary Materials: The following supporting information can be downloaded at: https://www.mdpi.com/article/10.3390/biom13030566/s1, Figure S1: The structure analysis procedure of 30S at 1 mM Mg^{2+}; Figure S2: Quality of the map for 30S at 1 mM Mg^{2+}; Figure S3: Overall maps of 30S under 2.5 and 10 mM Mg^{2+} conditions; Figure S4: A low concentration of Mg^{2+} induces a large rotation of the head in all directions; Figure S5: 30S under 1 mM Mg^{2+} showed different movements of h27; Figure S6: 30S at 1 mM Mg^{2+} showed a diverse S12 occupancy range; Figure S7: Movement of h17 does not affect the location of h16; Figure S8. A low magnesium concentration disassociates the ribosomal proteins.

Author Contributions: Conceptualization, T.Y. and F.Z.; funding acquisition, F.Z.; investigation, T.Y., J.J., Q.Y. and X.L.; project administration, T.Y. and F.Z.; resources, T.Y.; supervision, F.Z.; validation, T.Y. and F.Z.; visualization, T.Y. and F.Z.; manuscript—writing, original draft, T.Y. and J.J.; manuscript—writing, review and editing, T.Y. and F.Z. All authors have read and agreed to the published version of the manuscript.

Funding: This research was funded by the Guangdong Basic and Applied Basic Research Foundation (grant no. 2021A1515010805), the National Natural Science Foundation of China (grant no. 32171200), and the Shenzhen Science and Technology Program (grant no. JCYJ20220530115210023).

Data Availability Statement: Electron microscopy maps were deposited in the Electron Microscopy Data Bank under accession codes EMD-34987, EMD-34985, and EMD-34986 for 30S at 1, 2.5, and 10 mM Mg^{2+}, respectively, with EMD-34988 for $h17_{flexible}$, EMD-34989 for $h17_{out}$, EMD-34990 for $h17_{in-2}$, and EMD-34991 for $h17_{in}$.

Acknowledgments: We thank the Cryo-EM Center at the Southern University of Science and Technology for the cryoEM access, training, and cryoEM data collection, and the core research facility at the Southern University of Science and Technology for the ultracentrifugation.

Conflicts of Interest: The authors declare no conflict of interest.

References

1. Matsarskaia, O.; Roosen-Runge, F.; Schreiber, F. Multivalent ions and biomolecules: Attempting a comprehensive perspective. *Chemphyschem* **2020**, *21*, 1742–1767. [CrossRef] [PubMed]
2. Gesteland, R.F. Unfolding of Escherichia coli ribosomes by removal of magnesium. *J. Mol. Biol.* **1966**, *18*, 356–371. [CrossRef] [PubMed]
3. Klein, D.J.; Moore, P.B.; Steitz, T.A. The contribution of metal ions to the structural stability of the large ribosomal subunit. *Rna* **2004**, *10*, 1366–1379. [CrossRef] [PubMed]
4. Watson, Z.L.; Ward, F.R.; Méheust, R.; Ad, O.; Schepartz, A.; Banfield, J.F.; Cate, J.H. Structure of the bacterial ribosome at 2 Å resolution. *eLife* **2020**, *9*, e60482. [CrossRef]
5. Akanuma, G. Diverse relationships between metal ions and the ribosome. *Biosci. Biotechnol. Biochem.* **2021**, *85*, 1582–1593. [CrossRef] [PubMed]
6. Akanuma, G.; Yamazaki, K.; Yagishi, Y.; Iizuka, Y.; Ishizuka, M.; Kawamura, F.; Kato-Yamada, Y. Magnesium Suppresses Defects in the Formation of 70S Ribosomes as Well as in Sporulation Caused by Lack of Several Individual Ribosomal Proteins. *J. Bacteriol.* **2018**, *200*, e00212-18. [CrossRef]
7. Gavrilova, L.P.; Ivanov, D.A.; Spirin, A.S. Studies on the structure of ribosomes. 3. Stepwise unfolding of the 50 s particles without loss of ribosomal protein. *J. Mol. Biol.* **1966**, *16*, 473–489. [CrossRef]

8. Hori, N.; Denesyuk, N.A.; Thirumalai, D. Shape changes and cooperativity in the folding of the central domain of the 16S ribosomal RNA. *Proc. Natl. Acad. Sci. USA* **2021**, *118*, e2020837118. [CrossRef]
9. St John, A.C.; Goldberg, A.L. Effects of starvation for potassium and other inorganic ions on protein degradation and ribonucleic acid synthesis in Escherichia coli. *J. Bacteriol.* **1980**, *143*, 1223–1233. [CrossRef]
10. Selmer, M.; Dunham, C.M.; Murphy, F.V.t.; Weixlbaumer, A.; Petry, S.; Kelley, A.C.; Weir, J.R.; Ramakrishnan, V. Structure of the 70S ribosome complexed with mRNA and tRNA. *Science* **2006**, *313*, 1935–1942. [CrossRef]
11. Konevega, A.L.; Soboleva, N.G.; Makhno, V.I.; Semenkov, Y.P.; Wintermeyer, W.; Rodnina, M.V.; Katunin, V.I. Purine bases at position 37 of tRNA stabilize codon-anticodon interaction in the ribosomal A site by stacking and Mg^{2+}-dependent interactions. *RNA* **2004**, *10*, 90–101. [CrossRef] [PubMed]
12. Akanuma, G.; Kobayashi, A.; Suzuki, S.; Kawamura, F.; Shiwa, Y.; Watanabe, S.; Yoshikawa, H.; Hanai, R.; Ishizuka, M. Defect in the formation of 70S ribosomes caused by lack of ribosomal protein L34 can be suppressed by magnesium. *J. Bacteriol.* **2014**, *196*, 3820–3830. [CrossRef] [PubMed]
13. Failmezger, J.; Nitschel, R.; Sánchez-Kopper, A.; Kraml, M.; Siemann-Herzberg, M. Site-Specific Cleavage of Ribosomal RNA in *Escherichia coli*-Based Cell-Free Protein Synthesis Systems. *PLoS ONE* **2016**, *11*, e0168764. [CrossRef]
14. Görisch, H.; Goss, D.J.; Parkhurst, L.J. Kinetics of ribosome dissociation and subunit association studied in a light-scattering stopped-flow apparatus. *Biochemistry* **1976**, *15*, 5743–5753. [CrossRef]
15. Nierhaus, K.H. Mg^{2+}, K^+, and the ribosome. *J. Bacteriol.* **2014**, *196*, 3817–3819. [CrossRef] [PubMed]
16. Alatossava, T.; Jütte, H.; Kuhn, A.; Kellenberger, E. Manipulation of intracellular magnesium content in polymyxin B nonapeptide-sensitized Escherichia coli by ionophore A23187. *J. Bacteriol.* **1985**, *162*, 413–419. [CrossRef]
17. Johansson, M.; Zhang, J.; Ehrenberg, M. Genetic code translation displays a linear trade-off between efficiency and accuracy of tRNA selection. *Proc. Natl. Acad. Sci. USA* **2012**, *109*, 131–136. [CrossRef]
18. Zhang, J.; Ieong, K.W.; Johansson, M.; Ehrenberg, M. Accuracy of initial codon selection by aminoacyl-tRNAs on the mRNA-programmed bacterial ribosome. *Proc. Natl. Acad. Sci. USA* **2015**, *112*, 9602–9607. [CrossRef]
19. Pape, T.; Wintermeyer, W.; Rodnina, M. Induced fit in initial selection and proofreading of aminoacyl-tRNA on the ribosome. *EMBO J.* **1999**, *18*, 3800–3807. [CrossRef]
20. Schluenzen, F.; Tocilj, A.; Zarivach, R.; Harms, J.; Gluehmann, M.; Janell, D.; Bashan, A.; Bartels, H.; Agmon, I.; Franceschi, F.; et al. Structure of functionally activated small ribosomal subunit at 3.3 angstroms resolution. *Cell* **2000**, *102*, 615–623. [CrossRef]
21. Talkington, M.W.; Siuzdak, G.; Williamson, J.R. An assembly landscape for the 30S ribosomal subunit. *Nature* **2005**, *438*, 628–632. [CrossRef] [PubMed]
22. Adilakshmi, T.; Bellur, D.L.; Woodson, S.A. Concurrent nucleation of 16S folding and induced fit in 30S ribosome assembly. *Nature* **2008**, *455*, 1268–1272. [CrossRef] [PubMed]
23. Dutca, L.M.; Culver, G.M. Assembly of the 5' and 3' minor domains of 16S ribosomal RNA as monitored by tethered probing from ribosomal protein S20. *J. Mol. Biol.* **2008**, *376*, 92–108. [CrossRef] [PubMed]
24. Mulder, A.M.; Yoshioka, C.; Beck, A.H.; Bunner, A.E.; Milligan, R.A.; Potter, C.S.; Carragher, B.; Williamson, J.R. Visualizing ribosome biogenesis: Parallel assembly pathways for the 30S subunit. *Science* **2010**, *330*, 673–677. [CrossRef]
25. Zheng, S.Q.; Palovcak, E.; Armache, J.P.; Verba, K.A.; Cheng, Y.; Agard, D.A. MotionCor2: Anisotropic correction of beam-induced motion for improved cryo-electron microscopy. *Nat. Methods* **2017**, *14*, 331–332. [CrossRef]
26. Rohou, A.; Grigorieff, N. CTFFIND4: Fast and accurate defocus estimation from electron micrographs. *J. Struct. Biol.* **2015**, *192*, 216–221. [CrossRef]
27. Zivanov, J.; Nakane, T.; Forsberg, B.O.; Kimanius, D.; Hagen, W.J.; Lindahl, E.; Scheres, S.H. New tools for automated high-resolution cryo-EM structure determination in RELION-3. *eLife* **2018**, *7*, e42166. [CrossRef]
28. Emsley, P.; Lohkamp, B.; Scott, W.G.; Cowtan, K. Features and development of Coot. *Acta Cryst. D Biol. Cryst.* **2010**, *66*, 486–501. [CrossRef]
29. Pettersen, E.F.; Goddard, T.D.; Huang, C.C.; Couch, G.S.; Greenblatt, D.M.; Meng, E.C.; Ferrin, T.E. UCSF Chimera—A visualization system for exploratory research and analysis. *J. Comput. Chem.* **2004**, *25*, 1605–1612. [CrossRef]
30. Kinman, L.F.; Powell, B.M.; Zhong, E.D.; Berger, B.; Davis, J.H. Uncovering structural ensembles from single-particle cryo-EM data using cryoDRGN. *Nat. Protoc.* **2023**, *18*, 319–339. [CrossRef]
31. King, T.C.; Rucinsky, T.; Schlessinger, D.; Milanovich, F. Escherichia coli ribosome unfolding in low Mg^{2+} solutions observed by laser Raman spectroscopy and electron microscopy. *Nucleic Acids Res.* **1981**, *9*, 647–661. [CrossRef] [PubMed]
32. Spahn, C.M.; Gomez-Lorenzo, M.G.; Grassucci, R.A.; Jørgensen, R.; Andersen, G.R.; Beckmann, R.; Penczek, P.A.; Ballesta, J.P.; Frank, J. Domain movements of elongation factor eEF2 and the eukaryotic 80S ribosome facilitate tRNA translocation. *EMBO J.* **2004**, *23*, 1008–1019. [CrossRef] [PubMed]
33. Mohan, S.; Donohue, J.P.; Noller, H.F. Molecular mechanics of 30S subunit head rotation. *Proc. Natl. Acad. Sci. USA* **2014**, *111*, 13325–13330. [CrossRef]
34. Guo, Z.; Noller, H.F. Rotation of the head of the 30S ribosomal subunit during mRNA translocation. *Proc. Natl. Acad. Sci. USA* **2012**, *109*, 20391–20394. [CrossRef]
35. Maksimova, E.; Kravchenko, O.; Korepanov, A.; Stolboushkina, E. Protein Assistants of Small Ribosomal Subunit Biogenesis in Bacteria. *Microorganisms* **2022**, *10*, 747. [CrossRef] [PubMed]

36. Zhang, L.; Sato, N.S.; Watanabe, K.; Suzuki, T. Functional genetic selection of the decoding center in E. coli 16S rRNA. *Nucleic Acids Res Suppl.* **2003**, *3*, 319–320. [CrossRef]
37. Smith, T.F.; Lee, J.C.; Gutell, R.R.; Hartman, H. The origin and evolution of the ribosome. *Biol. Direct* **2008**, *3*, 16. [CrossRef] [PubMed]
38. Ben-Shem, A.; Jenner, L.; Yusupova, G.; Yusupov, M. Crystal structure of the eukaryotic ribosome. *Science* **2010**, *330*, 1203–1209. [CrossRef] [PubMed]
39. Agarwal, D.; Gregory, S.T.; O'Connor, M. Error-prone and error-restrictive mutations affecting ribosomal protein S12. *J. Mol. Biol.* **2011**, *410*, 1–9. [CrossRef]
40. Schuwirth, B.S.; Borovinskaya, M.A.; Hau, C.W.; Zhang, W.; Vila-Sanjurjo, A.; Holton, J.M.; Cate, J.H. Structures of the bacterial ribosome at 3.5 A resolution. *Science* **2005**, *310*, 827–834. [CrossRef]
41. Drygin, D.; Zimmermann, R.A. Magnesium ions mediate contacts between phosphoryl oxygens at positions 2122 and 2176 of the 23S rRNA and ribosomal protein L1. *Rna* **2000**, *6*, 1714–1726. [CrossRef] [PubMed]
42. Maguire, M.E.; Cowan, J.A. Magnesium chemistry and biochemistry. *Biometals* **2002**, *15*, 203–210. [CrossRef] [PubMed]
43. Shayan, R.; Rinaldi, D.; Larburu, N.; Plassart, L.; Balor, S.; Bouyssié, D.; Lebaron, S.; Marcoux, J.; Gleizes, P.E.; Plisson-Chastang, C. Good Vibrations: Structural Remodeling of Maturing Yeast Pre-40S Ribosomal Particles Followed by Cryo-Electron Microscopy. *Molecules* **2020**, *25*, 1125. [CrossRef] [PubMed]
44. Barandun, J.; Hunziker, M.; Klinge, S. Assembly and structure of the SSU processome-a nucleolar precursor of the small ribosomal subunit. *Curr. Opin. Struct. Biol.* **2018**, *49*, 85–93. [CrossRef]
45. Schedlbauer, A.; Iturrioz, I.; Ochoa-Lizarralde, B.; Diercks, T.; Lopez-Alonso, J.P.; Lavin, J.L.; Kaminishi, T.; Capuni, R.; Dhimole, N.; de Astigarraga, E.; et al. A conserved rRNA switch is central to decoding site maturation on the small ribosomal subunit. *Sci. Adv.* **2021**, *7*, eabf7547. [CrossRef]
46. Gabashvili, I.S.; Agrawal, R.K.; Spahn, C.M.; Grassucci, R.A.; Svergun, D.I.; Frank, J.; Penczek, P. Solution structure of the E. coli 70S ribosome at 11.5 A resolution. *Cell* **2000**, *100*, 537–549. [CrossRef]
47. Yusupov, M.M.; Yusupova, G.Z.; Baucom, A.; Lieberman, K.; Earnest, T.N.; Cate, J.H.; Noller, H.F. Crystal structure of the ribosome at 5.5 A resolution. *Science* **2001**, *292*, 883–896. [CrossRef]
48. Fourmy, D.; Yoshizawa, S.; Puglisi, J.D. Paromomycin binding induces a local conformational change in the A-site of 16S rRNA. *J. Mol. Biol.* **1998**, *277*, 333–345. [CrossRef]
49. Culver, G.M.; Kirthi, N. Assembly of the 30S Ribosomal Subunit. *EcoSal Plus* **2008**, *3*. [CrossRef]
50. Mizushima, S.; Nomura, M. Assembly mapping of 30S ribosomal proteins from E. coli. *Nature* **1970**, *226*, 1214. [CrossRef]
51. Demirci, H.; Wang, L.; Murphy, F.V.t.; Murphy, E.L.; Carr, J.F.; Blanchard, S.C.; Jogl, G.; Dahlberg, A.E.; Gregory, S.T. The central role of protein S12 in organizing the structure of the decoding site of the ribosome. *RNA* **2013**, *19*, 1791–1801. [CrossRef] [PubMed]
52. Ogle, J.M.; Murphy, F.V.; Tarry, M.J.; Ramakrishnan, V. Selection of tRNA by the ribosome requires a transition from an open to a closed form. *Cell* **2002**, *111*, 721–732. [CrossRef] [PubMed]
53. Moncany, M.L.; Kellenberger, E. High magnesium content of Escherichia coli B. *Experientia* **1981**, *37*, 846–847. [CrossRef]
54. Gall, A.R.; Datsenko, K.A.; Figueroa-Bossi, N.; Bossi, L.; Masuda, I.; Hou, Y.M.; Csonka, L.N. Mg^{2+} regulates transcription of mgtA in Salmonella Typhimurium via translation of proline codons during synthesis of the MgtL peptide. *Proc. Natl. Acad. Sci. USA* **2016**, *113*, 15096–15101. [CrossRef]
55. Rozov, A.; Khusainov, I.; El Omari, K.; Duman, R.; Mykhaylyk, V.; Yusupov, M.; Westhof, E.; Wagner, A.; Yusupova, G. Importance of potassium ions for ribosome structure and function revealed by long-wavelength X-ray diffraction. *Nat. Commun.* **2019**, *10*, 2519. [CrossRef]
56. Hury, J.; Nagaswamy, U.; Larios-Sanz, M.; Fox, G.E. Ribosome origins: The relative age of 23S rRNA Domains. *Orig. Life Evol. Biosph.* **2006**, *36*, 421–429. [CrossRef] [PubMed]
57. Hsiao, C.; Mohan, S.; Kalahar, B.K.; Williams, L.D. Peeling the onion: Ribosomes are ancient molecular fossils. *Mol. Biol. Evol.* **2009**, *26*, 2415–2425. [CrossRef] [PubMed]
58. Agmon, I. The dimeric proto-ribosome: Structural details and possible implications on the origin of life. *Int. J. Mol. Sci.* **2009**, *10*, 2921–2934. [CrossRef]
59. Davidovich, C.; Belousoff, M.; Wekselman, I.; Shapira, T.; Krupkin, M.; Zimmerman, E.; Bashan, A.; Yonath, A. The Proto-Ribosome: An ancient nano-machine for peptide bond formation. *Isr. J. Chem.* **2010**, *50*, 29–35. [CrossRef]
60. Rivas, M.; Fox, G.E. Further Characterization of the Pseudo-Symmetrical Ribosomal Region. *Life* **2020**, *10*, 201. [CrossRef]
61. Wang, W.; Li, W.; Ge, X.; Yan, K.; Mandava, C.S.; Sanyal, S.; Gao, N. Loss of a single methylation in 23S rRNA delays 50S assembly at multiple late stages and impairs translation initiation and elongation. *Proc. Natl. Acad. Sci. USA* **2020**, *117*, 15609–15619. [CrossRef] [PubMed]
62. Nikolay, R.; Hilal, T.; Schmidt, S.; Qin, B.; Schwefel, D.; Vieira-Vieira, C.H.; Mielke, T.; Bürger, J.; Loerke, J.; Amikura, K.; et al. Snapshots of native pre-50S ribosomes reveal a biogenesis factor network and evolutionary specialization. *Mol. Cell* **2021**, *81*, 1200–1215.e1209. [CrossRef] [PubMed]

Disclaimer/Publisher's Note: The statements, opinions and data contained in all publications are solely those of the individual author(s) and contributor(s) and not of MDPI and/or the editor(s). MDPI and/or the editor(s) disclaim responsibility for any injury to people or property resulting from any ideas, methods, instructions or products referred to in the content.

Article

Differential Participation of Plant Ribosomal Proteins from the Small Ribosomal Subunit in Protein Translation under Stress

Zainab Fakih, Mélodie B. Plourde and Hugo Germain *

Department of Chemistry, Biochemistry and Physics and Groupe de Recherche en Biologie Végétale, Université du Québec à Trois-Rivières, Trois-Rivières, QC G9A 5H9, Canada; zainab.fakih@uqtr.ca (Z.F.); melodie.bplourde@uqtr.ca (M.B.P.)
* Correspondence: hugo.germain@uqtr.ca

Abstract: Upon exposure to biotic and abiotic stress, plants have developed strategies to adapt to the challenges imposed by these unfavorable conditions. The energetically demanding translation process is one of the main elements regulated to reduce energy consumption and to selectively synthesize proteins involved in the establishment of an adequate response. Emerging data have shown that ribosomes remodel to adapt to stresses. In *Arabidopsis thaliana*, ribosomes consist of approximately eighty-one distinct ribosomal proteins (RPs), each of which is encoded by two to seven genes. Recent research has revealed that a mutation in a given single RP in plants can not only affect the functions of the RP itself but can also influence the properties of the ribosome, which could bring about changes in the translation to varying degrees. However, a pending question is whether some RPs enable ribosomes to preferentially translate specific mRNAs. To reveal the role of ribosomal proteins from the small subunit (RPS) in a specific translation, we developed a novel approach to visualize the effect of RPS silencing on the translation of a reporter mRNA (GFP) combined to the 5′UTR of different housekeeping and defense genes. The silencing of genes encoding for *NbRPSaA*, *NbRPS5A*, and *NbRPS24A* in *Nicotiana benthamiana* decreased the translation of defense genes. The *NbRACK1A*-silenced plant showed compromised translations of specific antioxidant enzymes. However, the translations of all tested genes were affected in *NbRPS27D*-silenced plants. These findings suggest that some RPS may be potentially involved in the control of protein translation.

Keywords: *Nicotiana benthamiana*; *Arabidopsis thaliana*; translation regulation; ribosomal proteins from the small subunit (RPS); VIGS; 5′untranslated regions; transient expression; plant defense

Citation: Fakih, Z.; Plourde, M.B.; Germain, H. Differential Participation of Plant Ribosomal Proteins from the Small Ribosomal Subunit in Protein Translation under Stress. *Biomolecules* **2023**, *13*, 1160. https://doi.org/10.3390/biom13071160

Academic Editors: Brigitte Pertschy and Ingrid Zierler

Received: 6 June 2023
Revised: 12 July 2023
Accepted: 13 July 2023
Published: 21 July 2023

Copyright: © 2023 by the authors. Licensee MDPI, Basel, Switzerland. This article is an open access article distributed under the terms and conditions of the Creative Commons Attribution (CC BY) license (https:// creativecommons.org/licenses/by/ 4.0/).

1. Introduction

As sessile beings, plants have developed various strategies to overcome the range of challenging conditions they are exposed to. These responses are built on finely tuned gene expressions, which, in turn, lead to protein level variations. Changes in protein level depend on the regulation of multiple factors, such as transcription, mRNA structure, stability, transport, storage, protein synthesis, and degradation [1,2]. Among them, the translation process is one of the main elements that finely modulates protein accumulation under both biotic and abiotic stress situations; its regulation reduces energy consumption and allows for the selective synthesis of proteins involved in the proper establishment of an appropriate response [3,4]. Many examples of global translational inhibition and the preferential production of key proteins that are critical for adapting to environmental conditions are known [5–8]. A general decrease in global translation levels is observed in plants under conditions of sucrose starvation [9,10] and those acting in response to cold stress [11]. Furthermore, the overall translation activity in plants is higher in the light than in the dark; this is correlated with the higher energetic status of the plant cells under light conditions [12].

Protein synthesis is mediated by ribosomes and ribosomal-associated proteins. Ribosome assembly occurs within the nucleolus and requires the coordinated production and transport of four rRNAs (5S, 5.8S, 18S, and 28S) and eighty-one ribosomal proteins (RPs) [13]. The eukaryotic ribosome, termed the 80S ribosome, consists of two ribonucleoprotein subunits; the 40S small subunit binds the mRNA and provides the decoding site, which is formed by the 18S rRNA and thirty-three small ribosomal proteins (RPS). The 60S large subunit, which is composed of the 5S, 5.8S, and 23S rRNAs and 48 large ribosomal proteins (RPL), catalyzes the formation of peptide bonds [14–16]. All of these RPs are present in a single copy in each ribosome, except for the RPs forming a flexible lateral stalk on the large subunit [17,18]. In the model plant, *Arabidopsis thaliana*, each RP can be encoded by two to seven different members of the small families [15]. Thus, the 81 RP families may produce up to 10^{34} different potential ribosome structural conformations that could theoretically serve as a source of translation heterogeneity [3]. Although each RP gene has multiple paralogs, their expressions appear to be differentially regulated by various environmental cues and treatments with signaling molecules [19–25]. This differential expression between gene families, as well as within specific ribosomal gene families, opens vast possibilities for the functional role of these RPs in stress conditions. Furthermore, ribosome composition has, to date, been examined in several mass spectrometric studies, which have identified different r-protein paralogs within ribosomes that act in response to different stimuli, showing that ribosome composition may also be dynamic [3,26–28]. This heterogeneity can constitute specialized ribosomes that may regulate mRNA translation and control protein synthesis. Thus, specialized ribosomes are defined as a functional subpopulation of ribosomes that appear, for example, after an altered condition; they work to constrain translation to specific mRNAs and to shape the acclimated proteome [29].

The differential expression between the RP genes and the ribosomal composition implies a diversified functional relevance regarding RPs [30]. This is consistent with accumulating evidence that emphasizes the RP involvement in several ribosome functions, as well as roles away from the ribosome, such as DNA repair, histone binding, transcription-factors activity, and cell-cycle regulation [31–33]. For instance, mutations in some RPs influence the integrity of ribosomes, in structure and in function. The mutational analysis of several prokaryotic RPs has highlighted their importance in a variety of ribosomal processes. *RPS12* was shown to be required for tRNA decoding in the ribosomal A site [34] and the *RPS4* and *RPS5* mutations showed ribosome translational inaccuracy [35]; whereas, *RPSa*, *RPS7*, and *RPS11* are essential for mRNA binding [36]. Furthermore, in mammalian cells, the binding of the RACK1 (Receptor for Activated C-Kinase 1) to ribosomes is essential for the full translation of capped mRNAs and the efficient recruitment of eukaryotic initiation factor 4E (eIF4E) [37]. In *Nicotiana benthamiana* and *Arabidopsis thaliana*, QM/RPL10A plays a transcriptional role in regulating translational mechanisms and defense-associated genes [38]; also, *RPS27B* is involved in the degradation of damaged RNAs (induced by genotoxic treatments) [39].

The involvement and specific constitution of the protein-translation machinery in plant defense is poorly studied. Some reports have shown that the deficiency and mutation of ribosome proteins themselves are associated with disease responses in plants. The silencing of *RPL12* and *RPL19* in *N. benthamiana* and *A. thaliana* showed compromised nonhost disease resistance against multiple bacterial pathogens [40]; the silencing of *RPL10* in *N. benthamiana* and *A. thaliana* showed compromised disease resistance against the nonhost pathogen *Pseudomonas syringae* pv. *tomato* T1 [38]; and the silencing of *RPS6* in *N. benthamiana* affected the accumulation of the Cucumber mosaic virus, Turnip mosaic virus (TuMV), and Potato virus A (PVA), but not the Turnip crinkle virus and Tobacco mosaic virus [41].

Despite these studies, a systematic understanding of the functional role of RPs in the context of plant defense is still lacking. In the present study, using previously published nuclear proteomes of plants under stress, we identified several RPS that accumulated in the nuclei (the site of ribosome biogenesis) of stressed plants. We hypothesize that the accumulated RPS paralogs generate ribosomes that shape the cellular translatome and plant defense responses. To address the role of the identified RPS in a specific translation, we first developed a translation assay in which we tested the production of the green fluorescent protein (GFP) fused to different 5′UTR corresponding to known defense genes, or housekeeping genes, in the leaves of RPS-silenced and control plants. We found that three tested proteins (*RPSaA*, *RPS5A*, and *RPS24A*) are involved in the efficient translation of some defense proteins. In contrast, the protein *RPS27D* is involved in the general translational activity of the ribosome; whereas, *RACK1A* is involved in the efficient translation of several antioxidant enzymes. Our technical approach defines a suitable methodological strategy for testing ribosomal protein requirements for the translation of specific groups of mRNAs. Moreover, this suggests that RPS paralogs play a crucial role in translational control.

2. Materials and Methods

2.1. Plant Growth and Stress Treatments

Seeds of *N. benthamiana* were vernalized for 48 h at 4 °C and plants were grown in soil (AgroMix) at 23 °C and 60% relative humidity with a 14 h/10 h light/dark cycle in a growth chamber.

For 2,6-dichloroisonicotinic acid (INA) treatment, 3-week-old plants were sprayed to imminent runoff with an aqueous solution of 0.65 mM INA containing 0.05% Sylgard 309 surfactant; whereas, the mock treatment consisted of only the Sylgard 309 aqueous solution. Leaf tissues were harvested 24 h after being sprayed with INA, as previously described [42]. INA was used to induce plant defense as it was shown to induce a response similar to those of salicylic acid and pathogen infection [43]. For cold stress treatments, 3-week-old plants were placed at 4 °C for 6 h [44].

For the biotic stress experiments, we used the bacterial pathogen *P. fluorescens* EtHAn (Effector-to-Host Analyzer) strain, which allowed for the development of the PTI response in *N. benthamiana* [45]. The bacterial suspension of *P. fluorescens* EtHAn at OD_{600} = 0.2 in 10 mM of $MgCl_2$ was infiltrated into the abaxial side of 3-week-old *N. benthamiana* leaves; tissue was collected 7 h post-inoculation. Leaf samples of infiltrated plants, with 10 mM of $MgCl_2$ grown under similar conditions, were used as a control to normalize the expression. All of the samples were collected in the form of three biological replicates after each time interval and were immediately frozen in liquid nitrogen and stored at −70 °C.

2.2. Differential Gene Expression Analysis

A gene expression analysis in a *N. benthamiana* plant was performed on RNA extracted from the frozen tissue using the Genezol Total RNA kit (Geneaid), following the manufacturer's instructions. The RNA quality was assessed by agarose gel electrophoresis and quantified by spectrophotometry. In total, 1 µg of each sample was used as the template for first-strand cDNA synthesis using the M-MuLV Reverse Transcriptase (New England Biolabs, Whitby, ON, Canada). Quantitative PCR amplification was performed on a CFX Connect detection system (Bio-Rad Laboratories, Mississauga, ON, Canada) using gene-specific primers and the SYBR Green PCR Master Mix (Bioline, Toronto, ON, Canada). The primers used were designed using the Primer 3 software; they were designed in such a way that they targeted a region that is completely absent of all other paralogous genes and is unique. This selection was performed using the VIGS Tool from the Sol Genomics Network (https://vigs.solgenomics.net/, accessed on 1 March 2018) (Figure S1). The specificity of the primers was then verified by using the Primer-Blast tool at NCBI. In total, a 100 ng cDNA template and 0.4 µM of each primer (listed in Supplementary Table S1) were used in a final volume of 20 µL. The amplification protocol included an initial denaturation at 95 °C for 2 min, with 40 cycles at 95 °C for 5 s, a primer-specific annealing temperature for 10 s, and an extension at 72 °C for 5 s. This was followed by constructing a melt curve at the end to estimate the amplification specificity of each gene. The data were analyzed with CFX Maestro qPCR software. *PP2A* and *UBQ1* (Polyubiquitin 1) were used as reference genes for normalization under INA conditions and those of a *P. fluorescens* EtHAn infection [46]. *ACT 2* and *UBQ1* were considered suitable genes to normalize with for the cold treatment [47]. The mean values of the relative fold change were calculated as per the $^{\Delta\Delta Ct}$

method [48]. RPS genes in each condition were defined as differentially expressed only if the expression value of the gene was more than 1.5-fold the control and had a *p*-value of less than 0.05 compared to the control.

The expression of the identified RPS genes in *Arabidopsis* was analyzed using the Genevestigator tool (https://genevestigator.com/, accessed on 6 February 2018) with the *Arabidopsis* Gene Chip platforms (ATH1: 22k array). The perturbation tool of the Genevestigator software was used to estimate the levels of gene expression as a heat map under different conditions. Data were presented as absolute $\log_2$ values of fold change compared with that of the control samples.

2.3. Virus-Induced Gene Silencing (VIGS)

The pBINTRA6 and pTV00 vectors were used for silencing in the *N. benthamiana*. The pTV00::*NbRPSaA*, pTV00::*NbRPS5A*, pTV00::*NbRPS27D*, pTV00::*NbRPS24A*, and pTV00::*NbRACK1A* constructs were developed and used for VIGS, as described [49]. In order to select VIGS silencing sections that were specific to a single paralog of the targeted protein, we used the VIGS Tool from the Sol Genomics Network (https://vigs.solgenomics.net/, accessed on 1 March 2018); we were able to design VIGS fragments unique to the 3′UTR of each targeted gene that was absent from the other paralogs (Figure S1). Table S2 provides a list of all of the paralogs of the investigated RPS. PCR was used to amplify the desired fragments with specific primers (Table S3) using genomic DNA prepared from the plant tissues. The amplified fragments of the RPS genes and the pTV00 vectors were digested by the restriction enzymes *Kpn*I and *Hin*dIII, according to the manufacturer's instructions; the purified products of the RPS sequence were inserted into the pTV00 vectors using T4 DNA ligase (NEB, England). The vectors were then transformed into competent cells of the *E. coli* strain DH5α. The selected positive clones with the correct sequence were used to transform the *Agrobacterium tumefaciens* strain of GV3101 electrocompetent cells. Plant infiltration was performed, as described previously [49]. The *Agrobacterium* strains of GV3101 containing pTV::*NbRPSaA*, pTV::*NbRPS5A*, pTV::*NbRPS27D*, pTV::*NbRPS24A*, or pTV::*NbRACK1A* and those of C58C1 containing pBINTRA6 were grown at 28 °C in a liquid Luria-Bertani medium including antibiotics (50 µg mL^{-1} kanamycin and 50 µg mL^{-1} rifampicin). After 24 h, the cells were harvested by centrifugation and resuspended in the infiltration buffer (10 mM of $MgCl_2$ with 200 µM of acetosyringone and 10 mM of MES, pH 5.6) to a final optical density, at 600 nm, of approximately 0.5 and were agitated for 2 h (28 °C) before mixing in a 1:1 ratio. The *Agrobacterium* mix, containing either pBINTRA6 or pTV-*NbRPS* vectors, was infiltrated using a needleless 1-mL syringe that was inserted into the lower leaves of 2-week-old *N. benthamiana* plants [50]. As a control, the empty cloning vector pTV was used to distinguish the nonspecific phenotypic effects of VIGS.

2.4. Quantitative RT-PCR

Leaf tissue was collected 3 weeks after TRV inoculation to test the downregulation of ribosomal protein-encoding gene transcripts in *N. benthamiana*-silenced plants. The total RNA was extracted from silenced and mock-infiltrated plants and the first-strand cDNA was synthesized with oligo(dT_{15}) primers using M-MuLV Reverse Transcriptase (New England Biolabs, Whitby, ON, Canada), according to the manufacturer's instructions. The RT-qPCR was performed using the CFX Connect detection system (Bio-Rad Laboratories, Mississauga, ON, Canada). *ACT 1* and *EF1α* were used to normalize the transcript levels [51]. Each sample was run in triplicate and repeated six times from two pooled biological replicates of silenced and non-silenced plants. The average of the six experiments was calculated and the results were graphed, with the corresponding standard deviations indicated with bars in the figures. The primers used in this study are listed in Table S4.

2.5. 5′UTR Chimeras and Plasmid Construction

The Cauliflower mosaic virus (CaMV) 35S promoter (p35S) and 5′UTR fusion constructs were assembled by PCR stitching. Briefly, two rounds of a PCR were carried out. In the first round, two separate PCRs were performed: one amplified the p35S from the *pB7FWG2* vector using specific primers listed in Supplementary Table S5; the other amplified the 5' upstream region of 5 defense genes, or 3 housekeeping genes, from the *N. benthamiana* genomic DNA using gene-specific primers (Table S5). The selection of these genes and the categorization of housekeeping and defense genes were made following a literature review. For instance, the *PP2A*, *F-BOX*, and *GAPDH* genes were consistently reported as housekeeping genes within the context of different viral infections in *Nicotiana benthamiana* [51–53] and under different conditions in other species [54–57]. Furthermore, the catalase, peroxidase, and ascorbate peroxidase proteins play a crucial role in overcoming various stress conditions and work as part of the antioxidant defense system [58]. In addition, the NPR1 (NONEXPRESSOR OF PR1) protein functions as a master regulator of plant hormone salicylic acid (SA)-signaling and plays an essential role in promoting defense responses [59]. Finally, the MAPK3 protein is implicated in stomatal development, biotic stress responses, and abiotic stress responses and is required for the complete "priming" of plants [60]. The mRNAs encoding these proteins showed a status indicating a higher translational efficiency in response to stress [61–63]. The 5'UTRs of these genes were identified using the Sol Genomics *N. benthamiana* draft genome (https://solgenomics.net/organism/Nicotiana_benthamiana/genome, accessed on 3 June 2023). In the second round, the products of these two PCRs, which overlapped at one end, were subsequently mixed and amplified.

Amplified fragments containing the p35S promoter and the 5'UTR were used to generate expression vectors, having different 5'UTRs linked to the reporter gene GFP. Amplicons were inserted into the pDONR221 vector (Invitrogen, part of Thermo Fisher Scientific, Waltham, MA, USA) via BP recombination reactions and then into the plant-expression vector PBGWFS7 via LR recombination reactions using Gateway technology [64].

2.6. Leaf-Infiltration Method

For transient GFP protein expression, constructs were introduced into the *A. tumefaciens* strain GV3101 by electroporation and were delivered into the leaf cells of silenced and non-silenced *N. benthamiana* (5-week-old) using the agroinfiltration method, as previously described [65]. Briefly, recombinant bacterial strains were grown overnight in a liquid Luria-Bertani medium with spectinomycin (50 mg/L); then, they were harvested and resuspended into an infiltration buffer (10 mM of $MgCl_2$ and 150 µM of acetosyringone) to obtain a 0.5 unit of optical density at 600 nm. One hour after resuspension, leaves were infiltrated on their abaxial side. To minimize leaf-to-leaf variation, each leaf was infiltrated with a vector containing the 5'UTR of two housekeeping genes (*F-box* and *PP2A*) as normalization controls, alongside vectors containing the 5'UTRs to be tested. Three independent infiltrations were made for each experiment and were compared using the Student's *t*-test. Ultimatley, $p < 0.05$ was represented with one star (*). The agro-infected leaves were collected at 5 days post-infiltration to be photographed and analyzed for GFP production by spectrofluorimetry.

2.7. Detection of GFP Fluorescence

Leaves producing GFP were photographed under UV illumination generated by a 100 W, hand-held, long-wave UV lamp (Model B-100, UVP, Upland, CA, USA). The GFP fluorescence intensity was quantified at an excitation of 485 nm and an emission of 538 nm using a Synergy H1 Microplate Reader, BioTek, as described by Diamos et al. [66]. GFP samples were prepared by a serial two-fold dilution with phosphate-buffered saline (PBS, 137 mM of NaCl, 2.6 mM of KCl, 10 mM of Na_2HPO_4, and 1.8 mM of KH_2PO_4, pH 7.4); 100 µL of each sample was added to black-wall 96-well plates (Thermo Fisher Scientific), in triplicate. All measurements were performed at room temperature and the reading of an extract from an uninfiltrated plant leaf was subtracted before graphing. A standard curve of fluorescence for the GFP concentration was generated by measuring the fluorescence of a dilution series of GFP (triplicate) in a 96-well plate in the plate reader.

2.8. Protein Extraction

Total protein extract was obtained by homogenizing agroinfiltrated leaf samples with a 1:5 ($w:v$) ice-cold extraction buffer (25 mM of sodium phosphate, pH 7.4, 100 mM of NaCl, 1 mM of EDTA, 0.2% Triton X-100, 10 mg/mL of sodium ascorbate, 10 mg/mL of leupeptin, and 0.3 mg/mL of phenylmethylsulfonyl fluoride) using a mortar and pestle. To enhance solubility, homogenized tissue was rotated at room temperature for 30 min. The crude plant extracts were clarified by centrifugation at $10,000 \times g$ for 10 min at 4 °C.

3. Results

3.1. Small Ribosomal Proteins Are Deregulated by Different Stresses in A. thaliana and N. benthamiana

Ribosome biogenesis represents a compendium of steps by which the ribosomes may become assembled, involving the import of most RPs into the nucleus and nucleolus and their association with rRNA to constitute the ribosomal subunits [13]. Hence, many studies have identified ribosomal proteins in the nuclei of various plants under stress [44,67–70]. In line with this idea, we analyzed the published datasets on the biotic and abiotic stress-responsive nuclear proteomes in various plant species to identify plant ribosomal proteins of the small subunit involved in disease resistance and selected RPS detected in the nuclei of stressed plants [44,67,68]. A previous proteomics analysis identified a subset of 11 RPS detected in the nuclei of elicited immunity in *Arabidopsis* plants following a chitosan elicitor treatment [44,67,68] (Figure S2a). In response to cold stress, eight RPS were overrepresented in the nuclear proteome of *Arabidopsis* [44] (Figure S2a). Furthermore, seven RPS had a significant change in abundance in the nucleus of a tomato (*Solanum lycopersicum*) during an infection caused by the oomycete pathogen *Phytophthora capsici* [68] (Figure S2a). From these three studies, a total of 15 different RPS displayed an increased nuclear abundance under various stress conditions.

The gene-expression patterns of these 15 RPS were evaluated under stress conditions in *Arabidopsis* using the Genevestigator application's compendium of microarray experiments. We evaluated the expression of these genes following cold stress (Figure S2b), elicitor treatment (Figure S2c), and biotic stress (Figure S2d). Genes that showed a strong induction in at least two conditions were selected as upregulated. Interestingly, six genes (*RPSaA*, *RPS10C*, *RPS12C*, *RPS19C*, *RPS27D*, and *RACK1A*) showed high levels of expression in response to all three stresses (Figure S2e).

To gain insights into the expression patterns of RPS genes in *N. benthamiana* plants in stress contexts and, also, to provide a comparative analysis of the RPS expression between the two model plants, we performed a quantitative reverse transcription qRT-PCR of the 15 RPS genes using *N. benthamiana* tissues with cold stress conditions (Figure 1a), an INA treatment, an analog of SA that induces plant defense [43] (Figure 1b), and infection with the bacteria *Pseudomonas fluorescens* EtHAn (Figure 1c). Five genes (*RPSaA*, *RPS5A*, *RPS24A*, *RPS27D*, and *RACK1A*) were highly regulated under the three stress treatments (Figure 1d). It is worth mentioning that the expression patterns of *RPSaA*, *RACK1A*, and *RPS27D* in *N. benthamiana* are consistent with the *Arabidopsis* data from the Genevestigator microarray database. We herein focus on these five RPS genes for further functional analyses in *N. benthamiana*.

3.2. RPSaA, RPS5A, and RPS24A Proteins Are Involved in the Translation of Defense Proteins Encoding mRNAs

To test whether the silencing of a specific *NbRPS* gene compromises defense genes' translations, direct measurements of chimeric reporter mRNA translational efficiencies were compared between RPS-silenced and mock-infiltrated plants. Each chimeric mRNA contained the 5'upstream region of either a defense gene or a housekeeping gene fused to the coding sequence of the green fluorescent protein (Table 1). All silenced plants showed more than a 50% down-regulation of the target transcripts (Figure 2a). Then, the different chimeric constructs were delivered into the *N. benthamiana* leaves of silenced and control plants by an *A. tumefaciens*-mediated transformation and the green fluorescence was monitored. To minimize leaf-to-leaf variation, each leaf was infiltrated with a vector containing the 5′UTR of two housekeeping genes, *F-box* and *PP2A*, as controls alongside vectors containing the 5′UTRs to be tested. No fluorescence was detected in the plant leaves infiltrated with empty vectors without any 5′UTRs (PBGWFS7 vector); whereas, a significant GFP fluorescence was observed with all of the 5′UTR-GFP chimeras in the control plants (Figure 2b). Using this system, we found that the GAPDH 5′UTR construct produced intense green fluorescence in both silenced and control plants; whereas, the constructs containing the 5′UTRs of catalase, peroxidase, ascorbate peroxidase, NPR1, and MAPK3 showed poor GFP signals in pTV::*NbRPSaA*, pTV::*NbRPS5A*, and pTV::*NbRPS24A* compared to the mock plant (Figure 2b). GFP fluorescence was quantified by spectrofluorimetry and was decreased by more than 50%; sometimes it was almost absent, particularly for the defense chimeric constructs in RPS-silenced plants (Figure 2c–e). These results indicated that *NbRPSaA*, *NbRPS5A*, and *NbRPS24A* are essential for the optimal translation of many defense genes in planta.

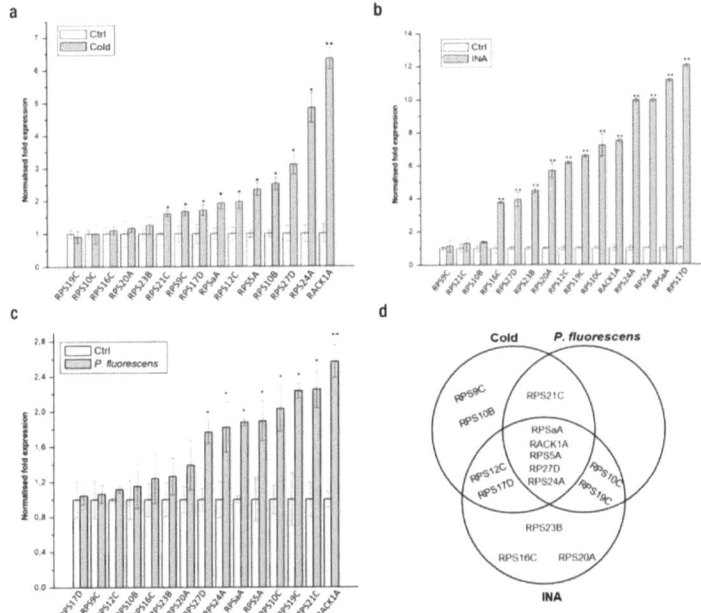

Figure 1. The mRNA levels of the ribosomal proteins of the small subunit are deregulated by different stresses in *N. benthamiana*. Relative expression of selected RPS genes in *N. benthamiana* following (**a**) cold stress, (**b**) INA treatment, and (**c**) *Pseudomonas fluorescens EtHAn* infection. * p-values < 0.05 and ** p-values < 0.01, Student's t-test. (**d**) Venn diagram of the deregulated RPS genes under the three conditions.

Table 1. List of the 5′UTRs used in this study.

Gene Symbol	Gene Name	Description	Reference
F-BOX	F-box protein	Normalizing gene	[52]
PP2A	Protein phosphatase 2A	Normalizing gene	[52]
GAPDH	Glyceraldehyde 3-phosphate dehydrogenase	Housekeeping gene	[52]
CAT	Catalase	ROS-scavenging enzymes	[71]
POX	Peroxidase	ROS-scavenging enzymes	[72]

Table 1. Cont.

Gene Symbol	Gene Name	Description	Reference
APX	Ascorbate peroxidase	ROS-scavenging enzymes	[72]
MAPK3	Mitogen-activated protein kinases 3	PAMP-triggered immunity (PTI)	[73]
NPR1	Nonexpressor of Pathogenesis-Related Genes1	Positive regulator of SAR	[74]

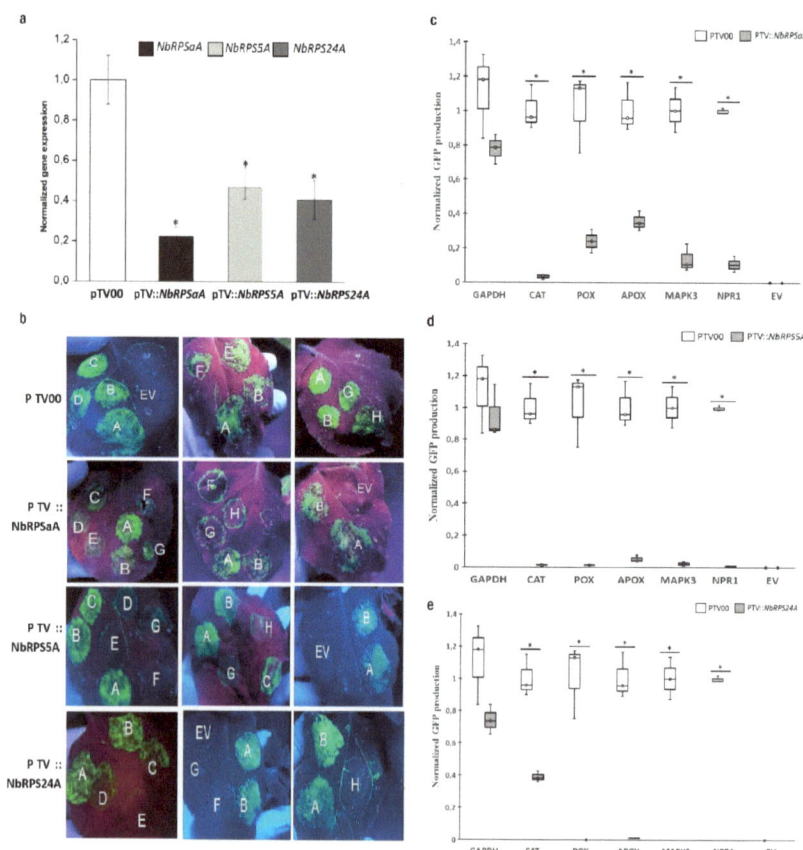

Figure 2. *RPSaA*, *RPS5A*, and *RPS24A* each have a role in the translation of defense proteins encoding mRNAs. (**a**) Relative expression levels of *NbRPSaA*, *NbRPS5A*, and *NbRPS24A* using quantitative RT-PCR analysis in the VIGS-treated *N. benthamiana* plants 21 days after agroinfiltration with TRV vectors. *ACTIN 1* and *EF1α* were used as internal references. Error bars represent the standard deviations of six independently infiltrated leaves from two biological replicates; asterisks (*) indicate significant differences based on the Student's *t*-test ($p < 0.05$). (**b**) GFP fluorescence in *N. benthamiana* leaves under UV light 5 days after infiltration (dpi), with *A. tumefaciens* carrying the p35S-5′UTR-GFP-expression cassettes (A: F-BOX, B: PP2A, C: GAPDH, D: CAT, E: POX, F: APX, G: NPR1, H: MAPK3, EV: Empty vector (PBGWFS7)). All leaves were infiltrated with the 5′ UTRs of two housekeeping gene vectors (A and B), in addition to the other vectors, as an internal control for leaf and plant variability. (**c–e**) Fluorimetric analysis of GFP accumulation. GFP fluorescence was quantified on ground tissue from three independently infiltrated leaves using a plate reader. Box plots show the replicate distributions in GFP concentration for each 5′UTR construct. The asterisks (*) represent significant differences between silenced and mock-infiltrated samples, based on the Student's *t*-test ($p < 0.05$).

3.3. RPS27D Is Required for Efficient Translation in N. benthamiana

The ribosomal protein S27 (RPS27), belongs to the 40S subunit and, through its zinc-finger-like motif, it acts as an RNA-binding protein and subsequently influences the transcription of many genes through transcript degradation [39]. *A. thaliana* and *N. benthamiana* both have four *RPS27* gene family members: A, B, C, and D. The amino acid similarity between *AtRPS27* and *NbRPS27* proteins is between 89.8 and 96.5% (Figure 3a) [26]. The alignment of the *S27* ribosomal protein sequences of different species (rice, barley, rat, and human) shows high conservation [39]. We sought to investigate the role of *NbRPS27D* in the translation of defense genes in *N. benthamiana* using the same TRV-mediated virus-induced gene-silencing approach to downregulate *NbRPS27D* expression. The silenced plants showed a more than 60% down-regulation of the target transcript compared to the control (Figure 3b). We then tested translational efficiency by analyzing the GFP accumulation from the agro-infiltration of chimeric mRNAs. *NbRPS27D* silencing resulted in a more than 50% decrease in GFP production in the zone of infiltration, with the vectors containing the 5′UTRs of defense genes (Figure 3c,d). Interestingly, a similar decrease was observed for the GFP vector containing the 5′UTR of the housekeeping gene, GAPDH (Figure 3c,d). The data presented here suggest that one paralog of ribosomal protein *S27* (*RPS27D*) may play a crucial role in the ribosome translational activity in *N. benthamiana*.

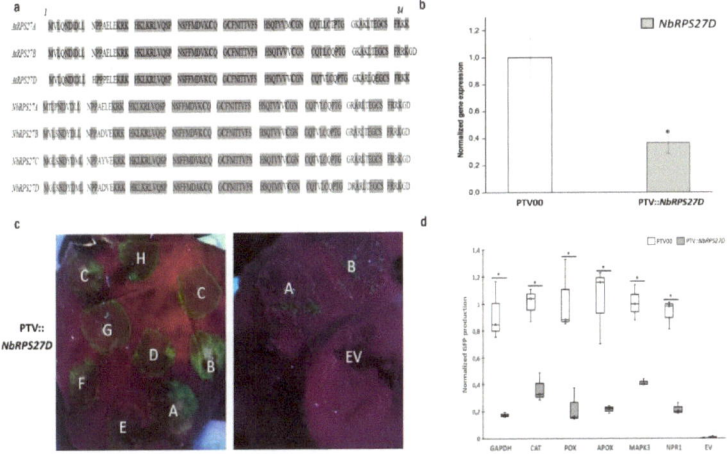

Figure 3. RPS27D is required for efficient translation in *N. benthamiana*. (**a**) Sequence alignment of *A. thaliana* and *N. benthamiana* ribosomal protein *S27* family members. Conserved amino acids in all the homologs are highlighted. Conserved cysteines forming a zinc finger are shown in bold. (**b**) Relative expression levels of *NbRPS27D* using quantitative RT-PCR analysis in the VIGS-treated *N. benthamiana* plants. ACTIN1 and EF1α were used as references. Error bars represent the standard deviations of six independently infiltrated leaves from two biological replicates; the asterisk (∗) indicates a significant difference based on the Student's *t*-test ($p < 0.05$). (**c**) GFP fluorescence in the *N. benthamiana* leaves of pTV::*NbRPS27D* plants under UV light 5 days after infiltration, with *A. tumefaciens* carrying the p35S-5′UTR-GFP-expression cassettes (A: F-BOX, B: PP2A, C: GAPDH, D: CAT, E: POX, F: APX, G: NPR1, H: MAPK3, EV: Empty vector (PBGWFS7)). All leaves were infiltrated with the 5′ UTRs of two housekeeping gene vectors (A and B), in addition to the other vectors, as an internal control for leaf and plant variability. (**d**) Fluorimetric analysis of GFP accumulation. GFP fluorescence was quantified on ground tissue from three independently infiltrated leaves using a plate reader. Box plots show the replicate distributions in GFP concentration for each 5′UTR construct. The asterisks (∗) represent significant differences between silenced and mock-infiltrated samples based on the Student's *t*-test ($p < 0.05$).

3.4. RACK1A Is Required for the Efficient Translation of Several Antioxidant Enzymes

RACK1 was originally isolated as a receptor for activated C-kinase 1. In addition to its signaling roles, it interacts with the ribosomal machinery, several cell surface receptors, and nuclear proteins [75]. The most stable and consistent interaction of RACK1 is the one it has with the ribosome. Indeed, RACK1 is found at the surface exposed region of the 40S ribosomal subunit, next to the mRNA exit channel [14,76,77]. It is known that RACK1 specifically modulates translational efficiency in various model systems [37,78,79]; however, its role in the efficient translation of mRNA subsets in the context of defense in planta is not well characterized. *N. benthamiana* has five RACK1 homologs [26] and *A. thaliana* has three [74]. *AtRACK1A* and *NbRACK1A* share 82% of their amino acid identities (Figure 4a). Since *RACK1A* is the paralog that was previously detected in the nucleus of stressed plants (Figure S2a), we silenced *NbRACK1A* to investigate its role in the translation of defense genes. Additionally, qRT-PCR analyses confirmed a down-regulation of more than 70% of the targeted transcript in the silenced plants compared to the control (Figure 4b). *NbRACK1A* silencing caused an important decrease in the GFP fluorescence in leaves infiltrated with the vectors containing the 5′UTRs of peroxidase, ascorbate peroxidase, and catalase (Figure 4c). Similarly, the quantification data indicate that the GFP production under these 5′UTRs was decreased in the pTV::*NbRACK1A* plants compared to control plants (Figure 4d). By contrast, *NbRACK1A* silencing had no effect on GFP production in the areas infiltrated with the other 5′UTRs (Figure 4c,d). These results suggest that *NbRACK1A* silencing in *N. benthamiana* compromises the translation of several antioxidant enzymes.

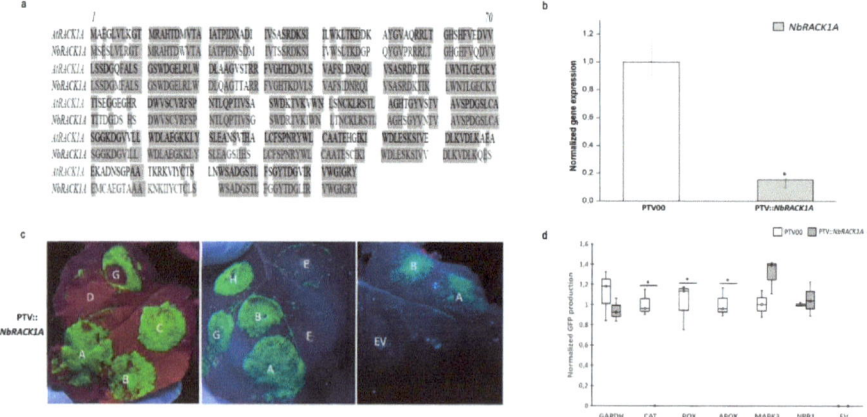

Figure 4. RACK1A is required for the efficient translation of several antioxidant enzymes. (**a**) Sequence alignment of *AtRACK1A* and *NbRACK1A*. (**b**) Relative expression levels of *NbRACK1A* using quantitative RT-PCR analysis in the VIGS-treated *N. benthamiana* plants. ACTIN1 and EF1α were used as references. Error bars represent the standard deviations of six independently infiltrated leaves from two biological replicates; the asterisk (*) indicates a significant difference based on the Student's *t*-test ($p < 0.05$). (**c**) GFP fluorescence in the *N. benthamiana* leaves of pTV::*NbRACK1A* plants under UV light 5 days after infiltration, with *A. tumefaciens* carrying the p35S-5′UTR-GFP-expression cassettes (A: F-BOX, B: PP2A, C: GAPDH, D: CAT, E: POX, F: APX, G: NPR1, H: MAPK3, EV: Empty vector (PBGWFS7)). All leaves were infiltrated with the 5′ UTRs of two housekeeping gene vectors (A and B), in addition to the other vectors, as an internal control for leaf and plant variability. (**d**) Fluorimetric analysis of GFP accumulation. GFP fluorescence was quantified on ground tissue from three independently infiltrated leaves using a plate reader. Box plots show the replicate distributions in GFP concentration for each 5′UTR construct. The asterisks (*) represent significant differences between silenced and mock-infiltrated samples based on the Student's *t*-test ($p < 0.05$).

4. Discussion and Conclusions

Plants' responses to stress vary according to stress-type and the outcome is mainly specific to a particular stress [80,81]. Recently developed technologies, such as ribosome profiling and quantitative proteomics, have shown that many stresses inhibit protein synthesis in cells [2,5,6,82]. Protein synthesis accounts for a large proportion of the energy budget of a cell and, thus, requires tight regulation [83]. However, a severe reduction in translation can be harmful during stress as it is precisely the time when cells require new protein synthesis in order to repair damage and adapt to the new environment [84]. Thus, selective translation regulation may allow cells to react to adverse conditions more effectively. As translation modulation is a fast response to environmental signals, the ribosome could be a player in this adaptation. It has been shown that translational regulation mainly takes place during the initiation steps [85]. In plants, the initiation of a translation requires initiation factors, mRNAs, tRNAs, and ribosomes. It involves numerous protein–RNA and protein–protein interactions. Briefly, the 40S subunit of the ribosome directly binds to mRNAs in a way that is dependent on the mRNAs' structures. After mRNA binding and scanning to the AUG start codon in a favorable context, the 60S subunit is recruited to form an 80S initiation complex capable of entry into the elongation phase [86]. During the initiation process, mRNA recruitment to the 40S ribosomal subunit is thought to be the rate-limiting step and is often modulated. The translation efficiency is determined by structural features in the 5' untranslated region (5'UTR) of the mRNA. These features not only determine how well an mRNA is translated but also whether specific ribosomal proteins and other proteins can interact with it [87].

Recent studies on *Arabidopsis* ribosomes revealed that numerous r-proteins are represented by two or more gene family members and most members of each family are expressed [29,88]; meanwhile, r-proteins are generally found as a single copy per ribosome [17,89]. However, the expression of each RP gene appears to be differentially regulated by different conditions. In line with this idea, changes in the expression patterns of these 15 RPS genes in *Arabidopsis* and *Nicotiana* were compared in order to gain insights into the regulation of the response to stress. We have shown differential expression under the three stress treatments of these seven (*RPSaA, RPS10C, RPS12C, RPS19C, RPS21C, RPS27D,* and *RACK1A*), five (*RPSaA, RPS5A, RPS24A, RPS27D,* and *RACK1A*), and three (*RPSaA, RACK1A,* and *RPS27D*) RPS genes in *Arabidopsis, Nicotiana,* and both plants, respectively. Similarly, several r-protein genes have been found to be upregulated under different stimuli in plants [19–21].These results indicate the fundamental stress-specific reprogramming of RP gene transcription under stress conditions.

With that in mind, it is interesting to speculate that these RPS endow ribosomes with the capacity for preferential mRNA selection for translation. To substantiate this hypothesis, using the VIGS method, we prepared plants with reduced levels of *RPSaA, RPS5A, RPS24A, RPS27D,* and *RACK1A* mRNAs and tested the translation efficiency of specific groups of selected mRNAs. *NbRPSaA-, NbRPS5A-,* and *NbRPS24A*-silenced *N. benthamiana* plants showed varying extents of compromised translation when compared to control plants. The green fluorescence intensity from the GAPDH 5'UTR-GFP chimera was similar in *RPSaA-, RPS5A-,* and *RPS24A*-silenced leaves and non-silenced control plants. However, the silencing of these three RPS genes resulted in a dramatic reduction in the translations of catalase, peroxidase, ascorbate peroxidase, NPR1, and MAPK3 5'UTR-GFP chimeras. Interestingly, previous studies have reported that *RPSaA, RPS5A,* and *RPS24A* have roles in the regulation of reactive oxygen species (ROS)-mediated systemic signaling. *RPS5A* may play an important role in dark treatment by participating in the autophagy regulatory process, which is triggered to degrade excessive ROS to help protect cells [90]. Then, it was shown that *RPSaA* and *RPS24A* had lower degradation rates and were more stable after oxidative stress [91]. Thus, the transcript levels of *RPSaA, RPS5A,* and *RPS24A* increased significantly in vanilla infected with *Fusarium oxysporum* f. sp. *vanilla* [20]; additionally, the expression of *RPS5A* was also induced by *Xanthomonas oryzae* pv. *oryzae* and *Rhizoctonia solani*, rice pathogens that, respectively, cause very serious bacterial leaf blight and sheath blight diseases [19]. Moreover, the expression of *RPSaA*, *RPS5A*, and *RPS24A* was upregulated under the three stress treatments in *Nicotiana* in our study. A differential accumulation of *RPSaA*, *RPS5A*, and *RPS24A* in the ribosomal apparatus was reported in *Arabidopsis* following treatment with the defense-inducing compound INA [28]. The data obtained in this study demonstrate that *RPSaA*, *RPS5A*, and *RPS24A* are required for the selective translation of defense genes to cope with unfavorable conditions.

Meanwhile, the data presented here suggest that *NbRPS27D* silencing leads to a general translation defect. As mentioned above, *RPS27* was shown to contain a conserved zinc finger motif, which may confirm its ability to interact with non-ribosomal components, especially mRNAs. In *Arabidopsis*, *RPS27B* has been said to act as a regulator of transcript stability in response to genotoxic treatments via the degradation of damaged RNAs [39]. Recent data have shown that mutations in *RPS27B* influence the integrity of ribosomes in both structure and function [90].

RACK1 is a highly conserved scaffold protein located at the surface exposed region of the 40S subunit, near the mRNA exit channel [92]. The recent identification of RACK1 as a core component of the small subunit of the ribosome suggests it possesses signaling functions, allowing for translation regulation in response to cell stimuli [93,94]. RACK1 regulates several signaling pathways by acting as a receptor for signaling proteins, such as the protein kinase C (PKC) family [95,96], and by controlling mRNA-specific translation [37,97]. Previous studies have reported on the role of *RACK1A* in plant immune signaling and hormone responses [98–100]. *RACK1A* has also been demonstrated to be a key regulator of reactive oxygen species (ROS)-mediated systemic signaling [99,101]. ROS are important and common messengers produced in response to various environmental stresses and are known to activate many MAPKs [102]. They are recognized as threshold-level signaling molecules that regulate adaptations to various biotic and abiotic stresses, i.e., the ROS level determines whether they will be defensive or destructive molecules, which is maintained by a balance between ROS-producing and ROS-scavenging pathways for normal cellular homeostasis [103]. Interestingly, RACK1 affects ROS levels and ROS levels also affect RACK1 gene expression [104]. It has been reported that the knockdown of endogenous RACK1 increases the intracellular ROS level following H_2O_2 stimulation in human hepatocellular carcinoma cells, leading to cell death promotion [105]. At the same time, Saelee et al. showed that *Penaeus monodon*-RACK1 protected shrimp cells from oxidative damage induced by H_2O_2 [106]. Núñez et al. have shown that RACK1 positively regulates the synthesis of cytoplasmic catalase, a detoxification enzyme induced by hydrogen peroxide treatment, and controls the cellular defense against the oxidative stress in the fission yeast *Schizosaccharomyces pombe* [97]. In contrast, rice RACK1 (*OsRACK1A*) has been shown to be involved in the immune response against pathogen attacks through enhanced reactive oxygen species (ROS) [99]. Our study is in line with *RACK1A* regulating the ROS in plants. The observation that *RACK1A* knockdown reduces the translation of catalase, APOX, and POX mRNAs is in agreement with RACK1's positive regulation of the detoxification enzyme synthesis induced by ROS. These enzymes are part of the antioxidant machinery, which helps to mitigate oxidative stress-induced damage. With respect to the role of RACK1 in signaling, accumulating evidence suggests that free RACK1 can act as a signaling molecule at a threshold level to enhance the production of ROS. Overall, it is noteworthy that the signal activated by free RACK1 is transient because, in the absence of ribosomal binding, the protein is unstable [37]. In contrast, RACK1 as a ribosomal protein controls the cellular defense against oxidative stress, positively regulating the translation of specific gene products involved in detoxification [97]. In conclusion, we propose that the association of *RACK1A* with the ribosome may indeed be regulated downstream of the ROS burst in order to modulate its translation functions; however, this possibility should be further investigated.

Our findings would argue that ribosomal proteins function in a modular fashion to decode genetic information in a context-dependent manner. The silencing of one r-protein, as was conducted in this study, could impact the stability and efficiency of the entire ribosome. In our case, the controls showed us that the overall translation efficiency was not impacted and, therefore, we believe that some variable ribosomal proteins additionally function in a coordinated manner to shape the translatome, which is adapted to different environmental cues in plants.

Overall, our study clearly demonstrates that some RPS are involved in the optimal translation regulation of many genes that are important for defense. However, our experimental design does not allow us to rule out whether the results presented herein are paralog-specific or if the effects could be true for all paralogs (or more than one paralog) of the same ribosomal protein. The findings of this study provide a novel strategy to assess translation efficiency that opens new and interesting avenues for research about the roles of ribosomal proteins during biotic and abiotic stress in *N. benthamiana*. Based on these data, we anticipate that some of the previously described biological functions of these RPS in plant immunity might be linked to their function as putative translational regulators. Future studies into the connection between the RP-mediated translation of defense proteins and the broader role of paralog specificity may provide a novel perspective on specialized ribosomes and translational control in plant disease.

Supplementary Materials: The following supporting information can be downloaded at: https://www.mdpi.com/article/10.3390/biom13071160/s1, Figure S1: Graphical representation of siRNA from targets (in blue) and off-targets (in red), suggesting the best construct predicted (in yellow) and allowing us to define a custom construct using the VIGS tool; Figure S2: (a) Venn diagram showing the deregulated ribosomal proteins detected in the nuclei of different plants under three types of stress collected from the previously published nuclear proteomes. Heat maps showing the differential expression patterns of the RPS genes deregulated in the nucleus during (b) cold treatment, (c) during elicitor treatments (d), and under biotic stresses. Relative expression ratios of the treatments versus controls are shown in green (down-regulated) and red (up-regulated). The scale on the top represents the $\log_2$ fold change value. The maximum value is displayed in dark red and the minimum value is displayed in light green. Images have been created and retrieved by Genevestigator v.3. using a meta-analysis tool and (e) the localization of deregulated RPS within the *Arabidopsis* 80S ribosome upon stress conditions. The visualization outlines mapped the upregulated RPS in response to cold stress, INA treatment, and biotic stress compared to control conditions. For the mapping, PyMOL visualization software was used to obtain a surface representation and to highlight proteins with significant changes. Red indicates RP families with increased transcripts following the three types of stress and purple indicates RP families with increased transcripts following either the biotic or elicitor treatment. Pink and blue represent RP families with increased transcript abundances following biotic and cold stress conditions, respectively; Table S1:List of the primer used to test the expression of the RPS in *N. benthamiana*; Table S2: *N. benthamiana* r-proteins, their size distribution (in amino acids, aa) and homologies to *A. thaliana*; Table S3: List of the primers used in the VIGS; Table S4: List of the primers used to test the silencing of RPS by qRT-PCR; Table S5: List of the primers used in P35s-5′UTR constructs for gateway clonning.

Author Contributions: Conceptualization, Z.F. and H.G.; methodology, Z.F. and H.G.; formal analysis, Z.F.; investigation, Z.F.; resources, H.G.; writing—original draft preparation, Z.F.; writing—review and editing, M.B.P. and H.G.; supervision, H.G.; project administration, H.G.; funding acquisition, H.G. All authors have read and agreed to the published version of the manuscript.

Funding: Canada Natural Sciences and Engineering Research Council Discovery program RGPIN 4002-2020.

Institutional Review Board Statement: Not applicable.

Informed Consent Statement: Not applicable.

Data Availability Statement: Data is contained within the article or Supplementary Material.

Acknowledgments: We are thankful to David Joly from University of Moncton for sharing the *Pseudomonas fluorescens* strain.

Conflicts of Interest: The authors declare no conflict of interest.

References

1. Echevarría-Zomeño, S.; Yángüez, E.; Fernández-Bautista, N.; Castro-Sanz, A.B.; Ferrando, A.; Castellano, M.M. Regulation of Translation Initiation under Biotic and Abiotic Stresses. *Int. J. Mol. Sci.* **2013**, *14*, 4670–4683. [CrossRef]
2. Fennoy, S.L.; Nong, T.; Bailey-Serres, J. Transcriptional and post-transcriptional processes regulate gene expression in oxygen-deprived roots of maize. *Plant J.* **1998**, *15*, 727–735. [CrossRef]
3. Hummel, M.; Cordewener, J.H.; de Groot, J.C.; Smeekens, S.; America, A.H.; Hanson, J. Dynamic protein composition of *Arabidopsis thaliana* cytosolic ribosomes in response to sucrose feeding as revealed by label free MSE proteomics. *Proteomics* **2012**, *12*, 1024–1038. [CrossRef]
4. Groppo, R.; Palmenberg, A.C. Cardiovirus 2A protein associates with 40S but not 80S ribosome subunits during infection. *J. Virol.* **2007**, *81*, 13067–13074. [CrossRef]
5. Bailey-Serres, J.; Sorenson, R.; Juntawong, P. Getting the message across: Cytoplasmic ribonucleoprotein complexes. *Trends Plant Sci.* **2009**, *14*, 443–453. [CrossRef]
6. Bailey-Serres, J. Selective translation of cytoplasmic mRNAs in plants. *Trends Plant Sci.* **1999**, *4*, 142–148. [CrossRef] [PubMed]
7. Liu, L.; Simon, M.C. Regulation of transcription and translation by hypoxia. *Cancer Biol.* **2004**, *3*, 492–497. [CrossRef] [PubMed]
8. Braunstein, S.; Karpisheva, K.; Pola, C.; Goldberg, J.; Hochman, T.; Yee, H.; Cangiarella, J.; Arju, R.; Formenti, S.C.; Schneider, R.J. A hypoxia-controlled cap-dependent to cap-independent translation switch in breast cancer. *Mol. Cell* **2007**, *28*, 501–512. [CrossRef]
9. Gamm, M.; Peviani, A.; Honsel, A.; Snel, B.; Smeekens, S.; Hanson, J. Increased sucrose levels mediate selective mRNA translation in *Arabidopsis*. *BMC Plant Biol.* **2014**, *14*, 306. [CrossRef] [PubMed]
10. Nicolaí, M.; Roncato, M.A.; Canoy, A.S.; Rouquié, D.; Sarda, X.; Freyssinet, G.; Robaglia, C. Large-Scale Analysis of mRNA Translation States during Sucrose Starvation in *Arabidopsis* Cells Identifies Cell Proliferation and Chromatin Structure as Targets of Translational Control. *Plant Physiol.* **2006**, *141*, 663–673.
11. Wang, L.; Li, H.; Zhao, C.; Li, S.; Kong, L.; Wu, W.; Kong, W.; Liu, Y.; Wei, Y.; Zhu, J.K.; et al. The inhibition of protein translation mediated by AtGCN1 is essential for cold tolerance in *Arabidopsis thaliana*. *Plant Cell Environ.* **2017**, *40*, 56–68. [CrossRef] [PubMed]
12. Juntawong, P.; Bailey-Serres, J. Dynamic Light Regulation of Translation Status in *Arabidopsis thaliana*. *Front. Plant Sci.* **2012**, *3*, 66. [CrossRef] [PubMed]
13. Sáez-Vásquez, J.; Delseny, M. Ribosome Biogenesis in Plants: From Functional 45S Ribosomal DNA Organization to Ribosome Assembly Factors. *Plant Cell* **2019**, *31*, 1945–1967. [CrossRef]
14. Chang, I.F.; Szick-Miranda, K.; Pan, S.; Bailey-Serres, J. Proteomic characterization of evolutionarily conserved and variable proteins of *Arabidopsis* cytosolic ribosomes. *Plant Physiol.* **2005**, *137*, 848–862. [CrossRef] [PubMed]
15. Barakat, A.; Szick-Miranda, K.; Chang, I.F.; Guyot, R.; Blanc, G.; Cooke, R.; Delseny, M.; Bailey-Serres, J. The organization of cytoplasmic ribosomal protein genes in the *Arabidopsis* genome. *Plant Physiol.* **2001**, *127*, 398–415.
16. Carroll, A.J. The *Arabidopsis* Cytosolic Ribosomal Proteome: From form to Function. *Front. Plant Sci.* **2013**, *4*, 32. [CrossRef]
17. Ban, N.; Nissen, P.; Hansen, J.; Moore, P.B.; Steitz, T.A. The complete atomic structure of the large ribosomal subunit at 2.4 A resolution. *Science* **2000**, *289*, 905–920. [CrossRef]
18. Yusupova, G.; Yusupov, M. High-resolution structure of the eukaryotic 80S ribosome. *Annu. Rev. Biochem.* **2014**, *83*, 467–486. [CrossRef]
19. Saha, A.; Das, S.; Moin, M.; Dutta, M.; Bakshi, A.; Madhav, M.S.; Kirti, P.B. Genome-Wide Identification and Comprehensive Expression Profiling of Ribosomal Protein Small Subunit (RPS) Genes and their Comparative Analysis with the Large Subunit (RPL) Genes in Rice. *Front. Plant Sci.* **2017**, *8*, 1553. [CrossRef]
20. Solano de la Cruz, M.T.; Adame-García, J.; Gregorio-Jorge, J.; Jiménez-Jacinto, V.; Vega-Alvarado, L.; Iglesias-Andreu, L.; Escobar-Hernández, E.E.; Luna-Rodríguez, M. Increase in ribosomal proteins activity: Translational reprogramming in *Vanilla planifolia* Jacks., against *Fusarium* infection. *bioRxiv* **2019**. [CrossRef]
21. Moin, M.; Bakshi, A.; Saha, A.; Dutta, M.; Madhav, S.M.; Kirti, P.B. Rice Ribosomal Protein Large Subunit Genes and Their Spatio-temporal and Stress Regulation. *Front. Plant Sci.* **2016**, *7*, 1284. [CrossRef]
22. Ban, Z.; Yan, J.; Wang, Y.; Zhang, J.; Yuan, Q.; Li, L. Effects of postharvest application of chitosan-based layer-by-layer assemblies on regulation of ribosomal and defense proteins in strawberry fruit (*Fragaria* × *ananassa*). *Sci. Hortic.* **2018**, *240*, 293–302. [CrossRef]
23. Wang, J.; Lan, P.; Gao, H.; Zheng, L.; Li, W.; Schmidt, W. Expression changes of ribosomal proteins in phosphate- and iron-deficient *Arabidopsis* roots predict stress-specific alterations in ribosome composition. *BMC Genom.* **2013**, *14*, 783. [CrossRef] [PubMed]
24. Vemanna, R.S.; Bakade, R.; Bharti, P.; Kumar, M.K.P.; Sreeman, S.M.; Senthil-Kumar, M.; Makarla, U. Cross-Talk Signaling in Rice During Combined Drought and Bacterial Blight Stress. *Front. Plant Sci.* **2019**, *10*, 193. [CrossRef]
25. Guimaraes, J.C.; Zavolan, M. Patterns of ribosomal protein expression specify normal and malignant human cells. *Genome Biol.* **2016**, *17*, 236. [CrossRef]
26. Eskelin, K.; Varjosalo, M.; Ravantti, J.; Mäkinen, K. Ribosome profiles and riboproteomes of healthy and Potato virus A- and Agrobacterium-infected *Nicotiana benthamiana* plants. *Mol. Plant Pathol.* **2019**, *20*, 392–409. [CrossRef] [PubMed]
27. Reschke, M.; Clohessy, J.G.; Seitzer, N.; Goldstein, D.P.; Breitkopf, S.B.; Schmolze, D.B.; Ala, U.; Asara, J.M.; Beck, A.H.; Pandolfi, P.P. Characterization and analysis of the composition and dynamics of the mammalian riboproteome. *Cell Rep.* **2013**, *4*, 1276–1287. [CrossRef]

28. Fakih, Z.; Plourde, M.B.; Nkouankou, C.E.T.; Fourcassie, V.; Bourassa, S.; Droit, A.; Germain, H. Specific alterations in riboproteomes composition of isonicotinic acid treated arabidopsis seedlings. *Plant Mol. Biol.* **2023**, *111*, 379–392. [CrossRef] [PubMed]
29. Genuth, N.R.; Barna, M. Heterogeneity and specialized functions of translation machinery: From genes to organisms. *Nat. Rev. Genet.* **2018**, *19*, 431–452. [CrossRef]
30. Luan, Y.; Tang, N.; Yang, J.; Liu, S.; Cheng, C.; Wang, Y.; Chen, C.; Guo, Y.N.; Wang, H.; Zhao, W.; et al. Deficiency of ribosomal proteins reshapes the transcriptional and translational landscape in human cells. *Nucleic Acids Res.* **2022**, *50*, 6601–6617. [CrossRef]
31. Kim, J.; Chubatsu, L.S.; Admon, A.; Stahl, J.; Fellous, R.; Linn, S. Implication of mammalian ribosomal protein S3 in the processing of DNA damage. *J. Biol. Chem.* **1995**, *270*, 13620–13629. [CrossRef]
32. Zhang, Y.; Lu, H. Signaling to p53: Ribosomal proteins find their way. *Cancer Cell* **2009**, *16*, 369–377. [CrossRef] [PubMed]
33. Ni, J.Q.; Liu, L.P.; Hess, D.; Rietdorf, J.; Sun, F.L. Drosophila ribosomal proteins are associated with linker histone H1 and suppress gene transcription. *Genes Dev.* **2006**, *20*, 1959–1973. [CrossRef] [PubMed]
34. Funatsu, G.; Wittmann, H.G. Ribosomal proteins. 33. Location of amino-acid replacements in protein S12 isolated from Escherichia coli mutants resistant to streptomycin. *J. Mol. Biol.* **1972**, *68*, 547–550. [CrossRef]
35. Stöffler, G.; Deusser, E.; Wittmann, H.G.; Apirion, D. Ribosomal proteins: XIX. Altered S5 ribosomal protein in an Escherichia coli revertant from streptomycin dependence to independence. *Mol. Gen. Genet. MGG* **1971**, *111*, 334–341. [CrossRef] [PubMed]
36. Brodersen, D.E.; Nissen, P. The social life of ribosomal proteins. *FEBS J.* **2005**, *272*, 2098–2108. [CrossRef] [PubMed]
37. Gallo, S.; Ricciardi, S.; Manfrini, N.; Pesce, E.; Oliveto, S.; Calamita, P.; Mancino, M.; Maffioli, E.; Moro, M.; Crosti, M.; et al. RACK1 Specifically Regulates Translation through Its Binding to Ribosomes. *Mol. Cell Biol.* **2018**, *38*, e00230-18. [CrossRef]
38. Ramu, V.S.; Dawane, A.; Lee, S.; Oh, S.; Lee, H.K.; Sun, L.; Senthil-Kumar, M.; Mysore, K.S. Ribosomal protein QM/RPL10 positively regulates defence and protein translation mechanisms during nonhost disease resistance. *Mol. Plant Pathol.* **2020**, *21*, 1481–1494. [CrossRef]
39. Revenkova, E.; Masson, J.; Koncz, C.; Afsar, K.; Jakovleva, L.; Paszkowski, J. Involvement of *Arabidopsis thaliana* ribosomal protein S27 in mRNA degradation triggered by genotoxic stress. *Embo J.* **1999**, *18*, 490–499. [CrossRef]
40. Nagaraj, S.; Senthil-Kumar, M.; Ramu, V.S.; Wang, K.; Mysore, K.S. Plant Ribosomal Proteins, RPL12 and RPL19, Play a Role in Nonhost Disease Resistance against Bacterial Pathogens. *Front. Plant Sci.* **2015**, *6*, 1192. [CrossRef]
41. Rajamäki, M.L.; Xi, D.; Sikorskaite-Gudziuniene, S.; Valkonen, J.P.T.; Whitham, S.A. Differential Requirement of the Ribosomal Protein S6 and Ribosomal Protein S6 Kinase for Plant-Virus Accumulation and Interaction of S6 Kinase with Potyviral VPg. *Mol. Plant Microbe Interact.* **2017**, *30*, 374–384. [CrossRef] [PubMed]
42. Cheng, Y.T.; Germain, H.; Wiermer, M.; Bi, D.; Xu, F.; Garcia, A.V.; Wirthmueller, L.; Despres, C.; Parker, J.E.; Zhang, Y.; et al. Nuclear pore complex component MOS7/Nup88 is required for innate immunity and nuclear accumulation of defense regulators in *Arabidopsis*. *Plant Cell* **2009**, *21*, 2503–2516. [CrossRef]
43. Conrath, U.; Chen, Z.; Ricigliano, J.R.; Klessig, D.F. Two inducers of plant defense responses, 2,6-dichloroisonicotinec acid and salicylic acid, inhibit catalase activity in tobacco. *Proc. Natl. Acad. Sci. USA* **1995**, *92*, 7143–7147. [CrossRef]
44. Bae, M.S.; Cho, E.J.; Choi, E.Y.; Park, O.K. Analysis of the *Arabidopsis* nuclear proteome and its response to cold stress. *Plant J.* **2003**, *36*, 652–663. [CrossRef]
45. Badel, J.L.; Piquerez, S.J.; Greenshields, D.; Rallapalli, G.; Fabro, G.; Ishaque, N.; Jones, J.D. In planta effector competition assays detect Hyaloperonospora arabidopsidis effectors that contribute to virulence and localize to different plant subcellular compartments. *Mol. Plant Microbe Interact.* **2013**, *26*, 745–757. [CrossRef]
46. Lu, X.; Liu, Y.; Zhao, L.; Liu, Y.; Zhao, M. Selection of reliable reference genes for RT-qPCR during methyl jasmonate, salicylic acid and hydrogen peroxide treatments in *Ganoderma lucidum*. *World J. Microbiol. Biotechnol.* **2018**, *34*, 92. [CrossRef] [PubMed]
47. Kreps, J.A.; Wu, Y.; Chang, H.-S.; Zhu, T.; Wang, X.; Harper, J.F. Transcriptome changes for *Arabidopsis* in response to salt, osmotic, and cold stress. *Plant Physiol.* **2002**, *130*, 2129–2141. [CrossRef] [PubMed]
48. Livak, K.J.; Schmittgen, T.D. Analysis of relative gene expression data using real-time quantitative PCR and the $2^{-\Delta\Delta CT}$ Method. *Methods* **2001**, *25*, 402–408. [CrossRef]
49. Ratcliff, F.; Martin-Hernandez, A.M.; Baulcombe, D.C. Technical Advance. Tobacco rattle virus as a vector for analysis of gene function by silencing. *Plant J.* **2001**, *25*, 237–245. [CrossRef]
50. Senthil-Kumar, M.; Mysore, K.S. Virus-induced gene silencing can persist for more than 2 years and also be transmitted to progeny seedlings in *Nicotiana benthamiana* and tomato. *Plant Biotechnol. J.* **2011**, *9*, 797–806. [CrossRef]
51. Zhang, G.; Zhang, Z.; Wan, Q.; Zhou, H.; Jiao, M.; Zheng, H.; Lu, Y.; Rao, S.; Wu, G.; Chen, J.; et al. Selection and Validation of Reference Genes for RT-qPCR Analysis of Gene Expression in *Nicotiana benthamiana* upon Single Infections by 11 Positive-Sense Single-Stranded RNA Viruses from Four Genera. *Plants* **2023**, *12*, 857. [CrossRef] [PubMed]
52. Liu, D.; Shi, L.; Han, C.; Yu, J.; Li, D.; Zhang, Y. Validation of reference genes for gene expression studies in virus-infected *Nicotiana benthamiana* using quantitative real-time PCR. *PLoS ONE* **2012**, *7*, e46451. [CrossRef] [PubMed]
53. Baek, E.; Yoon, J.-Y.; Palukaitis, P. Validation of reference genes for quantifying changes in gene expression in virus-infected tobacco. *Virology* **2017**, *510*, 29–39. [CrossRef] [PubMed]
54. Chi, C.; Shen, Y.; Yin, L.; Ke, X.; Han, D.; Zuo, Y. Selection and Validation of Reference Genes for Gene Expression Analysis in *Vigna angularis* Using Quantitative Real-Time RT-PCR. *PLoS ONE* **2016**, *11*, e0168479. [CrossRef]

55. Lilly, S.T.; Drummond, R.S.; Pearson, M.N.; MacDiarmid, R.M. Identification and validation of reference genes for normalization of transcripts from virus-infected *Arabidopsis thaliana*. *Mol. Plant Microbe Interact.* **2011**, *24*, 294–304. [CrossRef]
56. Czechowski, T.; Stitt, M.; Altmann, T.; Udvardi, M.K.; Scheible, W.-R. Genome-wide identification and testing of superior reference genes for transcript normalization in *Arabidopsis*. *Plant Physiol.* **2005**, *139*, 5–17. [CrossRef]
57. Migocka, M.; Papierniak, A. Identification of suitable reference genes for studying gene expression in cucumber plants subjected to abiotic stress and growth regulators. *Mol. Breed.* **2011**, *28*, 343–357. [CrossRef]
58. Rajput, V.D.; Harish; Singh, R.K.; Verma, K.K.; Sharma, L.; Quiroz-Figueroa, F.R.; Meena, M.; Gour, V.S.; Minkina, T.; Sushkova, S.; et al. Recent developments in enzymatic antioxidant defence mechanism in plants with special reference to abiotic stress. *Biology* **2021**, *10*, 267. [CrossRef]
59. Chen, J.; Mohan, R.; Zhang, Y.; Li, M.; Chen, H.; Palmer, I.A.; Chang, M.; Qi, G.; Spoel, S.H.; Mengiste, T.; et al. NPR1 Promotes Its Own and Target Gene Expression in Plant Defense by Recruiting CDK8. *Plant Physiol.* **2019**, *181*, 289–304. [CrossRef]
60. Taj, G.; Agarwal, P.; Grant, M.; Kumar, A. MAPK machinery in plants: Recognition and response to different stresses through multiple signal transduction pathways. *Plant Signal. Behav.* **2010**, *5*, 1370–1378. [CrossRef]
61. Xu, G.; Greene, G.H.; Yoo, H.; Liu, L.; Marqués, J.; Motley, J.; Dong, X. Global translational reprogramming is a fundamental layer of immune regulation in plants. *Nature* **2017**, *545*, 487–490. [CrossRef] [PubMed]
62. Li, Y.; Li, Q.; Beuchat, G.; Zeng, H.; Zhang, C.; Chen, L.Q. Combined analyses of translatome and transcriptome in *Arabidopsis* reveal new players responding to magnesium deficiency. *J. Integr. Plant Biol.* **2021**, *63*, 2075–2092. [CrossRef] [PubMed]
63. Yangueez, E.; Castro-Sanz, A.B.; Fernandez-Bautista, N.; Oliveros, J.C.; Castellano, M.M. Analysis of genome-wide changes in the translatome of *Arabidopsis* seedlings subjected to heat stress. *PLoS ONE* **2013**, *8*, e71425. [CrossRef] [PubMed]
64. Karimi, M.; Inzé, D.; Depicker, A. GATEWAY™ vectors for *Agrobacterium*-mediated plant transformation. *Trends Plant Sci.* **2002**, *7*, 193–195. [CrossRef]
65. Sparkes, I.A.; Runions, J.; Kearns, A.; Hawes, C. Rapid, transient expression of fluorescent fusion proteins in tobacco plants and generation of stably transformed plants. *Nat. Protoc.* **2006**, *1*, 2019–2025. [CrossRef]
66. Diamos, A.G.; Rosenthal, S.H.; Mason, H.S. 5′ and 3′ Untranslated Regions Strongly Enhance Performance of Geminiviral Replicons in *Nicotiana benthamiana* Leaves. *Front. Plant Sci.* **2016**, *7*, 200. [CrossRef]
67. Fakih, Z.; Ahmed, M.B.; Letanneur, C.; Germain, H. An unbiased nuclear proteomics approach reveals novel nuclear protein components that participates in MAMP-triggered immunity. *Plant Signal. Behav.* **2016**, *11*, e1183067. [CrossRef]
68. Howden, A.J.M.; Stam, R.; Martinez Heredia, V.; Motion, G.B.; Ten Have, S.; Hodge, K.; Marques Monteiro Amaro, T.M.; Huitema, E. Quantitative analysis of the tomato nuclear proteome during *Phytophthora capsici* infection unveils regulators of immunity. *New Phytol.* **2017**, *215*, 309–322. [CrossRef]
69. Ayash, M.; Abukhalaf, M.; Thieme, D.; Proksch, C.; Heilmann, M.; Schattat, M.H.; Hoehenwarter, W. LC-MS Based Draft Map of the *Arabidopsis thaliana* Nuclear Proteome and Protein Import in Pattern Triggered Immunity. *Front. Plant Sci.* **2021**, *12*, 744103. [CrossRef]
70. Palm, D.; Simm, S.; Darm, K.; Weis, B.L.; Ruprecht, M.; Schleiff, E.; Scharf, C. Proteome distribution between nucleoplasm and nucleolus and its relation to ribosome biogenesis in *Arabidopsis thaliana*. *RNA Biol.* **2016**, *13*, 441–454. [CrossRef]
71. Diaz-Albiter, H.; Mitford, R.; Genta, F.A.; Sant'Anna, M.R.; Dillon, R.J. Reactive oxygen species scavenging by catalase is important for female *Lutzomyia longipalpis* fecundity and mortality. *PLoS ONE* **2011**, *6*, e17486. [CrossRef] [PubMed]
72. Das, K.; Roychoudhury, A. Reactive oxygen species (ROS) and response of antioxidants as ROS-scavengers during environmental stress in plants. *Front. Environ. Sci.* **2014**, *2*, 53. [CrossRef]
73. Chang, M.; Chen, H.; Liu, F.; Fu, Z.Q. PTI and ETI: Convergent pathways with diverse elicitors. *Trends Plant Sci.* **2022**, *27*, 113–115. [CrossRef] [PubMed]
74. Backer, R.; Naidoo, S.; van den Berg, N. The NONEXPRESSOR OF PATHOGENESIS-RELATED GENES 1 (NPR1) and Related Family: Mechanistic Insights in Plant Disease Resistance. *Front. Plant Sci.* **2019**, *10*, 102. [CrossRef]
75. Adams, D.R.; Ron, D.; Kiely, P.A. RACK1, A multifaceted scaffolding protein: Structure and function. *Cell Commun. Signal.* **2011**, *9*, 22. [CrossRef]
76. Rabl, J.; Leibundgut, M.; Ataide, S.F.; Haag, A.; Ban, N. Crystal structure of the eukaryotic 40S ribosomal subunit in complex with initiation factor 1. *Science* **2011**, *331*, 730–736. [CrossRef]
77. Giavalisco, P.; Wilson, D.; Kreitler, T.; Lehrach, H.; Klose, J.; Gobom, J.; Fucini, P. High heterogeneity within the ribosomal proteins of the *Arabidopsis thaliana* 80S ribosome. *Plant Mol. Biol.* **2005**, *57*, 577–591. [CrossRef]
78. Volta, V.; Beugnet, A.; Gallo, S.; Magri, L.; Brina, D.; Pesce, E.; Calamita, P.; Sanvito, F.; Biffo, S. RACK1 depletion in a mouse model causes lethality, pigmentation deficits and reduction in protein synthesis efficiency. *Cell Mol. Life Sci.* **2013**, *70*, 1439–1450. [CrossRef]
79. Thompson, M.K.; Rojas-Duran, M.F.; Gangaramani, P.; Gilbert, W.V. The ribosomal protein Asc1/RACK1 is required for efficient translation of short mRNAs. *eLife* **2016**, *5*, e11154. [CrossRef]
80. Georgieva, M.; Vassileva, V. Stress Management in Plants: Examining Provisional and Unique Dose-Dependent Responses. *Int. J. Mol. Sci.* **2023**, *24*, 5105. [CrossRef]
81. Zhang, H.; Zhao, Y.; Zhu, J.-K. Thriving under Stress: How Plants Balance Growth and the Stress Response. *Dev. Cell* **2020**, *55*, 529–543. [CrossRef]

82. Hummel, M.; Dobrenel, T.; Cordewener, J.J.; Davanture, M.; Meyer, C.; Smeekens, S.J.; Bailey-Serres, J.; America, T.A.; Hanson, J. Proteomic LC-MS analysis of *Arabidopsis* cytosolic ribosomes: Identification of ribosomal protein paralogs and re-annotation of the ribosomal protein genes. *J. Proteom.* **2015**, *128*, 436–449. [CrossRef]
83. Hershey, J.W.; Sonenberg, N.; Mathews, M.B. Principles of translational control: An overview. *Cold Spring Harb. Perspect. Biol.* **2012**, *4*, a011528. [CrossRef] [PubMed]
84. Shcherbik, N.; Pestov, D.G. The Impact of Oxidative Stress on Ribosomes: From Injury to Regulation. *Cells* **2019**, *8*, 1379. [CrossRef] [PubMed]
85. Sonenberg, N.; Hinnebusch, A.G. Regulation of translation initiation in eukaryotes: Mechanisms and biological targets. *Cell* **2009**, *136*, 731–745. [CrossRef]
86. Roy, B.; von Arnim, A.G. Translational Regulation of Cytoplasmic mRNAs. *Arab. Book* **2013**, *11*, e0165. [CrossRef] [PubMed]
87. Merchante, C.; Stepanova, A.N.; Alonso, J.M. Translation regulation in plants: An interesting past, an exciting present and a promising future. *Plant J.* **2017**, *90*, 628–653. [CrossRef] [PubMed]
88. Browning, K.S.; Bailey-Serres, J. Mechanism of cytoplasmic mRNA translation. *Arab. Book/Am. Soc. Plant Biol.* **2015**, *13*, e0176. [CrossRef]
89. Tal, M.; Weissman, I.; Silberstein, A. A new method for stoichiometric analysis of proteins in complex mixture—Reevaluation of the stoichiometry of *E. coli* ribosomal proteins. *J. Biochem. Biophys. Methods* **1990**, *21*, 247–266. [CrossRef]
90. Shen, J.J.; Chen, Q.S.; Li, Z.F.; Zheng, Q.X.; Xu, Y.L.; Zhou, H.N.; Mao, H.Y.; Shen, Q.; Liu, P.P. Proteomic and metabolomic analysis of *Nicotiana benthamiana* under dark stress. *FEBS Open Bio* **2022**, *12*, 231–249. [CrossRef]
91. Salih, K.; Duncan, O.; Li, L.; O'Leary, B.; Fenske, R.; Troesch, J.; Millar, A. Impact of oxidative stress on the function, abundance and turnover of the *Arabidopsis* 80S cytosolic ribosome. *Plant J.* **2020**, *103*, 128–139. [CrossRef]
92. Sengupta, J.; Nilsson, J.; Gursky, R.; Spahn, C.M.; Nissen, P.; Frank, J. Identification of the versatile scaffold protein RACK1 on the eukaryotic ribosome by cryo-EM. *Nat. Struct. Mol. Biol.* **2004**, *11*, 957–962. [CrossRef] [PubMed]
93. Link, A.J.; Eng, J.; Schieltz, D.M.; Carmack, E.; Mize, G.J.; Morris, D.R.; Garvik, B.M.; Yates, J.R., 3rd. Direct analysis of protein complexes using mass spectrometry. *Nat. Biotechnol.* **1999**, *17*, 676–682. [CrossRef] [PubMed]
94. Nilsson, J.; Sengupta, J.; Frank, J.; Nissen, P. Regulation of eukaryotic translation by the RACK1 protein: A platform for signalling molecules on the ribosome. *EMBO Rep.* **2004**, *5*, 1137–1141. [CrossRef] [PubMed]
95. Ron, D.; Chen, C.H.; Caldwell, J.; Jamieson, L.; Orr, E.; Mochly-Rosen, D. Cloning of an intracellular receptor for protein kinase C: A homolog of the beta subunit of G proteins. *Proc. Natl. Acad. Sci. USA* **1994**, *91*, 839–843. [CrossRef]
96. Su, J.; Xu, J.; Zhang, S. RACK1, scaffolding a heterotrimeric G protein and a MAPK cascade. *Trends Plant Sci.* **2015**, *20*, 405–407. [CrossRef] [PubMed]
97. Núñez, A.; Franco, A.; Madrid, M.; Soto, T.; Vicente, J.; Gacto, M.; Cansado, J. Role for RACK1 orthologue Cpc2 in the modulation of stress response in fission yeast. *Mol. Biol. Cell* **2009**, *20*, 3996–4009. [CrossRef]
98. Rahman, M.A.; Fennell, H.; Ullah, H. Receptor for Activated C Kinase1B (OsRACK1B) Impairs Fertility in Rice through NADPH-Dependent H_2O_2 Signaling Pathway. *Int. J. Mol. Sci.* **2022**, *23*, 8455. [CrossRef]
99. Nakashima, A.; Chen, L.; Thao, N.P.; Fujiwara, M.; Wong, H.L.; Kuwano, M.; Umemura, K.; Shirasu, K.; Kawasaki, T.; Shimamoto, K. RACK1 functions in rice innate immunity by interacting with the Rac1 immune complex. *Plant Cell* **2008**, *20*, 2265–2279. [CrossRef]
100. Chen, J.G.; Ullah, H.; Temple, B.; Liang, J.; Guo, J.; Alonso, J.M.; Ecker, J.R.; Jones, A.M. RACK1 mediates multiple hormone responsiveness and developmental processes in *Arabidopsis*. *J. Exp. Bot.* **2006**, *57*, 2697–2708. [CrossRef]
101. Shirasu, K.; Schulze-Lefert, P. Complex formation, promiscuity and multi-functionality: Protein interactions in disease-resistance pathways. *Trends Plant Sci.* **2003**, *8*, 252–258. [CrossRef] [PubMed]
102. Jalmi, S.; Sinha, A. ROS mediated MAPK signaling in abiotic and biotic stress- striking similarities and differences. *Front. Plant Sci.* **2015**, *6*, 769. [CrossRef] [PubMed]
103. Mittler, R.; Vanderauwera, S.; Gollery, M.; Van Breusegem, F. Reactive oxygen gene network of plants. *Trends Plant Sci.* **2004**, *9*, 490–498. [CrossRef] [PubMed]
104. Fennell, H.W.W.; Ullah, H.; van Wijnen, A.J.; Lewallen, E.A. *Arabidopsis thaliana* and *Oryza sativa* receptor for activated C kinase 1 (RACK1) mediated signaling pathway shows hypersensitivity to oxidative stress. *Plant Gene* **2021**, *27*, 100299. [CrossRef]
105. Zhou, S.; Cao, H.; Zhao, Y.; Li, X.; Zhang, J.; Hou, C.; Ma, Y.; Wang, Q. RACK1 promotes hepatocellular carcinoma cell survival via CBR1 by suppressing TNF-α-induced ROS generation. *Oncol. Lett.* **2016**, *12*, 5303–5308. [CrossRef]
106. Saelee, N.; Tonganunt-Srithaworn, M.; Wanna, W.; Phongdara, A. Receptor for Activated C Kinase-1 protein from *Penaeus monodon* (Pm-RACK1) participates in the shrimp antioxidant response. *Int. J. Biol. Macromol.* **2011**, *49*, 32–36. [CrossRef]

Disclaimer/Publisher's Note: The statements, opinions and data contained in all publications are solely those of the individual author(s) and contributor(s) and not of MDPI and/or the editor(s). MDPI and/or the editor(s) disclaim responsibility for any injury to people or property resulting from any ideas, methods, instructions or products referred to in the content.

MDPI AG
Grosspeteranlage 5
4052 Basel
Switzerland
Tel.: +41 61 683 77 34

Biomolecules Editorial Office
E-mail: biomolecules@mdpi.com
www.mdpi.com/journal/biomolecules

Disclaimer/Publisher's Note: The title and front matter of this reprint are at the discretion of the Guest Editors. The publisher is not responsible for their content or any associated concerns. The statements, opinions and data contained in all individual articles are solely those of the individual Editors and contributors and not of MDPI. MDPI disclaims responsibility for any injury to people or property resulting from any ideas, methods, instructions or products referred to in the content.

www.ingramcontent.com/pod-product-compliance
Lightning Source LLC
LaVergne TN
LVHW072338090526
838202LV00019B/2441